T0256390

THE ELECTRICAL ENGINEERING
AND APPLIED SIGNAL PROCESSING SERIES

MIMO System Technology for Wireless Communications

THE ELECTRICAL ENGINEERING
AND APPLIED SIGNAL PROCESSING SERIES
Edited by Alexander Poularikas

The Advanced Signal Processing Handbook:
Theory and Implementation for Radar, Sonar,
and Medical Imaging Real-Time Systems
Stergios Stergiopoulos

The Transform and Data Compression Handbook
K.R. Rao and P.C. Yip

Handbook of Multisensor Data Fusion
David Hall and James Llinas

Handbook of Neural Network Signal Processing
Yu Hen Hu and Jenq-Neng Hwang

Handbook of Antennas in Wireless Communications
Lal Chand Godara

Noise Reduction in Speech Applications
Gillian M. Davis

Signal Processing Noise
Vyacheslav P. Tuzlukov

Digital Signal Processing with Examples in MATLAB®
Samuel Stearns

Applications in Time-Frequency Signal Processing
Antonia Papandreou-Suppappola

The Digital Color Imaging Handbook
Gaurav Sharma

Pattern Recognition in Speech and Language Processing
Wu Chou and Biing-Hwang Juang

Propagation Handbook for Wireless Communication System Design
Robert K. Crane

Nonlinear Signal and Image Processing: Theory, Methods, and Applications
Kenneth E. Barner and Gonzalo R. Arce

Smart Antennas
Lal Chand Godara

Mobile Internet: Enabling Technologies and Services
Apostolis K. Salkintzis and Alexander Poularikas

Soft Computing with MATLAB®
Ali Zilouchian

Wireless Internet: Technologies and Applications
Apostolis K. Salkintzis and Alexander Poularikas

Signal and Image Processing in Navigational Systems
Vyacheslav P. Tuzlukov

Medical Image Analysis Methods
Lena Costaridou

MIMO System Technology for Wireless Communications
George Tsoulos

THE ELECTRICAL ENGINEERING
AND APPLIED SIGNAL PROCESSING SERIES

MIMO SYSTEM TECHNOLOGY FOR WIRELESS COMMUNICATIONS

Edited by
George Tsoulos

Taylor & Francis
Taylor & Francis Group
Boca Raton   London   New York

A CRC title, part of the Taylor & Francis imprint, a member of the
Taylor & Francis Group, the academic division of T&F Informa plc.

Published in 2006 by
CRC Press
Taylor & Francis Group
6000 Broken Sound Parkway NW, Suite 300
Boca Raton, FL 33487-2742

International Standard Book Number-10: 0-8493-4190-6 (Hardcover)
International Standard Book Number-13: 978-0-8493-4190-8 (Hardcover)
Library of Congress Card Number 2005034989

Library of Congress Cataloging-in-Publication Data

MIMO system technology for wireless communications / edited by George Tsoulos.
     p. cm.
  Includes bibliographical references and index.
  ISBN-13: 978-0-8493-4190-8 (0-8493-4190-6 : alk. paper)
  1. MIMO systems. I. Tsoulos, George V., 1968-

TK5103.2.M565 2006
621.384--dc22

2005034989

Taylor & Francis Group
is the Academic Division of Informa plc.

Visit the Taylor & Francis Web site at
http://www.taylorandfrancis.com

and the CRC Press Web site at
http://www.crcpress.com

Pyrsia: Optical Telegraph of Kleoxenos and Dimoklitos

According to the Greek historian Polyvios, *Pyrsia* (communication via optical signals using torches — Greek: *pyrsos*) was invented by Alexandrine engineers Kleoxenos and Dimoklitos (4th century B.C.) and was further improved by Polyvios. The operation was based on the following concept:

1. Separate the Greek alphabet into groups of letters, generating an appropriate matrix.
2. Combine two groups of big torches, visible from a considerable distance with the help of diopters, in order to depict the appropriate letter (e.g., left/right torches represented lines/columns, respectively).

In order to start transmission of a message, two torches were used from one end, and the other end acknowledged that it was ready to receive the message, using two torches as well. Then assuming, for example, a 5×5 grouping of the Greek alphabet, and that lines/columns were represented by the left/right torches, if the letter "Θ" were to be transmitted, it would be represented by two torches on the left and three on the right (see photo below). Obviously, transmission of long messages was achieved by repeating the above method [e.g., the word "ΠΥΡΣΙΑ" is (4,1), (4,5), (4,2), (4,3), (2,4), (1,1)].

Furthermore, different groupings of the alphabet (e.g., 8×3) could also be used along with other methods of encryption (e.g., letters from right to left), for additional security.

Several elements of modern communications (some discussed in this book) are evidenced in this ancient Greek telecommunication system.

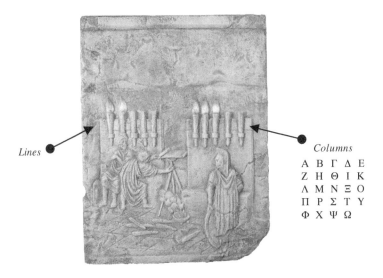

Lines

Columns

Α	Β	Γ	Δ	Ε
Ζ	Η	Θ	Ι	Κ
Λ	Μ	Ν	Ξ	Ο
Π	Ρ	Σ	Τ	Υ
Φ	Χ	Ψ	Ω	

Pyrsia: Optical Telegraph of Kleoxenos and Dimoklitos.
(Photograph from the Museum of Telecommunications of the Greek Organization of Telecommunications.)

Preface

Use of multiple antennas at both ends of wireless links is the result of the natural progression of more than four decades of evolution of adaptive antenna technology. Recent advances have demonstrated that multiple-input-multiple-output (MIMO) wireless systems can achieve impressive increases in overall system performance. The potential to provide the next major leap forward for wireless communications has led this technology to becoming the next frontier of wireless communications. As a result, it has received the attention not only of the international R&D community, but also of the wireless communications industry. This is evidenced from the international standardization efforts in the context of UMTS (e.g., 3GPP) and IEEE 802.11 (for wireless LANs) and also in the context of proposals for next generation (4G and beyond) wireless systems.

This book, *MIMO System Technology for Wireless Communications*, is a compilation of the work of several authors. The goal is not only to provide a complete reference source for readers interested in MIMO systems, but also to provide up-to-date information on several key issues related to the technology. For this reason, considerable effort was made to cover most of the elements of the technology and the related issues, some of which are not currently treated in the available textbooks. Following this direction, the following chapters are included in the book:

The chapters are organized so the reader builds upon the information provided and gradually reaches a point where more complex (system) issues are discussed. In this process, the reader is introduced to propagation modeling, theoretical and realistic performance analyses, space–time codes, different systems, implementation options and limitations (antenna arrays, channel knowledge, etc.), practical system development considerations, field trials, and network planning issues. Also, readers wishing to study further specific aspects of MIMO technology will find the references cited in each chapter particularly useful.*

George V. Tsoulos

---

* Another useful source of information on MIMO technology is a recent two-part special issue from the *IEEE Communications Magazine*:

"Adaptive antennas and MIMO systems for wireless communications — Part I," *IEEE Communications Magazine,* special issue, October 2004, G.V. Tsoulos (Guest Editor).

"Adaptive antennas and MIMO systems for wireless communications — Part II," *IEEE Communications Magazine,* special issue, December 2004, G.V. Tsoulos (Guest Editor).

Acknowledgments

I would like to express my gratitude to all the authors who contributed to this book; to Prof. Alex Poularikas, the editor of the Electrical Engineering and Applied Signal Processing Series of Taylor & Francis; and to the Taylor & Francis publishing staff for their support during this publication project.

The Editor

George Tsoulos graduated from the National Technical University of Athens, Department of Electrical and Computer Engineering, Greece, in 1992 and earned his Ph.D. from the University of Bristol, U.K., in 1997.

From 1994 until 1999 Dr. Tsoulos was a research associate and then a research fellow at the University of Bristol, working in the area of smart antennas for wireless communications. From 1999 until 2002 he was with the Global Technology Group of the PA Consulting Group, in Cambridge, U.K., where he worked for a range of leading companies across the world in the design and analysis of advanced wireless communication systems. In 2003 he joined the Institute of Communication and Computer Systems of the National Technical University of Athens (NTUA), in the context of the EC-funded research program, ENTER.

Dr. Tsoulos currently teaches at the University of Peloponnese, Department of Telecommunication Sciences and Technology, and the Greek Open University. He is also involved in smart antenna and MIMO research activities with the Department of Information Transmission Systems and Materials Technology at NTUA, and the Department of Informatics & Telecommunications, National & Kapodistrian University of Athens.

Contributors

G.E. Athanasiadou Department of Telecommunication Sciences and Technology, University of Peloponnese, Greece

Ernst Bonek Technische Universität Wien, Vienna, Austria

David Browne University of California, Los Angeles, California

Bedri Artug Cetiner Morehead State University, Space Science Center, Morehead, Kentucky

Babak Daneshrad University of California, Los Angeles, California

Mike Fitz University of California, Los Angeles, California

Andreas Forck Fraunhofer Institut for Telecommunications, Heinrich-Hertz-Institute, Berlin, Germany

Ajay Gumalla San Diego Research Center, San Diego, California

Jyri Hämäläinen Nokia Networks, Oulu, Finland

Thomas Haustein Fraunhofer Institute for Telecommunications, Heinrich-Hertz-Institute, Berlin, Germany

Robert W. Heath, Jr. The University of Texas at Austin, Austin, Texas

Christoph Juchems Institut für Angewandte Funksystemtechnik GmbH, Braunschweig, Germany

Volker Jungnickel Fraunhofer Institute for Telecommunications, Heinrich-Hertz-Institute, Berlin, Germany

Dimitra Kaklamani School of Electrical and Computer Engineering, National Technical University of Athens, Athens, Greece

Markku Kuusela Nokia Research Center, Nokia Group, Finland

Stephan Lang University of California, Los Angeles, California

Harry Lee San Diego Research Center, San Diego, California

David J. Love School of Electrical and Computer Engineering, Purdue University, West Lafayette, Indiana

Sergey Loyka   School of Information Technology and Engineering (SITE), University of Ottawa, Ottawa, Canada

Christoph Mecklenbräuker   ftw. Forschungszentrum Telekommunikation Wien, Vienna, Austria

Neelesh B. Mehta   Mitsubishi Electric Research Labs, Cambridge, Massachusetts

Andreas F. Molisch   Mitsubishi Electric Research Labs, Cambridge, Massachusetts *and* Department of Electroscience, Lund University, Lund, Sweden

Juan Mosig   Swiss Federal Institute of Technology, Lausanne, Switzerland

Thomas Neubauer   Symena, Vienna, Austria

Christian Oberli   Department of Electrical Engineering, Pontificia Universidad Católica de Chile, Santiago, Chile

Kari Pajukoski   Nokia Networks, Oulu, Finland

Christian B. Peel   Brigham Young University, Provo, Utah

Alexander D. Poularikas   University of Alabama, Huntsville, Alabama

Raghu Rao   Xilinx Inc., San Jose, California

Quentin H. Spencer   Distribution Control Systems, Inc., Hazelwood, Missouri

Thomas Svantesson   ArrayComm, Inc., San Jose, California

A. Lee Swindlehurst   Brigham Young University, Provo, Utah

Esa Tiirola   Nokia Networks, Oulu, Finland

George Tsoulos   Department of Telecommunication Sciences and Technology, University of Peloponnese, Greece

Jon W. Wallace   Brigham Young University, Provo, Utah

Antonis D. Valkanas   Intracom S.A., Athens, Greece

Risto Wichman   Helsinki University of Technology, Finland

Dimitra Zarbouti   School of Electrical and Computer Engineering, National Technical University of Athens, Athens, Georgia

Weijun Zhu   University of California, Los Angeles, California

Wolfgang Zirwas   Siemens AG, Munich, Germany

Contents

1

Spatio-Temporal Propagation Modeling

G.E. Athanasiadou

CONTENTS

1.1   Introduction

The evolution of wireless communications from analog to digital led to the enhancement of early propagation models, which provided information about power, in order to also consider time delay information. Further consideration of the space domain either with space diversity or smart antennas or, nowadays, MIMO systems has also pushed the evolution of propagation modeling toward more complex spatio-temporal considerations.

In this context, there is a plethora of radiowave propagation models, each developed and used for different applications. The right choice is critical for specific analyzes and depends on system and operational parameters such as the environment, speed, accuracy, cost and ease of use. Generally, experience has shown that for scenarios and parameters that are not very site specific, sufficient accuracy can be achieved at reasonable simulation speeds, with stochastic models. On the other hand, for more site-specific scenarios, more complex ray-tracing models that employ geographical databases are required to provide reasonable accuracy, but at the cost of increased run times.

This chapter starts with models that were developed in an attempt to describe propagation characteristics for space diversity and smart antenna applications. Then models developed to provide the necessary channel information for MIMO applications are discussed. Obviously, measurement campaigns played a key role in the development of these models, and hence, important results from such activities are reported for both cases.

Several references are cited throughout this chapter, but there are some good sources of information that the reader will find particularly useful, such as [1–5].

1.2   Directional Channel Modeling

Figure 1.1 shows that there are three different sources of scattering that affect signal propagation between the base station and the mobile:

1. Scatterers around the mobile station (MS): Similar height or higher than the mobile, hence, the received signal at the mobile usually arrives with wide angular spread.

2. Scatterers around the base station (BS): Generally, the energy arrives at the BS from identifiable clusters, which correspond to different propagation mechanisms (e.g., single reflections from high objects or from rooftop diffractions or street-guided propagation with multiple reflections from the building walls, etc.). For different operational

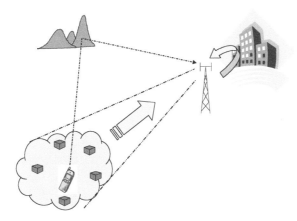

FIGURE 1.1
Scattering sources for radiowave propagation modeling.

scenarios there are different characteristics, e.g., in macrocells (BS at the same level or above the surrounding scatterers), multipath is contained within a relatively small angular spread with relatively low delay spread. In microcells (BS below rooftops), the angular spread is larger than in the macrocell case.

3. Multipath from remote scatterers is another possibility, e.g., in rural operational scenarios. It is usually contained within a very small angular spread but contributes large delay spread.

From Figure 1.1 we can see that each signal from the kth user experiences a different multipath environment, described by the amplitude ($\alpha_{i,k}$), phase ($\psi_{i,k}$), time delay ($\tau_{i,k}$), Doppler shift, and Angle-of-Arrival (AoA) components (time varying). A convenient way to characterize the radio channel is through its channel impulse response, which when modified to consider the AoA of the multipath components for an antenna array, produces the vector channel impulse response:

$$\mathbf{h}(t,\tau) = \sum_{i=1}^{K} \alpha_{i,k} \exp\left(-j\psi_{i,k}\right) \mathbf{a}\left(\varphi_{i,k},\theta_{i,k}\right)\delta\left(t-\tau_{i,k}\right)$$

where $\mathbf{a}(\varphi_{i,k},\theta_{i,k})$ is the complex array response vector of the receive antenna elements (x_m, y_m, z_m) for the ith multipath direction $(\varphi_{i,k}, \theta_{i,k})$ and operating frequency f:

$$\mathbf{a}\left(\varphi_{i,k},\theta_{i,k}\right) =$$

$$\left[1 \cdots \exp(j\frac{2\pi}{\lambda}(m-1)\left(x_m \cos\varphi_{i,k}\sin\theta_{i,k} + y_m \sin\varphi_{i,k}\sin\theta_{i,k} + z_m \cos\theta_{i,k}\right)\right]$$

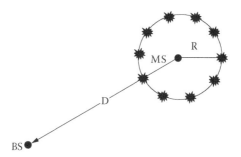

FIGURE 1.2
One-ring-of-scatterers model.

1.2.1 Ring of Scatterers [6]

In this model, the effective scatterers (each effective scatterer comprises a cluster of scatterers) are uniformly spaced on a circular ring around the mobile (Figure 1.2). In the one-ring-of-scatterers models, the BS is assumed to be elevated and therefore not obstructed by local scattering, while the MS is surrounded by scatterers and no Line-of-Sight (LOS) is assumed between BS and MS.

Based on Figure 1.2 and assuming that N scatterers are uniformly placed on a circle with radius R around the mobile at distance D from the base station, the discrete angle of arrivals is [7]:

$$\theta_i = \frac{R}{D}\sin\left(\frac{2\pi}{N}i\right), \quad i = 1, \cdots, N$$

The model was originally used to predict signal correlation as a function of antenna element spacing. Although correlation measurements at the BS and MS are consistent with a narrow/wide angular spread at the BS/MS, respectively, the power delay profile predicted by this model is not generally consistent with measurements. As a result there have been proposals (e.g., [8]) where additional rings of scatterers are added in an attempt to rectify this problem. Furthermore, because small scale fading requires consideration of Doppler shift, [8, 9] have proposed extensions that take into account this effect.

1.2.2 Discrete Uniform Distribution Model [10]

This is a model similar to Lee's ring of scatterers model [6]. Figure 1.3 shows that it considers scatterers evenly located within a narrow beamwidth centered around the direction of the mobile.

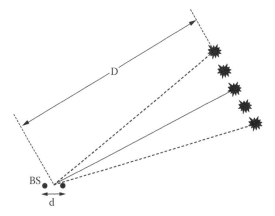

FIGURE 1.3
Discrete uniform distribution of scatterers.

Analysis performed by the same author suggests that, due to the fact that in practice the AoA is discrete, a continuous AoA distribution (reported as Gaussian for rural-suburban environments) will underestimate the correlation that exists between the antenna array elements.

1.2.3 Geometrically Based Single-Bounce (GBSB) Statistical Channel Models

This kind of model assumes that scatterers are placed in a region according to a spatial scatterer density function. From the location of each scatterer, the AoA, Time-of-Arrival (ToA), and signal amplitude can be determined along with the relevant probability density functions. In order to make the calculations easier, two important assumptions are usually made.

First, the signal undergoes only one reflection when it travels from the MS to the BS. Then, all scatterers confined within the scattering area are isotropic re-radiating elements, with random complex scattering coefficients (yet, in practice it is rather difficult to assign realistic scattering coefficients).

1.2.3.1 Geometrically Based Circular Model (Macrocell Model)

The idea behind this model [11–13] is shown in Figure 1.4. It assumes that the scatterers lie within radius R_m about the mobile. The joint ToA and AoA pdfs are reported in [2], where it is also shown that the circular model predicts a relatively high probability of multipath components with small excess delay along the line of sight, i.e., the model is more suitable for large cell environments, where all the multipath components lie within a small angular spread. The appropriate values for the radius of the scattering depend on the macrocellular type of environment (urban, dense urban, etc.) and the model can be "tuned" based on results from measurements.

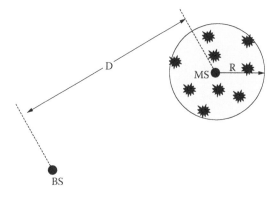

FIGURE 1.4
Geometrically based circular model.

1.2.3.2 *Geometrically Based Elliptical Model (Microcell Model)*

Figure 1.5 shows that this model places the base station and the mobile at the foci of an ellipse and distributes the scatterers uniformly within the ellipse. It was proposed in [14] for applications in microcellular environments, since in such environments antenna heights are relatively low, and hence, multipath scattering near the base station and the mobile is just as likely. The semi-major and semi-minor axes are calculated as a function of the maximum ToA to be considered, and this determines both the delay spread and angular spread of the channel.

According to [15], this model produces a high probability of scatterers with minimum excess delay along the LOS.

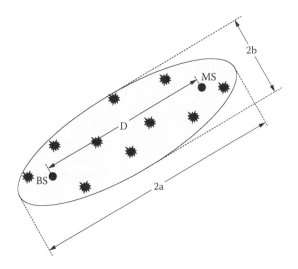

FIGURE 1.5
Geometrically based elliptical model.

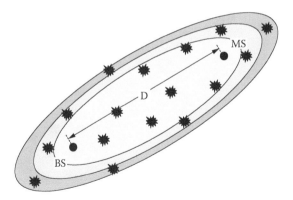

FIGURE 1.6
Elliptical subregions model.

1.2.3.3 Elliptical Subregions Model [16]

If the distribution of scatterers in elliptical subregions (corresponding to different ranges of excess delay) is considered, then we get the model shown in Figure 1.6.

This approach is similar to the Geometrical Based Elliptical Model proposed by [14], where the scatterers are uniformly distributed within the entire ellipse, while here the ellipse is subdivided into a number of elliptical subregions. A number of scatterers is then selected within each subregion employing a Poisson random variable, with its mean chosen to match the measured time characteristics. Furthermore, due to multiple reflection points of the scatterers, the multipath components arrive in clusters.

1.2.3.4 Elliptical Model with Dense Discrete Scatterers [17]

This is a wide-band spatial model, where diffuse scattering is modeled using dense discrete scatterers. The transmitter and receiver define the focal points of ellipses with constant propagation delay. The scatterers are distributed on ellipsoids. To simplify the model, only the intersection of the ellipsoid and the ground is used. A uniform distribution of scatterers is assumed, although the calculated distribution function for the scatterers around the mobile is not uniform (it has two spikes). Two channel models are defined — a rural macrocell model and an urban microcell model.

1.2.4 Gaussian Wide Sense Stationary Uncorrelated Scattering (GWSSUS)

This is a statistical model that makes assumptions about the received signal vector [18–20] (Figure 1.7). Although scatterers are grouped in clusters, spatio-temporal multipath is not resolvable within a cluster, i.e., the narrowband channel assumption is satisfied. Nevertheless, frequency-selective fading channels can be modeled by including multiple clusters.

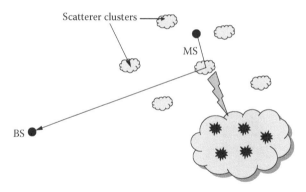

FIGURE 1.7
The GWSSUS model.

The steering vector(s), due to multipaths from the kth cluster, can be expressed in this case as the sum of all the contributions (K) from the scatterers within the kth cluster:

$$\mathbf{s}_k = \sum_{i=1}^{K} \alpha_{i,k} \exp\left(-j\psi_{i,k}\right) \mathbf{a}\left(\varphi_{i,k}\right)$$

where $\alpha_{i,k}$ is the amplitude, $\psi_{i,k}$ the phase and $\varphi_{i,k}$ the relative AoA of the ith multipath in the kth cluster, and $\mathbf{a}(\varphi)$ is the complex array response vector of the receive antenna elements for direction and operating frequency f.

If the number of scatterers in each φ cluster is sufficiently large, (≥ 10 from [18]), then this sum is Gaussian distributed (central limit theorem). Also, wide sense stationarity is assumed, which leads to the steering vector being multi-variate Gaussian distributed and, hence, described by its mean and covariance matrix, presented in [19].

A special case of the GWSSUS model is the Gaussian AoA model [21], where only one cluster is considered and the AoA statistics are assumed to be Gaussian distributed about the direction of the cluster (narrowband flat fading model).

1.2.5   A Stochastic Spatio-Temporal Propagation Model (SSTPM)

The GWSSUS channel model does not impose any conditions on the spatial distribution of the received power, hence, requiring additional information in order to be used in space–time studies. The GBSB channel models, on the other hand, do not provide information about the temporal evolution of the generated channel characteristics, resulting in consecutive snapshots being un-correlated, an un-realistic assumption.

In order to avoid these problems, [22] and [23] proposed a hybrid approach, which combined these two classes of channel models (the GBSB

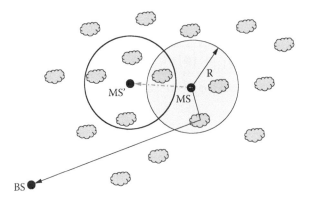

FIGURE 1.8
Stochastic spatio-temporal propagation model concept.

and GWSSUS) and further considered time variations associated with the movement of the mobile (non-stationary scenarios), as shown in Figure 1.8.

The scattering area in the presented model has a circular shape (although the general elliptical case can also be considered), and a uniform distribution of scatterers is assumed (given area density of scatterers, which depends on the type of environment, i.e., urban, suburban, rural). As shown in Figure 1.8, the mobile is located at the center of the circular scattering area, which moves together with the mobile. The scatterers are positioned at fixed locations throughout a large area (e.g., the cell), but active scatterers are only those within the circular scattering area. As a result, the number of multipaths is not fixed as the mobile moves but follows a random process (Poisson distributed). The expected number of scatterers depends on the scatterers' area density and the scattering area size, and hence, it depends on the type of the operational environment. The pdf of the "scatterers' lifetime" is also calculated from the authors, and is shown to depend on the size of the scattering area and the mobile speed.

The angular spread of the channel is described by the probability density function (pdf) of the angle of arrival and was derived in [24]:

$$f_\varphi(\varphi_k) = \begin{cases} \dfrac{2}{\pi}\cos(\varphi_k)\sqrt{\cos^2(\varphi_k)+F^2-1} & \text{for} -\varphi_{k\max} < \varphi_k < \varphi_{k\max} \\[2em] 0 & \text{elsewhere} \end{cases}$$

where $F = R/l_{BM}$, i.e., the cluster radius over the distance between the MS and the BS. The above expression is a special case of the formula derived in [24] for the elliptical case:

$$f_\varphi(\varphi) =$$

$$\begin{cases} \dfrac{1}{\Omega} \dfrac{\cos(\varphi)\sqrt{\cos^2(\varphi) - \left(\cos^2(\varphi) + E^{-2}\sin^2(\varphi)\right)\left(1 - F^2\right)}}{\left(E^2\cos^2(\varphi) + \sin^2(\varphi)\right)^2} & \text{for } -\varphi_{max} < \varphi < \varphi_{max} \\[2em] 0 & \text{elsewhere} \end{cases}$$

where a, b, d are the major and minor axes of the ellipse and the distance from the BS, respectively, $F = a/d$ and $E = b/a$, and Ω can be calculated from

$$\int_{-\varphi_{max}}^{\varphi_{max}} f_\varphi(\varphi)\, d\varphi = 1$$

As mentioned above, the consecutive channel characteristics have correlated temporal characteristics, hence allowing for the direct calculation of the correlation functions.

1.2.6 Extended Saleh-Valenzuela Model

In [25] extensive indoor measurements using a system that collects simultaneous time and angle of arrival data at 7 GHz have shown a clustering pattern in the time-angle multipath data. The model proposed for indoor environments employs the clustered "double Poisson" time-of-arrival model proposed by Saleh and Valenzuela [26], with statistical independence between time and angle. The mean angles of each cluster are distributed uniformly over all angles. The distribution of arrivals within clusters is approximately Laplacian, with standard deviations ranging from 22° to 26°.

1.2.7 Gaussian Scatter Density Model

The spatial characteristics of the radio channel are studied in [27], based on work presented in [28], with the Gaussian Scatter Density Model (GSDM). Starting from a Gaussian distribution of scatterers around a mobile station, expressions are provided for the pdfs of AoA, the power azimuth spectrum, the ToA, and the time delay spread, all as seen from a BS. Expressions are also provided for the rms delay spread, the rms angular spread and the spatial cross correlation function.

Also, with an appropriate choice of the standard deviation of the scattering region, the Gaussian density model is suitable for environments with small angular spreads (macrocells) or with large angular spreads (picocells). When the scattering width is small compared with the distance between the BS and the MS, it is shown that the pdf in the AoA reduces to the Gaussian function.

1.2.8   Gaussian AoA — Laplacian Power Azimuth Spectrum

A statistical model of azimuthal and temporal dispersion in mobile radio channels is presented in [29]. Based on field trial results, it is proposed that for typical urban environments, the power azimuth spectrum (PAS) can be modeled by a Laplacian function, while the power delay spectrum (PDS) can be modeled by a one-sided exponential decaying function.

Positive correlation between azimuth and delay spread also was observed from the measurements. Hence, propagation environments with high angular spread also have high delay spread, and vice versa. A significant increase in both angular and delay spread was also observed with the lowering of the BS antenna below rooftop height.

The pdf of the azimuth of the multipath rays was found to match a Gaussian function, while the pdf of their delays matched a one-sided exponential decay function.

For bad urban environments, a two-cluster model was proposed, with the PAS described by the sum of two Laplacian functions and the PDS by the sum of two exponentially decaying functions. The power azimuth-delay spectrum could not be expressed as a product of the PDS and PAS in this case, contrary to the typical urban scenario.

1.2.9   Semi-Elliptical Geometrical Model

In [30, 31] a geometric-based channel model with a semi-elliptical coverage area is applied in order to determine a new PAS model (called the secant square PAS model). This model is appropriate for environments where the BS is on a building with height close to that of the surrounding building scatterers, and the authors show that it is a better fit to the employed experimental results (from the European project TSUNAMI II) than the previously proposed Laplacian model for these scenarios.

1.2.10   Lognormal Distribution of Local Angular Spread (AS), Delay Spread (DS), and Shadow Fading (Macrocells)

In [32] the joint statistical behavior of the random variables describing the local AS, the local DS, and the shadow fading component is studied using measurements in macrocellular (including NLOS) scenarios. It is found that a log-normal distribution provides an accurate fit of the measured pdfs for all three parameters. Their spatial autocorrelation functions follow an exponential decay for typical and bad urban environments, and a double exponential decay in suburban environments. The decorrelation distance of the AS,

DS, and shadow fading is observed to be nearly identical within each environment class. The fact that the pdf and the spatial autocorrelation function of the three parameters are identical indicates that the propagation mechanisms leading to these effects are strongly related.

1.3 MIMO Propagation Modeling

The discussion up to now has considered single-input-multiple-output (SIMO) propagation models. This section focuses more on the multiple-input-multiple-output (MIMO) propagation models. Figure 1.9 shows a MIMO system with M transmit and N receive antenna elements. The general expression for the baseband signal in this case can be expressed as:

$$\mathbf{y}(t) = \mathbf{H}(t) * \mathbf{s}(t) + \mathbf{n}(t)$$

where $\mathbf{y}(t)$ is the received, $\mathbf{s}(t)$ the transmitted, $\mathbf{n}(t)$ the noise signal, * denotes convolution, and $\mathbf{H}(t)$ is the $M \times N$ channel matrix. If the signal bandwidth is narrow enough so that the channel can be considered approximately constant over frequency, then we get the narrowband MIMO channel matrix. Otherwise, we get the wideband MIMO channel matrix.

Based on a similar approach to that presented in the previous section, this section addresses two major categories of models:

- Deterministic
- Stochastic (parametric, geometric, correlation)

Some of the models presented in the following sections are based on propagation measurements ([1] and [5] include good surveys on space–time measurement campaigns).

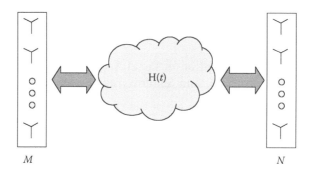

M N

FIGURE 1.9
MIMO system concept.

1.3.1 Deterministic Propagation Modeling with Ray Tracing

This modeling approach has evolved from SISO to SIMO and, more recently, MIMO scenarios, and hence, it is discussed here, in the MIMO propagation modeling section, since the last represents the more general approach.

Ray tracing is a technique based on Geometrical Optics (*GO*), an easily applied approximate method for estimating a high-frequency electromagnetic field [33]. The dissipating energy is considered to be radiating in infinitesimally small tubes, often called *rays*. These rays are normal to the surface of equal signal power, lie in the direction of propagation, and travel in straight lines, provided that the refractive index is constant. Their amplitude is governed by the conservation of energy flux in the ray tube. In *GO*, only direct, reflected and refracted rays are considered, and consequently abrupt transition areas occur, corresponding to the boundaries of the regions where these rays exist. The *Geometrical Theory of Diffraction (GTD)* [34] and its uniform extension, the *Uniform GTD (UTD)* [35, 36], complement the *GO* theory by introducing a new type of rays, known as the diffracted rays. The purpose of these rays is to remove the field discontinuities and to introduce proper field corrections, especially in the zero field areas predicted by *GO*.

The extended *Fermat principle* and the *principle of local field* are two basic concepts extensively used by the ray models [37]. While the original Fermat principle states that a *GO* ray follows the shortest path from a source point to a field point, the extended Fermat principle also includes the diffracted rays and states that these rays follow shortest path as well. The principle of the local field states that the high frequency boundary processes, such as reflection, refraction and diffraction, depend only on the electrical and geometrical properties of the scatterer in the immediate neighborhood of the point of interaction. The corresponding amplitude, phase and direction of a ray following reflections, refractions and diffractions can be calculated using a combination of Snell's laws, *UTD* and Maxwell's equations [33].

In a wireless communication system, the signal arriving at the receiving antenna consists of several multipath components, each of which is the result of the interaction of the transmitted waves with the surrounding environment. The application of *GO* and *UTD* to a given propagation problem requires that the given configuration is decomposed into simple geometrical configurations for which the reflection, transmission and diffraction coefficients can be calculated. All rays contributing significantly to the channel characterization at the examined position must be traced, and the complex impulse response $h(t)$ of the radio channel is then found as the sum of these contributions [38]. Here, the received signal is formed by N time delayed impulses (rays), each represented by an attenuated and phase-shifted version of the original transmitted impulse. For each ray, the model computes the amplitude A_n, the arrival time τ_n and phase ϑ_n. According to the objects encountered by the ith ray, its complex received field amplitude E_i (V/m) is given by:

$$E_i = E_0 \, f_{ti} \, f_{ri} \left\{ \prod_j R_j \prod_k T_k \prod_l A_l(s',s) D_l \right\} \frac{e^{-jkd}}{d}$$

where E_0 represents the reference field, f_{ti} and f_{ri} the transmitting and receiving antenna field radiation patterns in the direction of the ray, R_j the reflection coefficient for the jth reflector, T_k the wall transmission coefficient for the kth transmission, D_l the diffraction coefficient for the lth diffracting wedge and e^{-jkd} the propagation phase factor due to the path length d ($k = 2\pi/\lambda$, with λ the wavelength). The diffraction coefficients are also multiplied by a factor $A_l(s',s)$ which finds the correct spatial attenuation of the diffracted rays, given the $1/d$ dependence in the last term. An advantage of ray-tracing models over other propagation models is the ability to incorporate antenna radiation patterns and particularly to consider the effect of the radiation pattern on each ray individually.

In order to trace rays that are generated and launched from the transmitting antenna, two methods have been developed: the *imaging technique* and the *ray launching technique*. The *imaging technique* (e.g., [39, 40]) is based on the electromagnetic theory of images and works by generating an image table for each BS location, considering all the various wall reflection, transmission and diffraction permutations that are possible in a given area. The image information is then stored and used to compute the channel characteristics at each mobile location. In the *ray launching* approach (e.g., [41, 42]), rays are sent out at various angles and their paths are traced until a certain power threshold is reached. The number of rays considered and the distance from the transmitter to the receiver location determined the available spatial resolution and the accuracy of the model.

In the image-based models presented in [40] (for microcells), [43] (for indoor) and [44, 45] (for macrocells), the geometry of each ray is examined in three-dimensional (3D) space, and hence, both the azimuth and the elevation angles of arrival at the antennas are available. Moreover, the 3D antenna radiation patterns can be used and steered in any direction in space so that the channel can be examined for any antenna orientation. Note that the model works with the electromagnetic field of the rays and, hence, uses the radiation patterns of the field components. This feature, in conjunction with the fact that all reflections, transmissions and diffractions are computed using 3D vector mathematics, makes the models very useful in the study of different antenna polarizations and the examination of depolarization effects.

Since the field components can be calculated for each antenna element separately, as explained above, the MIMO channel matrix can be generated, as shown in Figure 1.10 and Figure 1.11. Figure 1.10 shows the geographical database of the area and examples of 2D-3D multipath visualization. It can be seen that ray tracing offers site-specific information for the radio channel characteristics and, hence, provides more accurate predictions. Figure 1.11 shows an example of the 3D impulse responses (amplitude-delay-AoA) for a MIMO scenario (between Tx element m and Rx element n). It shows polar plots for the AoA (azimuth) vs. power at the BS (a) and MS (b), the AoA (elevation) vs. power at the BS (c) and MS (d) and, finally, the ToA vs. power (e).

The application of ray-tracing models to study several aspects of propagation modeling has proven to be a popular method.

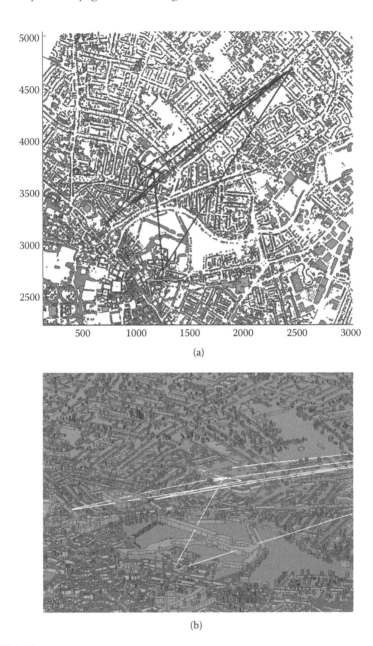

(a)

(b)

FIGURE 1.10
Map of the environment under investigation with 2D (a) and 3D (b) multipath examples.

In [46, 47] the spatial characteristics of microcellular environments are studied. Results showed that the signal is not uniformly distributed in the spatial domain, but instead is contained in a few narrow clusters. Although the number of clusters increases under NLOS conditions, for both LOS and NLOS positions, 90% of the power is contained within two clusters. Also, although the angular

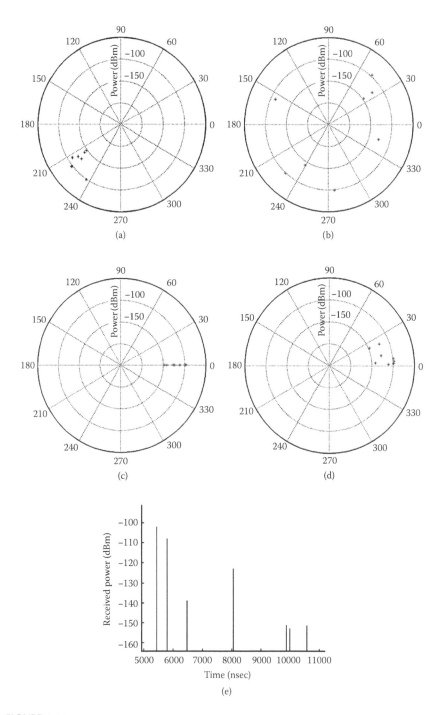

FIGURE 1.11
Example of ray-tracing 3D impulse responses for a MIMO scenario: AoA (azimuth) vs. power at the BS (a) and MS (b). AoA (elevation) vs. power at the BS (c) and MS (d). (e) ToA vs. power.

spread and the number of multipath rays almost doubles under NLOS conditions, there are still only two important clusters that contain most of the power with angular spread less than the beamwidth of an eight element array.

In [48] the spatial and temporal characteristics of 60GHz indoor channels were analyzed. Multipath components were resolved in time by using a sliding correlator with 10 ns resolution and in space by sweeping a directional antenna with 7° half power beamwidth in the azimuthal direction. Power delay profiles (PDPs) and power angle profiles (PAPs) were measured in various indoor and short-range outdoor environments. The measurement results confirm that the majority of the multipath components can be determined from image-based, ray-tracing techniques for LOS applications. For non-LOS propagation through walls, the metallic structure of composite walls must be considered. Also, statistical parameters of received power, AoA and ToA, were calculated from the measurements, which agreed well with the theoretical expectations.

Furthermore, ray-tracing models have also been used to produce the data sets required for the statistical evaluation of the parameters of stochastic models, as in [49]. The advantage of using deterministic predictions instead of field trial measurements is mainly that large data sets can be easily produced for many different test environments. Also, when using field trial results in order to produce a statistical model, the influence of the measurement antennas is included in the results and cannot be eliminated afterward. In [49], a stochastic model for the indoor mobile propagation channel is presented. The channel is described by multipath components, including 3D angles of arrival at the antennas. By relating the angle of arrival to the direct line between transmitter and receiver, a universal modeling approach, which is independent of the actual geometry, becomes possible. In each modeling step, path properties change according to the movement of the radio stations. The appearance and disappearance of multipath components are modeled by a genetic process.

1.3.2 Stochastic Propagation Modeling

There have been several such models proposed, analyzed and discussed the last few years; some of the most representative are briefly mentioned here:

- Wideband Directional Channel Model (WDCM), [50]. Geometrically based, parameterized for macro, microcells, circular-elliptical scattering area, single bounce.

- Geometrical MIMO channel model based on SISO parameters [51, 52]. Macrocellular broadband fixed wireless, multiple delay ellipses, circle of local scatterers around the MS, double bounces. To better represent such environments, two local rings are introduced in this model: a disc of exclusion, representing a scatterer-free area around the BS, and a smaller circular ring surrounding the CPE including a subset of the scatterers in the first ellipse (see Figure 1.12). The channel matrix is then calculated using a ray-based approach.

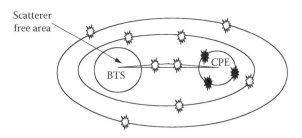

FIGURE 1.12
Physical combined model for macrocellular scenarios.

- Generic MIMO model [53]. Double scattering, far clusters, wave-guiding, guidelines to select the proper distribution of scatterers.
- Double Directional Channel Model (DDCM) [54, 55]. Parametric stochastic model, parameters through spatial scatterer distribution [1]. Tapped delay line with each multipath having complex characteristics (amplitude, delay, AoA).
- Indoor MIMO models:
  - EU IST METRA (Multi-Element Transmit Receive Antennas) project: A stochastic MIMO radio channel model for non-line-of-sight (NLOS) scenarios was proposed on the basis of the power correlation matrix of the MIMO radio channel [56].
  - EU IST SATURN (Smart Antenna Technology in Universal Broadband Wireless Networks) project: Based on the statistical characteristics of the measured data, both narrowband [57] and wideband [58] statistical models for NLOS MIMO propagation channels were developed.
- One (two) ring(s) MIMO channel models [59, 60]. Extension of [6] for MIMO channel modeling.
- One ring MIMO model with Von Mises angular distribution [61]. Von Mises angular pdf at the MS.
- Distributed scattering MIMO model [62]. Outdoor, narrowband, one group of scatterers near the BS and one near the MS.

1.3.2.1   The 3GPP MIMO Channel Model [63]*

A serious attempt to unify the propagation modeling approach for space–time models started a few years ago with the European Scientific Action COST259 [1]. This work received even more attention when adopted as the basis for spatial channel modeling in the context of the international standardization body of 3GPP. Naturally, it was influenced by many different

---

* 3GPP TSs and TRs are the property of ARIB, ATIS, ETSI, CCSA, TTA, and TTC who jointly own the copyright in them. They are subject to further modifications and are therefore provided to you "as is" for information purposes only. Further use is strictly prohibited.

proposals discussed in 3GPP (see technical documents for spatial channel model). This section describes in greater detail the parametric stochastic model that has been adopted recently by the 3GPP.

The combined 3GPP-3GPP2 spatial channel model ad-hoc group has specified parameters and methods for spatial channel modeling. The scope is the development of specifications for system-level evaluation with emphasis on the physical parameters, and link-level evaluation, defined only for calibration purposes (not for evaluation or comparisons).

As such, the following section presents the key characteristics of the spatial channel model that has been adopted for system-level simulation studies. (A detailed description is available in [63].)

1.3.2.1.1 *Spatial Channel Model for Simulations*

For an S element BS array and a U element MS array, the channel coefficients for one of N multipath components (note that these components are not necessarily resolvable in the time domain, since the time difference between successive paths may be less than a chip period) are given by an S-by-U matrix of complex amplitudes. The channel matrix for the nth multipath component ($n = 1, ..., N$), $\mathbf{H}_n(t)$, is a function of time, because the complex amplitude undergoes fast fading due to the movement of the MS.

The overall procedure for generating the channel matrices consists of three basic steps:

1. Specify an environment (either macro urban or macro suburban, or micro).
2. Obtain the simulation parameters associated with each environment.
3. Generate the channel coefficients based on these parameters.

Figure 1.13 shows the angular parameters used in the model. The following definitions are used:

Ω_{BS}	BS antenna array orientation, defined as the difference between the broadside of the BS array and the absolute North (**N**) reference direction
θ_{BS}	LOS AoD direction between the BS and MS, with respect to the broadside of the BS array
$\delta_{n,AoD}$	AoD for the nth path ($n = 1 ... N$) with respect to the LOS AoD θ_{BS}
$\Delta_{n,m,AoD}$	Offset for the mth sub-path ($m = 1 ... M$) of the nth path with respect to $\delta_{n,AoD}$
$\theta_{n,m,AoD}$	Absolute AoD for the mth sub-path of the nth path at the BS with respect to the BS broadside
Ω_{MS}	MS antenna array orientation, defined as the difference between the broadside of the MS array and the absolute North reference direction
θ_{MS}	Angle between the BS-MS LOS and the MS broadside
$\delta_{n,AoA}$	AoA for the nth path with respect to the LOS AoA θ_{MS}
$\Delta_{n,m,AoA}$	Offset for the mth sub-path of the nth path with respect to $\delta_{n,AoA}$
$\theta_{n,m,AoA}$	Absolute AoA for the mth sub-path of the nth path at the MS with respect to the MS broadside
v	MS velocity vector
θ_v	Angle of velocity vector with respect to the MS broadside: $\theta_v = arg(v)$

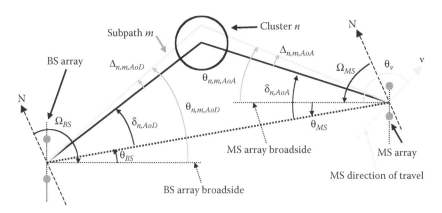

FIGURE 1.13
BS and MS angle parameters for the 3GPP model [63].

For the simulations some general assumptions are made (Table 1.1):

a. Uplink-downlink reciprocity, i.e., the AoD/AoA values are identical between the uplink and downlink.

b. For FDD systems, random sub-path phases between uplink and downlink are uncorrelated. For TDD systems, the phases are fully correlated.

c. Shadowing among different mobiles is uncorrelated.

d. The spatial channel model allows any type of antenna configuration (e.g., with size smaller than the shadowing coherence distance). In order to compare algorithms, reference antenna configurations based on uniform linear array with 0.5, 4 and 10 wavelength inter-element spacing are used.

e. The composite AS (angle spread), DS (delay spread) and SF(shadow fading) may be correlated parameters depending on the channel scenario, and are applied to all the sectors or antennas of a given base. The AS is composed of 6×20 sub-paths, and each has a precise angle of departure, which corresponds to an antenna gain from each BS antenna. The SF is a bulk parameter and is common among all the BS antennas or sectors.

f. The elevation spread is not modeled.

g. To allow comparisons of different antenna scenarios, the transmit power of a single antenna case is the same as the total transmit power of a multiple antenna case.

h. The generation of the channel coefficients assumes linear arrays (although the procedure can be generalized for other array configurations).

TABLE 1.1

Environment Parameters [63]

Channel Scenario	Suburban Macro	Urban Macro	Urban Micro
Number of paths (N)	6	6	6
Number of sub-paths (M) per path	20	20	20
Mean AS at BS	$E(\sigma_{AS}) = 5°$	$E(\sigma_{AS}) = 8°, 15°$	NLOS: $E(\sigma_{AS}) = 19°$
AS at BS as a lognormal random variable	$\mu_{AS} = 0.69$	$8° \; \mu_{AS} = 0.810$	N/A
$\sigma_{AS} = 10^{\circ}(\varepsilon_{AS}x + \mu_{AS}), x \sim \eta(0, 1)$	$\varepsilon_{AS} = 0.13$		$\varepsilon_{AS} = 0.34$
		$15° \; \mu_{AS} = 1.18$	$\varepsilon_{AS} = 0.210$
$r_{AS} = \sigma_{AoD}/\sigma_{AS}$	1.2	1.3	N/A
Per path AS at BS (fixed)	2°	2°	5° (LOS and NLOS)
BS per path AoD Distribution standard distribution	$\eta(0, \sigma^2_{AoD})$ where $\sigma_{AoD} = r_{AS}\sigma_{AS}$	$\eta(0, \sigma^2_{AoD})$ where $\sigma_{AoD} = r_{AS}\sigma_{AS}$	$U(-40°, 40°)$
Mean AS at MS	$E(\sigma_{AS, MS}) = 68°$	$E(\sigma_{AS, MS}) = 68°$	$E(\sigma_{AS, MS}) = 68°$
Per path AS at MS (fixed)	35°	35°	35°
MS Per path AoA Distribution	$\eta(0, \sigma^2_{AoA}(P_r))$	$\eta(0, \sigma^2_{AoA}(P_r))$	$\eta(0, \sigma^2_{AoA}(P_r))$
Delay spread as a lognormal random variable	$\mu_{DS} = -6.80$	$\mu_{DS} = -6.18$	N/A
$\sigma_{DS} = 10^{\circ}(\varepsilon_{DS}x + \mu_{DS}), x \sim \eta(0, 1)$	$\varepsilon_{DS} = 0.288$	$\varepsilon_{DS} = 0.18$	
Mean total RMS delay spread	$E(\sigma_{DS}) = 0.17 \; \mu s$	$E(\sigma_{DS}) = 0.65 \; \mu s$	$E(\sigma_{DS}) = 0.251 \; \mu s$ (output)
$r_{DS} = \sigma_{delays}/\sigma_{DS}$	1.4	1.7	N/A
Distribution for path delays			$U(0, 1.2 \; \mu s)$
Lognormal shadowing standard deviation, σ_{SF}	8dB	8dB	NLOS: 10 dB LOS: 4 dB
Pathloss model (dB), d is in meters	$31.5 + 35\log_{10}(d)$	$34.5 + 35\log_{10}(d)$	NLOS: $34.53 + 38\log_{10}(d)$ LOS: $30.18 + 6\log_{10}(d)$

The following are assumptions for the macrocell suburban and urban environments:

a. The macrocell pathloss is based on the modified COST231 Hata urban propagation model:

$$PL[dB] = \left(44.9 - 6.55\log_{10}\left(h_{BS}\right)\right)\log_{10}\left(\frac{d}{1000}\right) + 45.5 + \left(35.46 - 1.1h_{MS}\right)\log_{10}\left(f_c\right)$$

$$-13.82\log_{10}\left(h_{BS}\right) + 0.7h_{MS} + C$$

where h_{BS} and h_{MS} are the BS and MS antenna height (in meters), f_C is carrier frequency (in MHz), d is the distance between MS and BS (in meters) and C is a constant factor ($C = 0$ dB for suburban macro and $C = 3$ dB for urban micro).

b. Site-to-site SF corelation is 0.5.

The following are assumptions for the microcell environment:

a. The microcell NLOS pathloss is based on the COST 231 Walfish-Ikegami NLOS model with the following parameters: BS antenna height 12.5 m, building height 12 m, building to building distance 50 m, street width 25 m, MS antenna height 1.5 m, orientation 30 deg for all paths, and selection of metropolitan center. With these parameters, the equation simplifies to:

$$PL(dB) = -55.9 + 38*\log_{10}(d) + (24.5 + 1.5*f_c/925)*\log_{10}(f_c)$$

The resulting pathloss at 1900 MHz is: $PL(dB) = 34.53 + 38*\log_{10}(d)$, where d is in meters. The distance d is at least 20 m. A bulk log normal shadowing applying to all sub-paths has a standard deviation of 10 dB.

The microcell LOS pathloss is based on the COST 231 Walfish-Ikegami street canyon model with the same parameters as in the NLOS case. The pathloss is

$$PL(dB) = -35.4 + 26*\log_{10}(d) + 20*\log_{10}(f_c)$$

The resulting pathloss at 1900 MHz is $PL(dB) = 30.18 + 26*\log_{10}(d)$, where d is in meters. The distance d is at least 20 m. A bulk log normal shadowing applying to all sub-paths has a standard deviation of 4 dB.

b. Site-to-site correlation is $\zeta = 0.5$.

In [63], the algorithm that generates the model parameters is explained step by step for the different operational environments. These parameters are then used to generate the channel coefficients. For an S element linear BS array and a U element linear MS array, the channel coefficients for one of N multipath components are given by a U-by-S matrix of complex amplitudes. The channel matrix for the nth multipath component ($n = 1,...,N$) is denoted as $\mathbf{H}_n(t)$. The (u,s)th component ($s = 1,...,S$; $u = 1,...,U$) of $\mathbf{H}_n(t)$ is given by

$$h_{u,s,n}(t) = \sqrt{\frac{P_n \sigma_{SF}}{M}} \sum_{m=1}^{M} \left(\begin{array}{l} \sqrt{G_{BS}\left(\theta_{n,m,AoD}\right)} \exp\left(j\left[kd_s \sin\left(\theta_{n,m,AoD}\right) + \Phi_{n,m}\right]\right) \times \\ \sqrt{G_{MS}\left(\theta_{n,m,AoA}\right)} \exp\left(jkd_u \sin\left(\theta_{n,m,AoA}\right)\right) \times \\ \exp\left(jk\|\mathbf{v}\|\cos\left(\theta_{n,m,AoA} - \theta_v\right)t\right) \end{array} \right)$$

where

P_n	is the power of the nth path.
σ_{SF}	is the lognormal shadow fading, applied as a bulk parameter to the n paths for a given drop.
M	is the number of sub-paths per path.
$\theta_{n,m,AoD}$	is the AoD for the mth sub-path of the nth path.
$\theta_{n,m,AoA}$	is the AoA for the mth sub-path of the nth path.
$G_{BS}(\theta_{n,m,AoD})$	is the BS antenna gain of each array element.
$G_{MS}(\theta_{n,m,AoA})$	is the MS antenna gain of each array element.
k	is the wave number $2\pi/\lambda$ where λ is the carrier wavelength in meters.
d_s	is the distance in meters from BS antenna element s from the reference ($s = 1$) antenna. For the reference antenna $s = 1$, $d_1 = 0$.
d_u	is the distance in meters from MS antenna element u from the reference ($u = 1$) antenna. For the reference antenna $u = 1$, $d_1 = 0$.
$\Phi_{n,m}$	is the phase of the mth sub-path of the nth path.
$\|\mathbf{v}\|$	is the magnitude of the MS velocity vector.
θ_v	is the angle of the MS velocity vector.

The pathloss and the log normal shadowing is applied as bulk parameters to each of the sub-path components of the n path components of the channel.

Also in [63], a method of using polarized antennas in the above model is presented and the resulting channel characterization is given. Finally, the model includes options for far scatterer clusters, line of sight and urban canyon.

References

1. L.M. Correia, Ed. 2001. *Wireless Flexible Personalized Communications*, COST 259: European Cooperation in Mobile Radio Research, Chichester: John Wiley & Sons.
2. R.B. Ertel, P. Cardieri, K.W. Sowerby, T.S. Rappaport, and J.H. Reed. 1998. "Overview of spatial channel models for antenna array communication systems," *IEEE Personal Communications Magazine*, Vol. 5, No. 1, Feb., pp. 10–22.
3. K. Yu and B. Ottersten. 2002. "Models for MIMO propagation channels: a review," *Wireless Communications and Mobile Computing*, Vol. 2: pp. 653–666.
4. J.C. Liberti and T.S. Rappaport. 1999. *Smart Antennas for Wireless Communications*, Upper Saddle River, NJ: Prentice Hall.

5. L. Schumacher, L.T. Berger, J. Ramiro-Moreno, and T.B. Sorensen. 2004. "Propagation characterization and MIMO channel modeling for 3G," in *Adaptive Antenna Arrays*, S. Chandran, ed., Heidelberg: Springer-Verlag, pp. 377–393.

6. W.C.Y. Lee. 1973. "Effects on correlation between two mobile radio basestation antennas," *IEEE Transactions on Communications*, Vol. 21, No. 11, 1214–1224.

7. S.C. Swales, "Spectrum efficient cellular base station antenna architectures," Ph.D. Thesis, University of Bristol, U.K., 1990.

8. S.P. Stapleton, X. Carbo, and T. McKeen. 1994. "Tracking and diversity for a mobile communications base station array antenna," *Proc. IEEE VTC*, pp. 1695–1699.

9. S.P. Stapleton, X. Carbo, and T.McKeen, 1994. "Spatial channel simulator for phased arrays," *Proc. IEEE VTC*, pp. 1789–1792.

10. D. Asztely. 1996. "On antenna arrays in mobile communication systems: fast fading and GSM base station receiver algorithm," Ph.D. Dissertation, Royal Institute of Technology, Stockholm, Sweden, March.

11. P. Petrus, J.H. Reed, and T.S. Rappaport. 1997. "Effects of directional antennas at the base station on the Doppler spectrum," *IEEE Communications Letters*, Vol. 1, No. 2, March.

12. P. Petrus, J.H. Reed, and T.S. Rappaport. 1996. "Geometrically based statistical channel model for macrocellular mobile environment," *IEEE GLOBECOM*, pp. 1197–1201.

13. J. Fuhl, A. Molisch, and E. Bonek. 1998. "Unified channel model for mobile radio systems with smart antennas," *IEE Proc. — Radar, Sonar. Navig.* Vol. 145, No. 1, Feb., pp. 32–41.

14. J. Liberty and T. Rappaport. 1996. "A geometrically based model for line-of-sight multipath radio channels," *IEEE VTC*, pp. 844–848.

15. R.B. Ertel and J.H. Reed. 1999. "Angle and time of arrival statistics for circular and elliptical scattering models," *IEEE Journal on Selected Areas in Communications*, Vol. 17, Nov., pp. 1829–1840.

16. M. Lu, T. Lo, and J. Litva. 1997. "A physical spatio-temporal model of multipath propagation channels," *IEEE VTC*, pp. 180–184.

17. O. Norklit and J.B. Andersen. 1998. "Diffuse channel model and experimental results for array antennas in mobile environments," *IEEE Transactions on Antennas and Propagation*, Vol. 46, No. 6, June, pp. 834–840.

18. P. Zetterberg and B. Ottersten. 1994. "The spectrum efficiency of a basestation antenna array system for spatially selective transmission," *IEEE VTC*.

19. P. Zetterberg and P.L. Espensen. 1996. "A downlink beam steering technique for GSM/DCS1800/PCS1900," *IEEE PIMRC*, Taipei, Taiwan, October.

20. P. Mogensen, P. Zetterberg, H. Dam, P.L. Espensen, and F. Fredekirsen. 1997. "Algorithms and antenna array recommendations — part 1," ACTS TSUNAMI deliverable, January.

21. B. Ottersten. 1995. "Spatial Division Multiple Access (SDMA) in wireless Communications," Proc. Nordic Radio Symp.

22. R.J. Piechocki and G.V. Tsoulos. 1999. "Combined GWSSUS and GBSR channel model with temporal variations," Joint COST259/260 Workshop, April, Vienna, Austria.

23. R.J. Piechocki, J.P. McGeehan, and G.V. Tsoulos. 2001. "A new stochastic spatio-temporal propagation model (SSTPM) for mobile communications with antenna arrays," *IEEE Transactions on Communications*, Vol. 49, No. 5, May, pp. 855–862.

24. R.J. Piechocki, G.V. Tsoulos, and J.P. McGeehan. 1998. "Simple general formula for PDF of angle of arrival in large cell operational environment," *IEE Electronics Letters*, Vol. 34, September 3, pp. 1784–1785.

25. Q.H. Spencer, B.D. Jeffs, M.A. Jensen, and A.L. Swindlehurst. 2000. "Modeling the statistical time and angle of arrival characteristics of an indoor multipath channel," *IEEE Journal on Selected Areas in Communications*, Vol. 18, No. 3, March, pp. 347–360.

26. A.A. Saleh and R.A. Valenzuela. 1987. "A statistical model for indoor multipath propagation," *IEEE Journal on Selected Areas in Communications*, Vol. SAC-5, No. 2, Feb.

27. R. Janaswamy. 2002. "Angle and time of arrival statistics for the Gaussian scatter density model," *IEEE Transactions on Wireless Communications*, Vol. 1, July, pp. 488–497.

28. M.P. Lotter and P. van Rooyen. 1999. "Modeling spatial aspects of cellular CDMA/ SDMA systems," *IEEE Communications Letters*, Vol. 3, May, pp. 128–131.

29. K.I. Pedersen, P.E. Mogensen, and B.H. Fleury. 2000. "A stochastic model of the temporal and azimuthal dispersion seen at the base station in outdoor propagation environments," *IEEE Transactions on Vehicular Technology*, Vol. 49, No. 2, March, pp. 437–447.

30. S.A. Zekavat and C.R. Nassar. 2003. "Power-azimuth-spectrum modeling for antenna array systems: a geometric-based approach," *IEEE Transactions on Antennas and Propagation*, Vol. 51, No. 12, Dec., pp. 3292–3294.

31. S.A. Zekavat and C.R. Nassar. 2002. "Smart antenna arrays with oscillating beam patterns: characterization of transmit diversity using semi-elliptic-coverage geometric-based stochastic channel model," *IEEE Transactions on Communications*, Vol. 50, Oct., pp. 1549–1556.

32. A. Algans, K.I. Pedersen, and P.E. Mogensen. 2002. "Experimental analysis of the joint statistical properties of azimuth spread, delay spread, and shadow fading," *IEEE Journal on Selected Areas in Communications*, Vol. 20, No. 3, April, pp. 523–531.

33. C.A. Balanis. 1989. *Advanced Engineering Electromagnetics*, New York: John Wiley & Sons.

34. J.B. Keller. 1962. "Geometrical Theory of Diffraction," *Journal of the Optical Society of America*, Vol. 52, Feb., pp. 116–130.

35. R.G. Kouyoumjian and P.H. Pathak. 1974. "A uniform geometric theory of diffraction for an edge on a perfectly conducting surface," *IEEE Proceedings*, Vol. 62, No. 11, Nov., pp. 1448–1461.

36. R.J. Luebbers. 1984. "Finite conductivity uniform GTD versus knife edge diffraction in prediction of propagation path loss," *IEEE Trans. on Antennas & Propagation*, Vol. AP-32, No. 1, Jan., pp. 70–76.

37. H. Bach. 1977. *Modern Topics in Electromagnetics and Antennas*, London: Peter Peregrinus Ltd., Chap. 5.

38. G.L. Turin, et al. 1972. "A statistical model for urban multipath propagation," *IEEE Transactions on Vehicular Technology*, Vol. VT-21, Feb., pp. 1–9.

39. R.A. Valenzuela. 1993. "A ray tracing approach to predicting indoor wireless transmission," *IEEE VTC '93*, New Jersey, May 18–20, pp. 214–218.

40. G.E. Athanasiadou, A.R. Nix, and J.P. McGeehan. 2000. "A microcellular ray-tracing propagation model and evaluation of its narrowband and wideband predictions," *IEEE Journal on Selected Areas in Communications, Wireless Communications Series*, Vol. 18, No. 3, March, pp. 322–335.

41. S.Y. Seidel and T.S. Rappaport. 1994. "Site-specific propagation prediction for wireless in-building personal communications system design," *IEEE Transactions on Vehicular Technology*, Vol. 43, No. 4, Nov., pp. 1058–1066.

42. K.R. Schaubach and N.J. Davis IV. 1994. "Microcellular radio-channel propagation prediction," *IEEE Antennas and Propagation Magazine*, Aug., pp. 25–34.

43. G.E. Athanasiadou and A.R. Nix. 2000. "A novel 3D indoor ray-tracing propagation model: the path generator and evaluation of narrowband and wideband predictions," *IEEE Transactions on Vehicular Technology*, Vol. 49, No. 4, July, pp. 1152–1168.

44. G.E. Athanasiadou, I.J. Wassell, and C.L. Hong. 2004. "Deterministic propagation modeling and measurements for the broadband fixed wireless access channel," *IEEE VTCF 2004*, Los Angeles, Sept 26–29.

45. G.E. Athanasiadou and I.J. Wassell. 2005. "Comparisons of ray tracing predictions and field trial results for broadband fixed wireless access scenarios," *WSEAS Transactions on Communications*, Issue 8, Vol. 4, Aug., pp. 717–721.

46. G.V. Tsoulos, G.E. Athanasiadou, and M.A. Beach. 1998. "Adaptive antennas for microcellular and mixed cell environments with DS-CDMA." *Wireless Personal Communications Journal*, Special Issue on CDMA for Universal Personal Communications Systems, Kluwer Academic Publishers, Vol. 7, No. 2/3, Aug., pp. 147–169.

47. G.V. Tsoulos and G.E. Athanasiadou. 2002. "On the application of adaptive antennas to microcellular environments: radio channel characteristics and system performance," *IEEE Transactions on Vehicular Technology*, Vol. 51, No. 1, Jan., pp. 1–16.

48. H. Xu, V. Kukshya, and T.S. Rappaport. 2002. "Spatial and temporal characteristics of 60GHz indoor channels," *IEEE Journal on Selected Areas in Communications*, Vol. 20, No. 3, April, pp. 620–630.

49. T. Zwick, C. Fischer, D. Didascalou, and W. Wiesbeck. 2000. "A stochastic spatial channel model based on wave-propagation modeling," *IEEE Journal on Selected Areas in Communications*, Vol. 18, No. 1, Jan., pp. 6–15.

50. M. Marques and L. Correia. 2001. "A wideband directional channel model for UMTS microcells," *IEEE PIMRC*, pp. B-122–B-126.

51. C. Oestges, V. Erceg, and A. Paulraj. 2003. "A physical scattering model for MIMO macrocellular broadband wireless channels," *IEEE Journal on Selected Areas in Communications*, Vol. 21, No. 5, June, pp. 721–729.

52. C. Oestges, V. Erceg, and A. Paulraj. 2004. "Propagation modeling of MIMO multipolarized fixed wireless channels," *IEEE Transactions on Vehicular Technology*, Vol. 53, No. 3, May, pp. 644–654.

53. A. Molisch. 2002. "A generic model for MIMO wireless propagation channels," *IEEE ICC*, pp. 277–282.

54. M. Steinbauer, A. Molisch, and E. Bonek. 2001. "The double-directional channel model," *IEEE Antennas and Propagation Magazine*, Vol. 43, No. 4, pp. 51–63.

55. K. Kalliola, H. Laitinen, P. Vainikainen, M. Toeltsch, J. Laurila, and E. Bonek. 2003. "3-D double-directional radio channel characterization for urban macrocellular applications," *IEEE Transactions on Antennas and Propagation*, Vol. 51, No. 11, Nov., pp. 3122–3133.

56. K.I. Pedersen, J.B. Andersen, J.P. Kermoal, and P. Mogensen. 2000. "A stochastic multiple-input-multiple-output radio channel model for evaluation of space-time coding algorithms," *IEEE VTC*, Fall, pp. 893–897.

57. K. Yu, M. Bengtsson, B. Ottersten, D. McNamara, P. Karlsson, and M. Beach. 2001. "Second order statistics of NLOS indoor MIMO channels based on 5.2GHz measurements," *IEEE Globecom 2001*, Vol. 1, Nov., pp. 156–160.

58. K. Yu, M. Bengtsson, B. Ottersten, D. McNamara, P. Karlsson, and M. Beach. 2002. "A wideband statistical model for NLOS indoor MIMO channels," *IEEE VTC*, Spring, Vol. 1, pp. 370–374.

59. D.-S. Shiu, G.J. Foschini, M.J. Gans, and J.M. Kahn. 2000. "Fading correlation and its effect on the capacity of multielement antenna systems," *IEEE Transactions on Communications*, Vol. 48, No. 3, pp. 502–513.

60. D.-S. Shiu. 2000. *Wireless Communications Using Dual Antenna Arrays*, Norwell, MA: Kluwer Academic Publishers.

61. A. Abdi and M. Kaveh. 2002. "A space-time correlation model for multielement antenna systems in mobile fading channels," *IEEE Journal on Selected Areas in Communications*, Vol. 20, No. 3, pp. 550–560.

62. D. Gesbert, H. Bolcskei, D. Gore, and A. Paulraj. 2000. "MIMO wireless channels: capacity and performance," *Global Telecommunications Conference*, Vol. 2, Nov., pp. 1083–1088.

63. 3rd Generation Partnership Project, Technical Specification Group Radio Access Network. *Spatial channel model for Multiple Input Multiple Output (MIMO) simulations*, 3GPP TR 25.996 V6.1.0 (2003-09) Technical Report.

2

Theory and Practice of MIMO Wireless Communication Systems

Dimitra Zarbouti, George Tsoulos, and Dimitra Kaklamani

CONTENTS

Summary

This chapter introduces the principles of MIMO systems employing the necessary mathematical analysis to consider the achieved capacity performance. In this context, flat fading across time and frequency is considered, and the Rayleigh model is employed for describing the wireless channel. Furthermore, the case of spatial selective fading is examined by considering LOS propagation, with the Ricean model.

The mathematical representation of the MIMO system is performed through a complex matrix, which depends on the scenario considered each time (i.e., flat or selective spatial fading). The capacity achieved by the MIMO channel in all the above cases is studied with the use of the Shannon extended capacity formula. The capacity performance results, developed from the simulations performed, are related to the number of the multiple antenna elements that the Rx and the Tx are equipped with, the distance between them and the degree of correlation evidenced.

2.1   Shannon's Capacity Formula

Shannon's capacity formula approximated theoretically the maximum achievable transmission rate for a given channel with bandwidth B, transmitted signal power P and single side noise spectrum N_o, based on the assumption that the channel is white Gaussian (i.e., fading and interference effects are not considered explicitly).

$$C = B \cdot \log_2\left(1 + \frac{P}{N_o B}\right) \tag{2.1}$$

In practice, this is considered to be a SISO scenario (single input, single output) and Equation 2.1 gives an upper limit for the achieved error-free SISO transmission rate. If the transmission rate is less than C bits/sec (bps), then an appropriate coding scheme exists that could lead to reliable and error-free transmission. On the contrary, if the transmission rate is more than C bps, then the received signal, regardless of the robustness of the employed code, will involve bit errors.

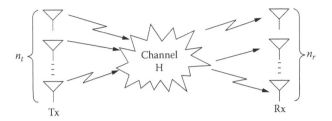

FIGURE 2.1
The MIMO channel.

2.2 Extended Capacity Formula for MIMO Channels

For the case of multiple antennas at both the receiver and the transmitter ends (Figure 2.1), the channel exhibits multiple inputs and multiple outputs and its capacity can be estimated by the extended Shannon's capacity formula, as described below.

2.2.1 General Capacity Formula

We consider an antenna array with n_t elements at the transmitter and an antenna array with n_r elements at the receiver. The impulse response of the channel between the jth transmitter element and the ith receiver element is denoted as $h_{i,j}(\tau,t)$. The MIMO channel can then be described by the $n_r \times n_t$ $\mathbf{H}(\tau,t)$ matrix:

$$\mathbf{H}(\tau,t) = \begin{bmatrix} h_{1,1}(\tau,t) & h_{1,2}(\tau,t) & \cdots & h_{1,n_t}(\tau,t) \\ h_{2,1}(\tau,t) & h_{2,2}(\tau,t) & \cdots & h_{2,n_t}(\tau,t) \\ \vdots & \vdots & \ddots & \vdots \\ h_{n_r,1}(\tau,t) & h_{n_r,2}(\tau,t) & \cdots & h_{M_R,n_t}(\tau,t) \end{bmatrix} \tag{2.2}$$

The matrix elements are complex numbers that correspond to the attenuation and phase shift that the wireless channel introduces to the signal reaching the receiver with delay τ. The input-output notation of the MIMO system can now be expressed by the following equation:

$$\mathbf{y}(t) = \mathbf{H}(\tau,t) \otimes \mathbf{s}(t) + \mathbf{u}(t) \tag{2.3}$$

where $\otimes$ denotes convolution, $\mathbf{s}(t)$ is a $n_t \times 1$ vector corresponding to the n_t transmitted signals, $\mathbf{y}(t)$ is a $n_r \times 1$ vector corresponding to the n_r received signals and $\mathbf{u}(t)$ is the additive white noise.

If we assume that the transmitted signal bandwidth is narrow enough that the channel response can be treated as flat across frequency, then the discrete-time description corresponding to Equation 2.3 is

$$\mathbf{r}_\tau = \mathbf{H}\mathbf{s}_\tau + \mathbf{u}_\tau \tag{2.4}$$

The capacity of a MIMO channel was proved in [1, 4] that can be estimated by the following equation:

$$C = \max_{tr(\mathbf{R}_{ss}) \leq p} \log_2 \left[\det\left(\mathbf{I} + \mathbf{H}\mathbf{R}_{ss}\mathbf{H}^H\right) \right] \tag{2.5}$$

where $\mathbf{H}$ is the $n_r \times n_t$ channel matrix, $\mathbf{R}_{ss}$ is the covariance matrix of the transmitted vector $\mathbf{s}$, $\mathbf{H}^H$ is the transpose conjugate of the $\mathbf{H}$ matrix and p is the maximum normalized transmit power. Equation 2.5 is the result of extended theoretical calculations, and its practical use is not obvious. Nevertheless, we can perform linear transformations at both the transmitter and receiver converting the MIMO channel to $n = \min(n_r, n_t)$ SISO subchannels (given that the channel is linear) and, hence, reach more insightful results. These transformations can be found in [1] and are briefly described in the following section.

2.2.2 Transformation of the MIMO Channel into *n* SISO Subchannels

Every matrix $\mathbf{H} \in \mathbb{C}^{n_r \times n_t}$ can be decomposed accordingly to its singular values. Suppose that for the aforementioned channel matrix this transformation is given by Equation 2.6.

$$\mathbf{H} = \mathbf{U}\mathbf{D}\mathbf{V}^H \tag{2.6}$$

where the matrices $\mathbf{U}$, $\mathbf{V}$ are unitaries of dimensions $n_r \times n_r$ and $n_t \times n_t$ accordingly, while $\mathbf{D}$ is a non-negative diagonal matrix of dimensions $n_r \times n_t$. The diagonal elements of matrix $\mathbf{D}$ are the singular values of the channel matrix $\mathbf{H}$. The algorithm of singular value decomposition that provides the above transformation can be found in [2].

The operations that lead to the linear transformation of the channel into $n = \min(n_r, n_t)$ SISO subchannels are described as follows: First, the transmitter multiplies the signal to be transmitted $\mathbf{x}_\tau$ with the matrix $\mathbf{V}$, the receiver multiplies the received signal $\mathbf{r}_\tau$ and noise with the conjugate transpose of the matrix $\mathbf{U}$. The above are presented in Equation 2.7 through Equation 2.9.

$$\mathbf{s}_\tau = \mathbf{V} \cdot \mathbf{x}_\tau \tag{2.7}$$

$$\mathbf{y}_\tau = \mathbf{U}^H \cdot \mathbf{r}_\tau \tag{2.8}$$

$$\mathbf{n}_\tau = \mathbf{U}^H \cdot \mathbf{u}_\tau \tag{2.9}$$

Substituting Equation 2.4 into Equation 2.8 gives:

$$\mathbf{y}_\tau = \mathbf{U}^H \cdot \mathbf{r}_\tau \Rightarrow$$

$$\mathbf{y}_\tau = \mathbf{U}^H \mathbf{H} \mathbf{s}_\tau + \mathbf{U}^H \mathbf{u}_\tau \overset{(2.7),(2.9)}{\Rightarrow}$$

$$\mathbf{y}_\tau = \mathbf{U}^H \mathbf{H} \mathbf{V} \mathbf{x}_\tau + \mathbf{n}_\tau \overset{(2.6)}{\Rightarrow}$$

$$\mathbf{y}_\tau = \mathbf{U}^H \mathbf{U} \mathbf{D} \mathbf{V}^H \mathbf{V} \mathbf{x}_\tau + \mathbf{n}_\tau$$

Since $\mathbf{U}$ and $\mathbf{V}$ are unitary matrices, they satisfy $\mathbf{U}^H \mathbf{U} = \mathbf{I}_{n_r}$, $\mathbf{V}^H \mathbf{V} = \mathbf{I}_{n_t}$ and hence:

$$\mathbf{y}_\tau = \mathbf{D} \mathbf{x}_\tau + \mathbf{n}_\tau \tag{2.10}$$

Each component of the received vector $\mathbf{y}_\tau$ can be written as:

$$\mathbf{y}_\tau^k = \varepsilon_k \mathbf{x}_\tau^k + \mathbf{n}_\tau^k \tag{2.11}$$

where ε_k are the singular values of matrix $\mathbf{H}$ according to the transformation that took place above. Equation 2.11 implies that the initial (n_r, n_t) MIMO system has been transformed into $n = \min(n_r, n_t)$ SISO subchannels, as illustrated in Figure 2.2.

The above analysis of a Multiple Antenna Element (MEA) system capacity is presented in [1]. It was proven in [1] that the total capacity of n SISO

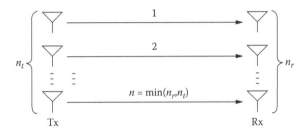

FIGURE 2.2
Conversion of the MIMO channel into n SISO subchannels.

subchannels is the sum of the individual capacities and as a result the total MIMO capacity is:

$$C = \sum_{k=1}^{n} \log_2\left(1 + p_k \varepsilon_k^2\right) \tag{2.12}$$

where p_k is the power allocated to the kth subchannel and ε_k^2 is its power gain. We notice that according to the singular value decomposition algorithm ε_k^2, $k = 1, 2, \ldots, n$ are the eigenvalues of the $\mathbf{HH}^H$ matrix, which are always non-negative. Furthermore, regardless of the power allocation algorithm used, p_k must satisfy

$$\sum_{k=1}^{n} p_k \leq p$$

because of the wanted power constraint.

At this point, there are two cases of particular interest that need further consideration: the knowledge (or not) by the transmitter of the Channel State Information (CSI). These are described in the following sections.

2.2.3   No CSI at the Transmitter

Considering Equation 2.12, we notice that the achieved capacity depends on the algorithm used for allocating power to each subchannel. The theoretical analysis assumes the channel state known at the receiver. This assumption stands correct since the receiver usually performs tracking methods in order to obtain CSI, however the same consideration does not apply to the transmitter.

When the channel is not known at the transmitter, the transmitting signal **s** is chosen to be statistically non-preferential, which implies that the n_t components of the transmitted signal are independent and equi-powered at the transmit antennas. Hence, the power allocated to each of the n_t subchannels is $p_k = p/n_t$. Applying the last expression to Equation 2.5 gives:

$$C = \log_2\left[\det\left(\mathbf{I} + \frac{p}{n_t}\mathbf{HH}^H\right)\right] \tag{2.13}$$

or

$$C = \sum_{k=1}^{n} \log_2\left(1 + \frac{p}{n_t}\varepsilon_k^2\right) \tag{2.14}$$

Equation 2.14 can be produced from Equation 2.13 as described in greater detail in Appendix 2A.

2.2.4 CSI at the Transmitter

In cases in which the transmitter has knowledge of the channel, it can perform optimum combining methods during the power allocation process. In that way, the SISO subchannel that contributes to the information transfer the most is supplied with more power.

One method to calculate the optimum power allocation to the n subchannels is to employ the waterpouring algorithm (a detailed discussion of this algorithm can be found in [3]).

Considering the assumption of CSI at the transmitter, we can proceed to the following capacity formula.

$$C = \sum_{k=1}^{n} \log_2 \left(1 + \frac{\gamma_k \cdot p}{n_t} \varepsilon_k^2 \right) \tag{2.15}$$

The difference between Equation 2.14 and Equation 2.15 is the coefficient γ_k that corresponds to the amount of power that is assigned to the kth subchannel. This coefficient is given by:

$$\gamma_k = E \left\{ |s_k|^2 \right\} \tag{2.16}$$

and satisfies the constraint

$$\sum_{k=1}^{n} \gamma_k = n_t.$$

The goal with the waterpouring algorithm is to find the optimum γ_k that maximizes the capacity given in Equation 2.15.

2.2.5 Channel Estimation at the Transmitter

As mentioned earlier, the CSI is not usually available at the transmitter. In order for the transmitter to obtain the CSI, two basic methods are used: the first is based on feedback and the second on the reciprocity principle.

In the first method the forward channel is calculated by the receiver and information is sent back to the transmitter through the reverse channel. This method does not function properly if the channel is changing fast. In that case, in order for the transmitter to get the right CSI, more frequent estimation and feedback are needed. As a result, the overhead for the reverse channel becomes prohibitive. According to the reciprocity principle, the forward and reverse channels are identical when the time, frequency and antenna locations are the same. Based on this principle the transmitter may use the CSI obtained by the reverse link for the forward link. The main problem with this method emerges when frequency duplex schemes are employed.

2.3   Remarks on the Extended Shannon Capacity Formula

In this section we will use mathematical tools in order to derive the theoretical upper and lower bounds of MIMO capacity. The algebraic expressions used, as well as the assumptions considered here, are summarized in Section 2.3.1.

In Section 2.3.3 we introduce the Effective Degrees of Freedom (EDoF), which we will use for the simulation justification in Section 2.7.

2.3.1   Bounds on MIMO Capacity

The lower and upper bounds of MIMO capacity were first derived in [1]. We proceed with a short description of those bounds. Four basic assumptions are considered in the following, summarized here for simplicity.

- The transmitter has no previous knowledge of the channel.
- The parallel subchannels produced by the decomposition of the MIMO channel are independent.
- The wireless channel is submitted to Rayleigh fading.
- The transmitter antenna array elements are less than the receiver's antenna elements ($n_t < n_r$).

In addition, we cite four mathematical expressions that will be used for deriving the wanted capacity bounds.

$$\det\left(DD^H + RR^H\right) \geq \prod_\ell \left(\left|D_{\ell,\ell}\right|^2 + \left|R_{\ell,\ell}\right|^2\right) \qquad (2.17)$$

where matrices D and R are diagonal and upper-triangular, respectively.

$$\det\left(I + AB\right) = \det\left(I + BA\right) \qquad (2.18)$$

$$\det\left(I + QAQ^H\right) = \det\left(I + A\right) \qquad (2.19)$$

where A, B are square matrices and Q is a unitary matrix.

$$\det(X) \le \prod_{\ell} X_{\ell,\ell} \qquad (2.20)$$

where X is a non-negative definite matrix.

Since the channel is submitted to Rayleigh fading, the channel matrix **H** is given by $\mathbf{H_W}$, which is referred to as spatially white matrix. The elements of $\mathbf{H_W}$ can be modeled as zero mean circularly symmetric complex Gaussian (ZMCSCG) random variables. The $\mathbf{H_W}$ has particular statistical properties that can be found in [3, 12].

We consider the transformation $\mathbf{H_W} = \mathbf{QR}$, where **Q** is a unitary and **R** is an upper-triangular matrix. This tranformation is referred to as Householder Transformation [5]. According to this transformation the elements of **R** above the main diagonal are statistically independent, while the magnitude of the main diagonal entries are chi-squared distributed with $2n_r$, $2(n_r - 2 + 1)$, ..., $2(n_r - n_t + 1)$ degrees of freedom.

Using Equation 2.13:

$$C = \log_2 \left[\det\left(\mathbf{I} + \frac{p}{n_t} \mathbf{HH}^H \right) \right] = \log_2 \left[\det\left(\mathbf{I} + \frac{p}{n_t} \mathbf{QRR}^H \mathbf{Q}^H \right) \right]^{(2.19)} \Rightarrow$$

$$C = \log_2 \left[\det\left(\mathbf{I} + \frac{p}{n_t} \mathbf{RR}^H \right) \right]^{(2.17)} \ge \log_2 \left[\prod_{\ell} \left(1 + \frac{p}{n_t} |R_{\ell,\ell}|^2 \right) \right] \Rightarrow$$

$$C \ge \sum_{\ell=1}^{n_t} \log_2 \left(1 + \frac{p}{n_t} |R_{\ell,\ell}|^2 \right) \qquad (2.21)$$

Equation 2.21 corresponds to the lower capacity bound and practically shows that this bound is defined by the sum of the capacities of n_t independent subchannels with power gains that follow the chi-square distribution with $2n_r$, $2n_r - 2$, ..., $2(n_r - n_t + 1)$ degrees of freedom.

In order to find the upper bound of the capacity, Equation 2.13 is used again:

$$C \le \sum_{\ell=1}^{n_t} \log_2 \left(1 + \frac{p}{n_t} \left(|R_{\ell,\ell}|^2 + \sum_{m=\ell+1}^{n_t} |R_{\ell,m}|^2 \right) \right) \qquad (2.22)$$

The upper bound of the capacity is the sum of the capacities of n_t independent subchannels, with power gains chi-squared distributed and with degrees of freedom $2(n_r + n_t - 1)$, $2(n_r + n_t - 3)$, ..., $2(n_r - n_t + 1)$. The difference of the mean values of the upper and lower bounds is less than 1b/s/Hz.

2.3.2 Capacity of Orthogonal Channels

It is interesting to study the case where the capacity of the MIMO channel is maximized. We consider the simple case of $n_r = n_t = n$, along with a fixed total power transfer through the SISO subchannels (i.e., $\sum_{k=1}^{n} \varepsilon_k^2 = a$, where a is a constant). The capacity in Equation 2.14 is concave in the variables ε_k^2 (k = 1, 2, ..., n) and, as a result, it is maximized when $\varepsilon_k^2 = \varepsilon_i^2 = a/n$, ($k$, i = 1, 2, ..., n). The last equation reveals that the $\mathbf{HH}^H$ matrix has n equal eigenvalues. Hence, H must be an orthogoal matrix, i.e., $\mathbf{HH}^H = \mathbf{H}^H\mathbf{H} = (a/n)\mathbf{I}_n$. Substituting $\mathbf{HH}^H = (a/n)\mathbf{I}_n$ into Equation 2.13:

$$C = \log_2\left[\det\left(\mathbf{I} + \frac{pa}{n^2}\mathbf{I}_n\right)\right] \Rightarrow C = n\cdot\log_2\left(1 + \frac{pa}{n^2}\right) \tag{2.23}$$

If $\left\|\mathbf{H}_{i,j}\right\|^2 = 1$, the matrix H satisfies $\mathbf{HH}^H = n\mathbf{I}_n$, hence, Equation 2.13 becomes:

$$C = \log_2\left[\det\left(\mathbf{I} + \frac{p}{n}n\mathbf{I}_n\right)\right] \Rightarrow C = n\cdot\log_2\left(1 + p\right) \tag{2.24}$$

The last equation indicates that the capacity of an orthogonal MIMO channel is n times the capacity of the SISO channel.

2.3.3 Effective Degrees of Freedom

Based on Equation 2.14, we can assume that in high SNR regime, capacity can increase linearly with n. Specifically, for high SNR regime ($\varepsilon_k^2 p/n) \gg 1$ Equation 2.14 becomes:

$$C \approx \sum_{k=1}^{n} \log_2\left(\frac{p}{n}\varepsilon_k^2\right) \tag{2.25}$$

However, this assumption is not always confirmed. For some subchannels ($\varepsilon_k^2 p/n$) is much smaller than one, and as a result the information transferred by these channels is nearly zero. This phenomenon is present in at least three cases:

- when the transmission is serviced through a low-powered device
- when there is a long-range communication application
- when there is strong fading correlation between subchannels

In the last case, the fading induced to a certain subchannel may cause the minimization of its corresponding ε_k^2.

In practice, the aforementioned cases are very likely to happen, therefore, the concept of EDoF is introduced. Intuitively, the EDoF value corresponds to the number of subchannels that actually contribute to the information transfer. In a more mathematical approach, the EDoF value indicates the non-zero singular values of the channel matrix **H**. A more detailed description of the EDoF can be found in [1].

2.4 Capacity of SIMO — MISO Channels

Single input, multiple output (SIMO) and multiple input, single output (MISO) channels are special cases of MIMO channels. In this paragraph we discuss the capacity formulas for the case of SIMO and MISO channels.

For a SIMO channel $n_t = 1$, so $n = \min(n_r, n_t) = 1$; hence, the CSI at the transmitter does not affect the SIMO channel capacity:

$$C_{SIMO} = \log_2\left(1 + p \cdot \varepsilon_1^2\right) \tag{2.26}$$

If we consider $|h_i|^2 = 1$ then $\varepsilon_1^2 = n_r$ (the proof can be found in Appendix 2B). Hence Equation 2.26 becomes:

$$C_{SIMO} = \log_2\left(1 + p \cdot n_r\right) \tag{2.27}$$

For a MISO channel $n_r = 1$ and $n = \min(n_r, n_t) = 1$. With no CSI at the transmitter, the capacity formula can be expressed as:

$$C_{MISO} = \log_2\left(1 + \frac{p}{n_t} \cdot \varepsilon_1^2\right) \tag{2.28}$$

If we make the same assumption as earlier and consider that $|h_i|^2 = 1$ then $\varepsilon_1^2 = n_t$ (the proof is cited in Appendix 2B); hence, Equation 2.28 becomes:

$$C_{MISO} = \log_2\left(1 + p\right) \tag{2.29}$$

Comparing Equation 2.29 and Equation 2.27 we can see that $C_{SIMO} > C_{MISO}$. This is because the transmitter, as opposed to the receiver, cannot exploit the antenna array gain since it has no CSI and, as a result, cannot retrieve the receiver's direction.

2.5   Stochastic Channels

In order to use the aforementioned capacity formulas, it is necessary to obtain the channel matrix expression. There are many spatial channel models that are used for this purpose ([6–11]).

The simulations that take place in Section 2.7 consider a stochastic channel approach. Specifically, the Rayleigh and the Rice models are used. The description of these models is cited below. Consequently, under the stochastic channel consideration, the capacity achieved becomes a random variable, and in order to study its behavior, we use stochastic quantities, as described below.

2.5.1   Ergodic Capacity

The ergodic capacity of a MIMO channel is the ensemble average of the information rate over the distribution of the elements of the channel matrix **H** [3], and it is given by:

$$\bar{C} = E\{I\} \tag{2.30}$$

When there is no CSI at the transmitter, we can substitute Equation 2.13 into Equation 2.30, so the ergodic capacity is given by

$$\bar{C} = E\left\{\log_2\left(\det\left(\mathbf{I}+\frac{p}{n_t}\mathbf{H}\mathbf{H}^H\right)\right)\right\} \tag{2.31}$$

Whereas with CSI at the transmitter we use Equation 2.15, and the ergodic capacity is given by:

$$\bar{C} = E\left\{\sum_{k=1}^{n}\log_2\left(\mathbf{I}+\frac{p}{n_t}\gamma_k\varepsilon_k^2\right)\right\} \tag{2.32}$$

Figure 2.3 illustrates the ergodic capacity for different antenna configurations as a function of the SNR, when the channel is unknown at the transmitter. As expected, the ergodic capacity increases with SNR. In addition, the ergodic capacity of a SIMO channel appears to be greater than the ergodic capacity of a MISO channel. The reason for this behavior, as previously explained, lies in the fact that the transmitter cannot exploit the antenna array gain since it has no CSI.

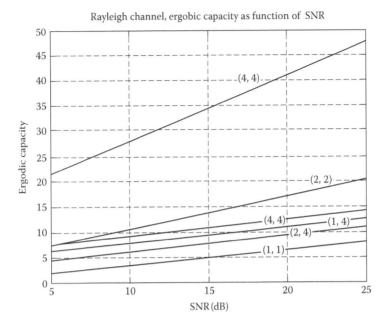

FIGURE 2.3
Ergodic capacity as a function of SNR and number of elements.

2.5.2 Outage Capacity

The outage capacity quantifies the level of capacity performance guaranteed with a certain level of reliability [3, 12]. For $q\%$ outage capacity $C_{out,q}$, indicates the maximum capacity level the system can achieve with probability $(100 - q)\%$. For stochastic channels we can observe that there is always a possibility of outage for a given MIMO system realization, regardless of the wanted rate. Hence, there is a tradeoff between the system's outage capacity and the achieved information rate.

2.6 MIMO Capacity with Rice and Rayleigh Channels

In this section we discuss the capacity expressions for the cases of Rayleigh and Ricean channels, as well as when spatial fading correlation is induced to the signal due to the limited distance between the array elements (the transmitter is considered blind for the discussion below, i.e., does not have CSI).

When the wireless environment is characterized by strong multipath activity, then the number of paths between the transmitter and receiver allows the use of the central limit theorem [13] and the envelope of the received signal follows the Rayleigh distribution. However, in cases that the location of buildings leads to the street waveguide propagation phenomenon, and in areas near the base station where a line of sight (LOS) component may dominate, the Ricean distribution is more suitable. The receiver in that scenario "sees" a dominant signal component along with lower power components caused by multipath. The dominant component that reaches the receiver may not be the result of LOS propagation, e.g., the dominant component may be the mean value of strong multipaths caused by large scatterers. The Ricean *K*-factor of the channel is defined as the ratio of the powers of the dominant and the fading components [13].

$$K = \frac{A^2}{2\sigma^2} \tag{2.33}$$

Obviously, *K* = 0 indicates a Rayleigh fading channel while $K \to \infty$ indicates a non-fading one.

2.6.1 MIMO Channel Matrix for Rayleigh Propagation Conditions

The channel matrix **H** in Equation 2.31 depends on the channel model. Specifically, in cases where the wireless channel is submitted to Rayleigh fading and the array antennas do not introduce additional correlation to the transmitted/received signal, the channel matrix becomes spatially white.

The ergodic capacity formula under the assumption of Rayleigh channels and equal power allocation is (following the analysis in Section 2.5.1):

$$\bar{C} = E\left\{\log_2\left(\det\left(\mathbf{I} + \frac{p}{n_t}\mathbf{H}_W\mathbf{H}_W^H\right)\right)\right\} \tag{2.34}$$

Equation 2.34 is used for the simulations concerning the Rayleigh channel that will be shown in the next section.

Under the assumption of $n_r = n_t = n$, it would be interesting to study the case of $n \to \infty$ [3]. Using the strong law of large numbers [14] we get:

$$\frac{1}{n}\mathbf{H}_W(\mathbf{H}_W)^H \to \mathbf{I}_n \quad as \quad n \to \infty \tag{2.35}$$

Therefore, the Rayleigh channel capacity bound is given by:

$$C \xrightarrow{n \to \infty} \log_2\left[\det\left(\mathbf{I}_n + p\mathbf{I}_n\right)\right] = \log_2\left[\left(1+p\right)^n\right] = n \cdot \log_2\left(1+p\right) \Rightarrow$$

$$C \to n \cdot \log_2\left(1+p\right) \quad \text{when } n \to \infty \tag{2.36}$$

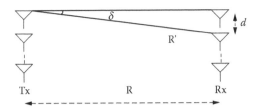

FIGURE 2.4
Geometry of a Tx and an Rx linear antenna array.

Considering the last expression, two things can be noticed:

- Capacity does not depend on the nature of the channel matrix, as it increases linearly with n for a fixed SNR.
- Every 3 dB increase of SNR corresponds to an n bits/sec/Hz increase in capacity.

2.6.2 MIMO Channel Matrix for Ricean Propagation Conditions

In the presence of a dominant component between the transmitter and the receiver, the wireless channel can be modeled as the sum of a constant and a variable component caused by scattering [3, 15].

$$\mathbf{H}_{\text{Rice}} = \sqrt{\frac{K}{K+1}} e^{j\phi_0} \mathbf{H}_{\text{LOS}} + \sqrt{\frac{1}{K+1}} \mathbf{H}_{\text{Rayleigh}} \qquad (2.37)$$

In Equation 2.37 $\mathbf{H}_{\text{Rice}}$ is the MIMO channel matrix, $\mathbf{H}_{\text{Rayleigh}}$ is the MIMO matrix corresponding to the variable component, $\mathbf{H}_{\text{LOS}}$ is the MIMO matrix corresponding to the constant signal component, K is the Ricean–K factor and ϕ_0 is the phase shift of the signal when propagating from a transmitting antenna element to the corresponding receiving antenna element.

The $\mathbf{H}_{\text{Rayleigh}}$ matrix is spatially white, and its structure was described earlier in this chapter. $\mathbf{H}_{\text{LOS}}$ can be derived by the procedure described in the following (for more details see [15]).

The general configuration of a multiple transmitting and receiving antenna array is illustrated in Figure 2.4.

In Figure 2.4, R is the distance between the transmitter and the receiver and d is the interelement distance. The matrix $\mathbf{H}_{\text{LOS}}$ is given by:

$$\mathbf{H}_{\text{LOS}} = \begin{bmatrix} 1 & e^{j\theta} & \cdots & e^{j(n_t-1)\theta} \\ e^{-j\theta} & 1 & \cdots & \vdots \\ \vdots & \vdots & \ddots & e^{j\theta} \\ e^{-j(n_r-1)\theta} & e^{-j(n_r-2)\theta} & \cdots & 1 \end{bmatrix} \qquad (2.38)$$

where θ is the angle corresponding to phase shift between the neighbor array elements.

In order to simplify the analysis, the distance between the receiving and the transmitting antennas is assumed substantially larger than the distance between the antenna elements. So, under the assumption of $R \gg d$, θ is minimized to the point that it can be omitted from the matrix in Equation 2.38 (see Appendix 2C for the proof). In that case $\mathbf{H}_{LOS}$ is given by an $n_r \times n_t$ matrix, with ones as elements (we refer to this matrix as $\mathbf{H}(1)$).

Also, it is obvious that ϕ_0 affects the contribution of $\mathbf{H}_{LOS}$ to $\mathbf{H}_{Rice}$. For reasons of simplicity it is assumed that

$$\phi_0 = \pi/4, \text{ so } e^{j\phi_0} = 1/\sqrt{2} + j1/\sqrt{2}$$

As a result, the real and the imaginary parts of the $\mathbf{H}_{Rice}$ elements are influenced in the same manner. After some manipulation, Equation 2.37 becomes:

$$\mathbf{H}_{Rice} = \sqrt{\frac{K}{K+1}}\left(\frac{1}{\sqrt{2}} + j\frac{1}{\sqrt{2}}\right)\mathbf{H}(1) + \sqrt{\frac{1}{K+1}}\mathbf{H}_w \qquad (2.39)$$

Equation 2.39 is used to produce the simulation results shown in the next section.

The assumptions made regarding the Ricean channel analysis can be summarized as follows:

- The dominant component is considered to be caused by LOS propagation.
- The distance between the transmitter and the receiver is considered substantially larger than the interelement distance.

Although these assumptions might not always be valid, the results indicate the effect of the dominant component on the MIMO system capacity, generally. In cases that the dominant signal component is caused by directional multipath propagation, this component is time varying, and hence, the above analysis cannot be applied.

However, the case $R \ll d$ can be found in multibase operations [3]. In these scenarios the transmit/receive antenna elements are cited in different base stations. The matrix that describes the constant component of the Ricean channel, in that case, is orthogonal. In Figure 2.5 is illustrated the ergodic capacity of a Ricean channel when the $\mathbf{H}_{LOS}$ is orthogonal and when $\mathbf{H}_{LOS} = \mathbf{H}(1)$.

Apparently, Figure 2.5 shows that the form of the $\mathbf{H}_{LOS}$ matrix, which represents the fixed channel component, influences the capacity for large values of K-factor. Specifically, the channel with the orthogonal $\mathbf{H}_{LOS}$ outperforms the channel with the degenerate $\mathbf{H}_{LOS}$ for increasing K.

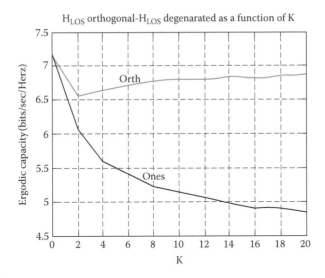

FIGURE 2.5
Ergodic capacity of a 2×2 Ricean channel when the $\mathbf{H}_{LOS}$ is orthogonal, and $\mathbf{H}_{LOS} = \mathbf{H}(1)$.

2.6.3 Channel Matrix with Spatial Fading Correlation

The Rayleigh channel assumes flat fading in the space, time, and frequency domains. However, the signal's components arriving at the receiver may experience correlation due to the limited distance of the antenna elements. As a result, the use of $\mathbf{H}_W$ as the channel matrix is inappropriate. In order to include the correlation effect the following equation is used [1, 3]:

$$\text{vec}\left(\mathbf{H}\right) = \mathbf{R}^{\frac{1}{2}}\text{vec}\left(\mathbf{H}_W\right) \tag{2.40}$$

where vec($\mathbf{H}$), denotes a vector* made by the columns of $\mathbf{H}$, and $\mathbf{R}$ is the covariance matrix of the channel of dimension $n_r n_t \times n_r n_t$.

$$\mathbf{R} = E\left\{\text{vec}\left(\mathbf{H}\right)\text{vec}\left(\mathbf{H}\right)^H\right\} \tag{2.41}$$

The analysis can be simplified with the use of the channel matrix of Equation 2.42.

$$\mathbf{H} = \mathbf{R}_R^{\frac{1}{2}}\mathbf{H}_W\mathbf{R}_T^{\frac{1}{2}} \tag{2.42}$$

---

* If $\mathbf{H} = [\mathbf{h}_1\ \mathbf{h}_2\ \cdots\ \mathbf{h}_{n_t}]$ is $n_r \times n_t$ then vec($\mathbf{H}$) = $[\mathbf{h}_1^T\ \mathbf{h}_2^T...\mathbf{h}_{n_t}^T]$ is $n_r\,n_t \times 1$.

where $\mathbf{R}_R$ is the reception correlation matrix and $\mathbf{R}_T$ is the transmission correlation matrix. Equation 2.42 is derived by 2.40 under the assumption that matrix $\mathbf{R}_R$ and $\mathbf{R}_T$ remain unchanged, regardless of the transmitting and receiving elements, respectively.

The correlation matrices $\mathbf{R}_T$ and $\mathbf{R}_R$ are calculated using two different models. The first (used for the simulations), calculates these matrices as a function of the distance between the receiving and transmitting elements [16]. A short description follows assuming that the $\mathbf{R}_T$, $\mathbf{R}_R$ matrices have the form:

$$\mathbf{R}_T = \begin{bmatrix} 1 & r_T & r_T^4 & \cdots & r_T^{(n_t-1)^2} \\ r_t & 1 & r_T & \ddots & \vdots \\ r_T^4 & r_T & 1 & \ddots & r_T^4 \\ \vdots & \ddots & \ddots & \ddots & r_T \\ r_T^{(n_t-1)^2} & \cdots & r_T^4 & r_T & 1 \end{bmatrix}$$

$$\mathbf{R}_R = \begin{bmatrix} 1 & r_R & r_R^4 & \cdots & r_R^{(n_r-1)^2} \\ r_R & 1 & r_R & \ddots & \vdots \\ r_R^4 & r_R & 1 & \ddots & r_R^4 \\ \vdots & \ddots & \ddots & \ddots & r_R \\ r_R^{(n_r-1)^2} & \cdots & r_R^4 & r_R & 1 \end{bmatrix}$$

$$(2.43)$$

where r is the fading correlation between two adjacent antenna elements and it is approximated by:

$$r(d) \approx \exp\left(-23 \cdot \Delta^2 \cdot d^2\right) \tag{2.44}$$

Δ is the angular spread and d is the distance in wavelengths between the antenna elements (Figure 2.6). In order to simplify the procedure we can make the following assumptions concerning the model:

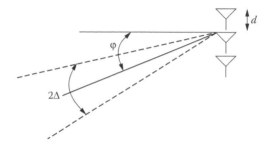

FIGURE 2.6
The mean angle of arrival (φ) and the angle spread (Δ) of an incoming multipath signal.

- For very small $r(d)$ the higher order terms of the above matrices can be omitted. Hence, the correlation matrices take the form of triagonal matrices.

- Moreover, if the same interelement distance for both the transmitter and the receiver is considered ($d_r = d_r$ and $r_T = r_R = r$), we can use a single parameter model that simplifies the capacity calculations.

The final form of the matrices used in simulations are:

$$\mathbf{R}_T = \begin{bmatrix} 1 & r & 0 & \cdots & 0 \\ r & 1 & r & \ddots & \vdots \\ 0 & r & 1 & \ddots & 0 \\ \vdots & \ddots & \ddots & \ddots & r \\ 0 & \cdots & 0 & r & 1 \end{bmatrix} \quad \mathbf{R}_R = \begin{bmatrix} 1 & r & 0 & \cdots & 0 \\ r & 1 & r & \ddots & \vdots \\ 0 & r & 1 & \ddots & 0 \\ \vdots & \ddots & \ddots & \ddots & r \\ 0 & \cdots & 0 & r & 1 \end{bmatrix} \quad (2.45)$$

The second model is described in [17], where the analysis from [18] is adopted and calculates the correlation matrices via the following formula:

$$R_{ik} = J_0[z(i-k)] \quad (2.46)$$

where J_0 is the zero-order Bessel function of the first kind, $z = 2\pi d/\lambda$, and R_{ik} is the signal correlation coefficient between the ith and the kth antenna array element.

For very small Δ and $\varphi = 0$ Equation 2.46 can be approximated by:

$$R_{ik} \approx \frac{\sin(z(i-k)\Delta)}{z(i-k)\Delta} \quad (2.47)$$

The largest value of Δ that maintains the validity of Equation 2.47 is $\pi/4$.

2.7   Simulations

In this section we present simulation results for the three capacity cases presented in Section 2.6. The Cumulative Distribution Function (CDF) of the capacity is produced for the following scenarios:

- Rayleigh channel without spatial fading correlation
- Rice channel with the dominant component caused by LOS propagation

- Rice channel with the dominant component caused by weak multipath
- Rayleigh channel with spatial fading correlation

2.7.1 MIMO Capacity for a Rayleigh Channel without Spatial Fading Correlation

The capacity formula used for the simulations of this section is presented in Equation 2.13. The channel matrix that it is used for capacity calculations is $\mathbf{H}_W$. This matrix is full rank and its elements are independent variables that follow a Gaussian distribution. As a result, the MIMO channel is transformed into exactly $n = rank(\mathbf{H}_W) = \min(n_r, n_t)$ SISO subchannels.

Figure 2.7 indicates that increasing the number of antenna elements leads to a capacity increase. Especially, we notice that the large capacity increase involves array antennas at both the transmitter and the receiver. For example, the (8,1) MIMO channel supports lower capacity gain than the (2,2) MIMO channel. This is justified by the MIMO system transformation concept mentioned earlier. Specifically, the (8,1) channel gives $n = 1$ while (2,2) gives $n = 2$. The result can be justified considering the fact that the independent SISO subchannels that resulted from the MIMO system transformation are responsible for the information transfer.

Also, Figure 2.7 indicates that the presence of an antenna array at the receiver is more important than the presence of the same antenna array at the transmitter. For example, we notice again that the channel (4,1) presents

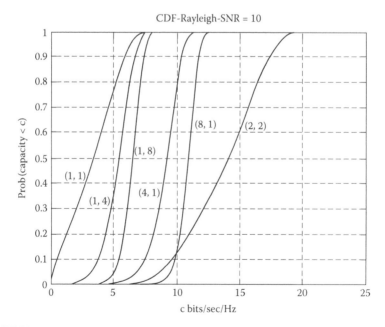

FIGURE 2.7
MIMO capacity for a Rayleigh channel with different antenna array elements (SNR=10 dB).

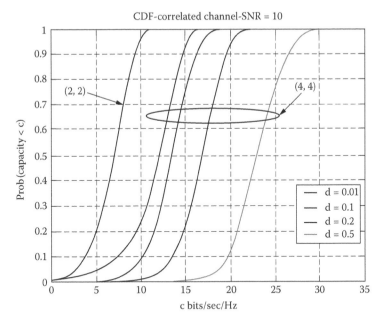

FIGURE 2.8
CDFs of capacity for the Rayleigh MIMO channel with spatial fading correlation.

better capacity behavior than the channel (1,4). The explanation for this lies in the assumption that the transmitter does not have CSI, and as a result it "equi-powers" the elements regardless of the channel. On the contrary, the receiver is considered to possess this information, and as a result it may use its antenna array for optimum combining based on CSI.

2.7.2 MIMO Capacity for a Rayleigh Channel with Spatial Fading Correlation

In this case the channel matrix is given by Equation 2.42. Figure 2.8 illustrates the capacity for different antenna configurations and interelement spacing distances, d in wavelengths.

Figure 2.8 proves the great effect of correlation to the MIMO channel capacity. We can easily notice that the uncorrelated channel* ($d = 0.5 \lambda$) offers high capacity performance in comparison to rest cases. Specifically, the ergodic capacity of the (4,4) uncorrelated channel is about 13 bits/sec/Hz greater than the fully correlated one. So as the distance between the antenna elements decreases, the capacity decreases too. The reason lies in the correlation increase with the decrease of the interelement distance. Correlation between the transmitted and received signals decreases the

---

* We consider the scenario where half wavelength interelement distance introduces low enough correlation that the fades can be considered independent [1].

independent propagation paths and, as a result, decreases the information transferred.

Also, we notice that the (4,4) MIMO channel achieves higher capacity compared to the (2,2) channel, under any correlation conditions.

2.7.3   MIMO Capacity for a Ricean Channel

The channel matrix that it is used for capacity calculations is given by Equation 2.39. Figure 2.9 illustrates the CDFs of capacity for different antenna configurations and for a constant Ricean factor equal to $K = 4$. Figure 2.10 illustrates the CDFs of capacity for a (2,2) MIMO channel for different Ricean factors. The CDF for $K = 0$ represents the case of Rayleigh fading channel.

Apparently, Figure 2.9 indicates that the use of array antennas at both the transmitter and receiver improves substantially the capacity. The same result arose in the case of the simple Rayleigh channel studied in the previous section.

Figure 2.10 proves that the presence of a fixed component can cause great damage to the MIMO channel capacity performance. We easily see that the channel capacity decreases when the Ricean factor increases. The value of K corresponds to the strength of the dominant component. As K increases, the dominant component appears stronger and the correlation coefficient increases too. As mentioned before, correlation leads to the limitation of the independent paths that transfer information and, hence, to lower capacity gains.

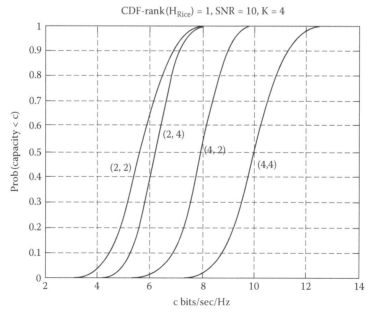

FIGURE 2.9
CDFs of capacity for the Ricean MIMO channel with K = 4.

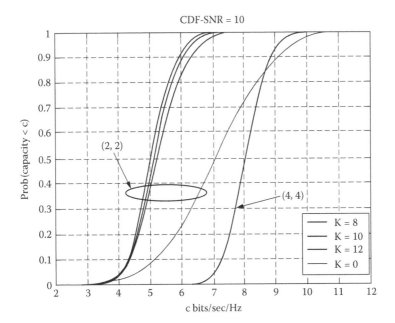

FIGURE 2.10
CDFs of capacity for a (2,2) channel for different K-factors.

Finally, Figure 2.10 shows that the capacity decrease, due to the fixed component reaching the receiver, can be easily compensated by the use of more elements. Specifically, we notice that the (4,4) channel with $K = 12$ outperforms the (2,2) Rayleigh fading channel ($K = 0$).

Appendix 2A

Capacity of a MIMO Channel Using the Singular Values of the Channel Matrix H

In Equation 2.13 we presented the MIMO channel capacity with no CSI at the transmitter. Equation 2.13 can be transformed into Equation 2.14 as follows:
  According to singular value decomposition

$$\mathbf{H} = \mathbf{UDV}^{H} \qquad (2A.1)$$

$\mathbf{D}$ is a diagonal matrix with elements the singular values of $\mathbf{H}$. The singular values of a complex matrix are always non-negative and equal to the square root of the eigenvalues of the positive semi-hermitian matrix $\mathbf{HH}^H$. Let ε_k be the kth singular value of $\mathbf{H}$, hence, ε_k^2 will be the kth eigenvalue of $\mathbf{HH}^H$.

Using transformation Equation 2A.1 for the $\mathbf{HH}^H$ matrix we get:

$$\mathbf{HH}^H = \mathbf{UDV}^H \cdot \left(\mathbf{UDV}^H\right)^H = \mathbf{UDV}^H\mathbf{VD}^H\mathbf{U}^H = \mathbf{UDD}^H\mathbf{U}^H \Rightarrow$$

$$\mathbf{HH}^H = \mathbf{UD}^2\mathbf{U}^H \tag{2A.2}$$

$$= \mathbf{UD}^2\mathbf{U}^H$$

We replace Equation 2A.2 in Equation 2.13:

$$C = \log_2\left[\det\left(\mathbf{I}+\frac{p}{n}\mathbf{UD}^2\mathbf{U}^H\right)\right],$$

we use the equality $\det\left[\mathbf{I}+\mathbf{AB}\right]=\det\left[\mathbf{I}+\mathbf{BA}\right]$

$$C = \log_2\left[\det\left(\mathbf{I}+\frac{p}{n}\mathbf{D}^2\right)\right] = \log_2\left[(1+\frac{p}{n}\varepsilon_1^2)(1+\frac{p}{n}\varepsilon_2^2)\cdots(1+\frac{p}{n}\varepsilon_n^2)\right] \Rightarrow \tag{2A.3}$$

$$C = \sum_{k=1}^{n}\log_2\left(1+\frac{p}{n}\varepsilon_k^2\right)$$

Appendix 2B

Proof That $\varepsilon_1^2 = n_r$ When $n_t = 1$ and $\varepsilon_1^2 = n_t$ When $n_r = 1$, When $|h_{i,j}| = 1$

We use the known algebraic equality $\sum_{k=1}^{n_r}\varepsilon_k^2 = Tr\left(\mathbf{HH}^H\right)$:

$$\sum_{k=1}^{n_r}\varepsilon_k^2 = Tr\left(\mathbf{HH}^H\right) = \sum_{i=1}^{n_r}\sum_{j=1}^{n_t}h_{i,j}\cdot\overline{h}_{i,j} = \sum_{i=1}^{n_r}\sum_{j=1}^{n_t}|h_{i,j}|^2 \Rightarrow$$

$$\sum_{k=1}^{n_r}\varepsilon_k^2 = \sum_{i=1}^{n_r}\sum_{j=1}^{n_t}|h_{i,j}|^2 \tag{2B.1}$$

When $n_t = 1$, Equation 2B.1 gives:

$$\sum_{k=1}^{n_r} \varepsilon_k^2 = \sum_{i=1}^{n_r}\sum_{j=1}^{1} |h_{i,j}|^2 \Rightarrow \varepsilon_1^2 + \varepsilon_2^2 + \ldots + \varepsilon_{n_r}^2 = n_r \qquad (2B.2)$$

However, since $n_t = 1$, $rank(\mathbf{HH}^H) = 1$ and as a result there is only one $\varepsilon_k^2 \neq 0$. Hence, Equation 2B.2 takes the form:

$$\varepsilon_1^2 = n_r \qquad (2B.3)$$

When $n_r = 1$, Equation 2B.1 gives:

$$\sum_{k=1}^{n_r} \varepsilon_k^2 = \sum_{i=1}^{1}\sum_{j=1}^{n_t} |h_{i,j}|^2 \Rightarrow \varepsilon_1^2 = n_t \qquad (2B.4)$$

Appendix 2C

Minimization of Angular Shift When R ≫ d

In accordance with Figure 2.4, we denote θ the angular shift between the neighbor elements; this angle is given by the equation:

$$\theta = 2\pi \frac{R - R'}{c} f = 2\pi \frac{R - R'}{\lambda} \qquad (2C.1)$$

where λ is the wavelength of the transmitted signal.
From the figure above we have:

$$\left. \begin{array}{l} R = R'\cos\delta \\ R' = \dfrac{d}{\sin\delta} \end{array} \right\} \Rightarrow \begin{array}{l} R = \dfrac{d\cos\delta}{\sin\delta} \\ R' = \dfrac{d}{\sin\delta} \end{array} \qquad (2C.2)$$

substituting Equation 2C.2 into Equation 2C.1:

$$\theta = 2\pi\frac{R-R'}{\lambda} = -\frac{2\pi}{\lambda}\cdot\left(\frac{d}{\sin\delta} - \frac{d\cos\delta}{\sin\delta}\right) = -\frac{2\pi d}{\lambda}\cdot\frac{1-\cos\delta}{\sin\delta} = -\frac{2\pi d}{\lambda}\cdot\frac{2\sin^2\left(\delta/2\right)}{\sin\delta} \Rightarrow$$

$$\theta = -\frac{4\pi d}{\lambda}\cdot\frac{\sin^2\left(\delta/2\right)}{\sin\delta} \tag{2C.3}$$

Assuming that $R \gg d$ or $d/R \ll 1$ then δ approaches zero and we can use the approximation:

$$\sin\delta = \frac{d}{R'} \approx \delta \tag{2C.4}$$

Finally, Equation 2C.3 is given by:

$$\theta = -\frac{4\pi d}{\lambda}\cdot\frac{\delta^2}{4\delta} = -\frac{\pi d\delta}{\lambda} \approx -\frac{\pi d^2}{\lambda R'} \ll 1 \tag{2C.5}$$

Equation 2C.5 proves that under the aforementioned assumptions θ may be omitted from the matrix (Equation 2.38).

References

1. D.S. Shiu, J. Foschini, J. Gans, and J.M Kahn. 2000. "Fading correlation and its effect on the capacity of multielement antenna system," *IEEE Transactions on Communications*, 48, 502, 2000.
2. http://www.cs.ut.ee/~toomas_1/linalg/lin2/node14.html
3. A. Paulraj, R. Nabar, and D. Gore. 2003. *Introduction to Space-Time Wireless Communications*, Cambridge: Cambridge University Press, Chap. 4.
4. I.E. Telatar, "Capacity of multi-antenna Gaussian channels," *Eur. Trans. Telecom.*, Vol. 10, No. 6, Dec. 1999.
5. http://rkb.home.cern.ch/rkb/AN16pp/node123.html
6. B.R. Ertel and P. Cardieri. 1998. "Overview of spatial channel models for antenna array communication systems," *IEEE Personal Communication Magazine*, Vol. 5, Feb. 1998, pp. 10–22.
7. G.V. Tsoulos and G.E. Athanasiadou. 2002. "On the application of adaptive antennas to microcellular environments: radio channel characteristics and system performance," *IEEE Trans. Veh. Technol.*, Vol. 51, No. 1, Jan. 2002, pp. 1–16.
8. G.E. Athanasiadou, A.R. Nix, and J.P. McGeehan. 2000. "A microcellular ray-tracing propagation model and evaluation of its narrowband and wideband predictions," *IEEE J. Select. Areas Commun.*, Vol. 18, Mar. 2000, pp. 322–335.

9. G.E. Athanasiadou and A.R. Nix. 2000. "A novel 3D indoor ray-tracing propagation model: the path generator and evaluation of narrowband and wideband predictions," *IEEE Trans. Veh. Technol.*, Vol. 49, July 2000, pp. 1152–1168.

10. 3GPP TR 25.996 V6.1.0 (www.3gpp.org).

11. L. Correia, ed. 2001. *Wireless Flexible Personalised Communications*, New York: Wiley.

12. G.J. Foschini. 1996. "Layered space-time architecture for wireless communications in a fading environment when using multi-element antennas," *Bell Labs Tech. J.*, Autumn 1996, pp. 41–59.

13. J.G. Proakis. 1983. *Digital Communications*, 4th ed., New York: McGraw-Hill, Chap. 2.

14. A. Papoulis. 1984. *Probability, Random Variables, and Stochastic Processes*, New York: McGraw-Hill.

15. M.A. Khalighi, J.-M. Brossier, G. Jourdain, and K. Raoof. 2001. "On capacity of Ricean MIMO channels," in *Proc. 12th IEEE Int. Symp. Personal, Indoor and Mobile Radio Communications*, A-150 to A-154, Vol. 1, 2001.

16. A. Zelst and J.S. Hammerschmidt. 2002. "A single coefficient spatial correlation models for multiple-input multiple output (MIMO) radio channels," in *Proc. URSI XXVIIth General Assembly*, 2002.

17. S. Loyka and G.V. Tsoulos. 2002. "Estimating MIMO system performance using the Correlation matrix approach," *IEEE Commun. Letters*, 6, 19, 2002.

18. J. Salz and J.H. Winters. 1994. "Effect of fading correlation on adaptive arrays in digital mobile radio," *IEEE Trans. Veh. Technol.*, Vol. 43, Nov. 1994, pp. 1049–1057.

3

Information Theory and Electromagnetism: Are They Related?

Sergey Loyka and Juan Mosig

CONTENTS

3.1 Introduction

Multi-antenna systems have recently emerged as a highly efficient strategy for wireless communications in rich multipath channels [1–4]. However, it is also well recognized that the wireless propagation channel has a profound impact on MIMO system performance [3–8]. In ideal conditions (uncorrelated high rank channel) the MIMO capacity scales roughly linearly as the number of Tx/Rx antennas. The effect of channel correlation is to decrease the capacity and, at some point, this is the dominant effect. This effect is highly dependent on the scenario considered. Many practically important scenarios have been studied and some design guidelines have been proposed as well.

Here we analyze the effect of propagation channel from a completely different perspective [9–14]. Electromagnetic waves are used as the primary carrier of information. The basic electromagnetism laws, which control the electromagnetic field behavior, are expressed as Maxwell equations [15,16]. Hence, we ask the question: What is the impact, if any, of Maxwell equations on the notion of information in general and on channel capacity in particular? In other words, do the laws of electromagnetism impose any limitations on the achievable channel capacity? Below, we concentrate on this last question and try to answer it. We are not targeting particular scenarios; rather, we are going to look at fundamental limits that hold in any scenario. Analyzing MIMO channel capacity allows one, in our opinion, to come very close to answering this question.

Our approach is a three-fold one [13]. First, we employ the channel correlation argument and introduce the concept of an *ideal scattering* to demonstrate that the minimum antenna spacing is limited to about half a wave length for *any* channel (i.e., locating antennas closer to each other will result in a capacity decrease because of correlation).

Second, we use the plane wave spectrum expansion of a generic electromagnetic wave and the Nyquist sampling theorem in the spatial domain to show that the laws of electromagnetism in its general form (Maxwell equations) limit the antenna spacing to half a wavelength, $d_{min} = \lambda/2$, for linear antenna arrays but only asymptotically when the number of antennas $n \to \infty$. For a finite number of antennas, this limit is slightly less than $\lambda/2$ because a slight oversampling is required to reduce the truncation error when using the sampling series. In any case, the existence of the minimum spacing limits the number of antennas and the MIMO capacity for a given aperture size. It should be emphasized that this limitation is scenario-independent. It follows directly from Maxwell equations and is valid in any situation.

Third, we consider the MIMO capacity of waveguide and cavity channels and demonstrate that there is a final number of degrees of freedom in that environment also, which is dictated directly by Maxwell equations that can be exploited for MIMO communications. Electromagnetics and information theory can be nicely united in this case to produce insight that is not available by using either of these disciplines separately. In particular, it turns out that the traditional single-mode transmission, which is so popular in the electromagnetics community, is optimal only at a small signal-to-noise ratio (SNR).

3.2   MIMO Channel Capacity

We employ the celebrated Foschini-Telatar formula for the MIMO channel capacity [1,2], which is valid for a fixed linear $n \times n$ matrix channel with additive white Gaussian noise and when the transmitted signal vector is composed of statistically independent equal power components, each with a Gaussian distribution, and the receiver knows the channel,

$$[\text{bits/s/Hz}]. \tag{3.1}$$

Here, n is the number of transmit/receive antennas, ρ is the average SNR, $\mathbf{I}$ is a $n \times n$ identity matrix, $\mathbf{G}$ is the normalized channel matrix (the entries are complex channel gains from each Tx to each Rx antenna), $\text{tr}[\mathbf{GG}^+] = n$, which is considered to be frequency independent over the signal bandwidth, and $^+$ denotes transpose conjugate. For simplicity, we consider a $n \times n$ channel, but the results also hold true, sometimes with minor modifications, for a $n_R \times n_T$ channel, where n_R and n_T are the number of Tx and Rx antennas, respectively, $n_R \neq n_T$.

In an ideal case of orthogonal full-rank channel Equation 3.1 reduces to

$$C = n \log_2 \left(1 + \rho / n\right), \tag{3.2}$$

i.e., the capacity is maximum and scales roughly linearly with the number of antennas.

3.3   The Laws of Electromagnetism

It follows from Equation 3.1 that the MIMO channel capacity crucially depends on the propagation channel $\mathbf{G}$. Since electromagnetic (EM) waves are used as the carrier of information, the laws of electromagnetism must have an impact on the MIMO capacity. They ultimately determine the behavior of $\mathbf{G}$ in different scenarios. Hence, we outline the laws of electromagnetism from a MIMO system perspective. In their most general form, they are expressed as Maxwell equations with charge and current densities as the field sources [15,16]. Appropriate boundary conditions must be applied in order to solve them. We are interested in the application of Maxwell equations to find the channel matrix $\mathbf{G}$ in Equation 3.1. Since the Rx antennas are physically separated from the Tx ones, we assume that the physical support of our channel is a source-free space, which includes scatterers, where EM waves do propagate. In this case, Maxwell equations simplify to the system of two decoupled wave equations [15]:

$$\nabla^2 \mathbf{E} - \frac{1}{c^2} \frac{\partial^2 \mathbf{E}}{\partial t^2} = 0, \quad \nabla^2 \mathbf{H} - \frac{1}{c^2} \frac{\partial^2 \mathbf{H}}{\partial t^2} = 0, \tag{3.3}$$

where $\mathbf{E}$ and $\mathbf{H}$ are electric and magnetic field vectors, and c is the speed of light. There are six independent field components (or "polarizational degrees of freedom") associated with Equation 3.3 (three for electric and three for magnetic fields), which can be used for communication in rich-scattering environments. Only two of them survive in free space at the far-field region ("poor scattering"). Hence, in a generic scattering case the number of polarizational degrees of freedom varies between two and six, and each of them can be used for communication. Using the Fourier transform in time domain,

$$\phi(\mathbf{r}, \omega) = \int \phi(\mathbf{r}, t) e^{-j\omega t} dt. \tag{3.4}$$

Equation 3.3 can be expressed as [15]

$$\nabla^2 \phi(\mathbf{r}, \omega) + (\omega / c)^2 \phi(\mathbf{r}, \omega) = 0, \tag{3.5}$$

where ϕ denotes any of the components of **E** and **H**, **r** is a position vector and ω is the frequency. For a given frequency ω (i.e., narrowband assumption), Equation 3.5 is a second-order partial differential equation in **r**. It determines ϕ (for given boundary conditions, i.e., a Tx antenna configuration and scattering environment) and, ultimately, the channel matrix and the channel capacity. Note that Equation 3.5 is general as it does not require any significantly restrictive assumptions. The source-free region assumption seems to be quite natural (i.e., Tx and Rx antennas are physically separated), and the narrowband assumption is simplifying but not restrictive since Equation 3.5 can be solved for *any* frequency and, further, the capacity of a frequency-selective channel can be evaluated using well-known techniques.

Unfortunately, the link between Equation 3.5 and the channel matrix **G** is not explicit at all. A convenient way to study this link is to use the spatial domain Fourier transform, i.e., the plane-wave spectrum expansion,

$$\phi(\mathbf{k}, \omega) = \int \phi(\mathbf{r}, \omega) e^{j\mathbf{k} \cdot \mathbf{r}} d\mathbf{r}$$

$$\phi(\mathbf{r}, t) = \frac{1}{(2\pi)^4} \iint \phi(\mathbf{k}, \omega) e^{j(\omega t - \mathbf{k} \cdot \mathbf{r})} d\mathbf{k} d\omega, \tag{3.6}$$

where **k** is the wave vector. Using Equation 3.6, Equation 3.5 can be reduced to [15]

$$\left(|\mathbf{k}|^2 - (\omega/c)^2 \right) \phi(\mathbf{k}, \omega) = 0. \tag{3.7}$$

Hence, $|\mathbf{k}| = \omega/c$, and the electromagnetic field is represented in terms of its plane-wave spectrum $\phi(\mathbf{k}, \omega)$, which in turn is determined through given boundary conditions, i.e., scattering environment and Tx antenna configuration. In the next sections, we discuss limitations imposed by Equation 3.5 through Equation 3.7 on the MIMO channel capacity.

3.4   Spatial Capacity and Correlation

The channel capacity is defined as the maximum mutual information [17],

$$C = \max_{p(\mathbf{x})} \{ I(\mathbf{x}, \mathbf{y}) \}, \tag{3.8}$$

where $\mathbf{x}, \mathbf{y}$ are Tx and Rx vectors, and the maximum is taken over all possible transmitted vectors subject to the total power constraint, $P_x = \langle \mathbf{x}\mathbf{x}^+ \rangle \leq P_t$. Under some conditions (quasi-static frequency-flat channel with additive white Gaussian noise (AWGN), with perfect channel state information (CSI) at the receiver), this results in Equation 3.1. In order to study the impact of the electromagnetics laws on the channel capacity and, following the approach of [1,2], we define the *spatial capacity S* as the maximum mutual information between the Tx vector on one side and the pair of the Rx vector $\mathbf{y}$ and the channel $\mathbf{G}$ (assuming perfect CSI at the Rx end) on the other, the maximum being taken over both the Tx vector and EM field distributions,

$$S = \max_{p(\mathbf{x}),\, \mathbf{E}} \{ I(\mathbf{x}, \{ \mathbf{y}, \mathbf{G}(\mathbf{E}) \}) \},$$

$$\text{const.:}\ \langle \mathbf{x}^+\mathbf{x} \rangle \leq P_T,\ \ \nabla^2 \mathbf{E} - \frac{1}{c^2}\frac{\partial^2 \mathbf{E}}{\partial t^2} = 0,\ \ \mathbf{E} = \mathbf{E}_0 \forall \{\mathbf{r}, t\} \in B, \tag{3.9}$$

where, to be specific, we assume that the electric field $\mathbf{E}$ is used to transmit data ($\mathbf{H}$ field can be used in the same way), and the last constraint is due to the boundary condition B associated with the scattering environment. The first constraint is the classical power constraint and the second one is due to the wave equation. The channel matrix $\mathbf{G}$ is a function of $\mathbf{E}$ since the electric field is used to send data. The spatial capacity S is difficult to find in general, since the constraints include a partial differential equation with arbitrary boundary conditions.

One may consider a reduced version of this problem by defining a spatial MIMO capacity as a maximum of the conventional MIMO channel capacity (per unit bandwidth, i.e., in bits/s/Hz) over possible propagation channels (including Tx and Rx antenna locations and scatterers' distribution), subject to some possible constraints. In this case, we replace Equation 3.9 by

$$S = \max_{\mathbf{G}} \{ C(\mathbf{G}) \},\ \text{const.:}\ \mathbf{G} \in \mathcal{M}(\text{Maxwell}), \tag{3.10}$$

where the constraint $\mathcal{M}(\text{Maxwell})$ is due to the Maxwell (wave) equations and the capacity is maximized by changing $\mathbf{G}$ (within some limits), for example, by appropriate positioning of antennas. Unfortunately, the explicit form of the constraint M is not known. Additional constraints (due to a limited aperture, for example) may be also included. The aperture constraint was discussed in [25] by introducing the concept of intrinsic capacity, which is somewhat similar to our concept of spatial capacity. Note that the second definition (Equation 3.10) will give a spatial capacity, which is, in general, less than or equal to that in the first definition (Equation 3.9).

We have termed the maxima in Equation 3.9 and Equation 3.10 "spatial capacity" or "capacity of a given space." Since we have to vary the channel during this maximization, the name "channel capacity" seems to be inappropriate simply because the channel is not fixed. On the other hand, we vary the channel within some limits, i.e., within given space. Thus, the term "capacity of a given space," or "spatial capacity," seems to be appropriate.

The question arises: What is this maximum and what are the main factors that have an impact on the maximum? Using the ray-tracing (geometrical optics) arguments and the recent result on the MIMO capacity, we further demonstrate that there exists an optimal distribution of scatterers and of Tx/Rx antennas that provides the maximum possible capacity in a given region of space. Hence, the MIMO capacity per unit spatial volume can be defined in a fashion similar to the traditional definition of the channel capacity per unit bandwidth. This allows the temporal and spatial domains to enter into the analysis on equal footing and, hence, demonstrates explicitly the space–time symmetry of the capacity problem in the spirit of special relativity in physics.

In order to proceed further, we need some additional assumptions. Considering a specific scenario would not allow us to find a fundamental limit simply because the channel capacity would depend on too many specific parameters. For example, in outdoor environments the Tx and Rx ends of the system are usually located far away from each other. Hence, any MIMO capacity analysis (and optimization) must be carried out under the constraint that the Tx and Rx antennas cannot be located close to each other. However, there exists no fundamental limitation on the minimum distance between the Tx and Rx ends. Thus, this maximum capacity would not be a fundamental limit. In a similar way, a particular antenna design may limit the minimum distance between the antenna elements, but it is just a design constraint rather than a fundamental limit. Similarly, the antenna design has an effect on the signal correlation (due to the coupling effect, for example), but this effect is very design-specific and, hence, is not of a fundamental nature. In other words, the link between the wave equations, Equation 3.3 or Equation 3.7 and the channel matrix $\mathbf{G}$ is far from explicit, since too many facts depend on Tx and Rx antenna designs and on many other details.

We will instead consider a reduced version of this problem. In particular, we investigate the case when the Tx and Rx antenna elements are constrained to be located within given Tx and Rx antenna apertures. We are looking for such location of antenna elements (within the given apertures) and such distribution of scatterers that the MIMO capacity ("spatial capacity") is maximum. While this maximum may not be achievable in practice, it gives a good indication as to what the potential limits of MIMO technology are.

In order to avoid the effect of design-specific details, we adopt the following assumptions. First, we consider a limited antenna aperture size (1D, 2D or 3D) for both the Tx and Rx antennas. All the Tx (Rx) antenna elements must be located within the Tx (Rx) aperture. As it is well known, a rich scattering environment is required in order to achieve high MIMO capacity. Thus, second, the rich ("ideal") scattering assumption is adopted in its most abstract form. Specifically, it is assumed that there is an infinite number of randomly and uniformly located ideal scatterers (the scattering coefficient equals to unity), which form a uniform scattering medium in the entire space (including the space region considered) and which do not absorb EM field. This is the concept of "ideal scattering" (which cannot be better than that). Third, antenna array elements are considered to be ideal field sensors with no size and no coupling between the elements in the Rx (Tx) antenna array. Our goal is to find the maximum MIMO channel capacity in such a scenario (which posses no design-specific details) and the limits imposed by the electromagnetism laws. It should be emphasized that the effect of electromagnetism laws is already implicitly included in some of the assumptions above. In order to simplify analysis further, we use the ray (geometrical) optics approximation, which justifies the ideal scattering assumption above.

Knowing that the capacity increases with the number of antennas, we try to use as many antennas as possible. Is there any limit to it? Since antennas have no size (by the assumption above), the given apertures can accommodate the infinite number of antennas. However, if antennas are located close to each other the channel correlation increases and, consequently, the capacity decreases. A certain minimum distance between antennas must be respected in order to avoid capacity decrease, even in ideal rich scattering. Figure 3.2 demonstrates this effect for uniformly spaced linear array antennas for the scattering scenario depicted in Figure 3.1: if $d < d_{min}$, the effect of correlation is significant and the capacity is less than the maximum one [8]. While d_{min} depends on the scattering environment, i.e., the angular spread Δ of incoming multipath,

$$\frac{d_{min}}{\lambda} \approx \max\left\{\frac{1}{2\Delta}, 0.5\right\},$$

even in rich scattering (i.e., $\Delta = 360°$) $d_{min} \approx \lambda/2$, which is consistent with the Jakes model [22]. While the model above is a two-dimensional (2D) one, it can be extended to 3D applying it to both orthogonal planes and, due to the symmetry of the problem (no preferred direction), similar results should hold in 3D as well. Rigorous analysis shows that the correlation between adjacent elements in that case is $\sin(2\pi d/\lambda)/(2\pi d/\lambda)$ (with the first zero at $d_0 = \lambda/2$) and the same minimum spacing requirement holds true, $d_{min} = \lambda/2$.

We summarize the effect of d_{min} as follows. When we increase the number of antenna elements over a fixed aperture, the capacity at first increases. But at some point, due to aperture limitation, we have to decrease the distance

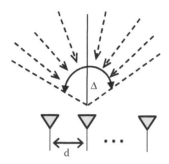

FIGURE 3.1
Incoming multipath signals arrive to a uniform linear antenna array of isotropic elements within $\pm\Delta/2$ of the broadside direction.

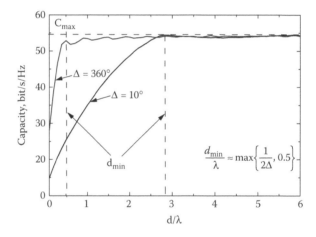

FIGURE 3.2
The average capacity vs. antenna spacing for the uniform linear array and the single-cluster multipath channel of Figure 3.1 with angular spread Δ at the Rx end (the Tx end is assumed to be uncorrelated); $n = 10$, SNR = 30 dB. When $\Delta = 360°$, $d_{min} \approx \lambda/2$.

between adjacent elements to accommodate new ones within the fixed aperture. When the element spacing decreases, the capacity increase slows down, and finally, when the element spacing is less than the minimum distance, $d_{min} \approx \lambda/2$, the capacity saturates. Hence, there is an optimal number of antennas, for which the full capacity is achieved with the minimum number of antenna elements (i.e., the minimum complexity). Figure 3.3 demonstrates this capacity saturation effect for a fixed-aperture antenna array using the model in [8]. Note the dual nature of capacity saturation: the capacity saturates with increasing the element spacing over d_{min} for a given number of elements n (1st type saturation — Figure 3.2) and also with increasing n over number N_{opt} for given aperture length L (2nd type saturation — Figure 3.3).

A similar capacity saturation argument has already been presented earlier in [9]. However, no appropriate model has been developed, and also the

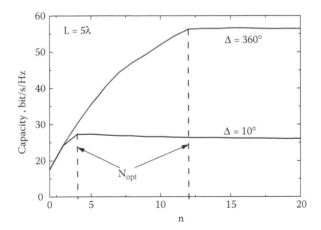

FIGURE 3.3
The average capacity vs. the number of elements n of a uniform linear array and a single-cluster multipath channel with the angular spread Δ at the Rx end. The Tx end is assumed to be uncorrelated, the number of Tx antenna elements is $n_T = 10$; the aperture length $L = 5\lambda$, SNR = 30 dB.

optimal number of antennas has not been evaluated. Using the model in [8] or the equivalent 3D model results in $d_{min} \approx \lambda/2$ for $\Delta = 360°$ (uniform scattering), and the optimal number N_{opt} of antennas for a given aperture size L is straightforward to evaluate (1D aperture, i.e., linear antenna array):

$$N_{opt} \approx 2L/\lambda + 1. \tag{3.11}$$

For an angular spread smaller than 360°, the optimal number of antennas decreases correspondingly,

$$N_{opt} \approx \frac{2L}{\lambda} \min\{\Delta, 1\} + 1.$$

This result is a spatial domain equivalent of the $2WT$ dimension theorem [36].

It should be emphasized that the 2nd type capacity saturation has been observed under the assumption of fixed average SNR at the receiver, which is equivalent to $tr\{\mathbf{G}\mathbf{G}^+\} = n_T$, where n_T is the number of Tx antennas, which is fixed. Since increasing n results in more power collected by the Rx antenna elements, which is equivalent to increasing the average SNR, it was speculated that the 2nd type saturation may not exist if the SNR increase is taken into account.* As we show later, this is not so (since the total power collected by the Rx array is limited for fixed L).

---

* A. Molisch, private communication.

While our analysis above was based on 1D antenna arrays, the similar saturation effects can be observed for 2D and 3D arrays as well. Additionally, the capacity saturation effect has been also noted for circular arrays [24]. Hence, this effect is not a consequence of a specific array geometry but rather a generic property of any array: capacity saturates as long as adjacent element spacing is about $\lambda/2$, regardless of the geometry.* Finally, we note that the results above are consistent with the diversity combining analysis, where the minimum spacing is about half a wavelength (for $\Delta = 360°$) as well [22], and with the earlier speculation in [1].

3.5   Spatial Sampling and MIMO Capacity

In the previous section, we argued that the channel correlation limits the minimum antenna spacing to half a wavelength (even in the case of "ideal" scattering). In this section, we demonstrate that the same limit can be obtained directly from the wave equations, Equation 3.3 or Equation 3.5, without reference to the channel correlation.

Let us start with the wave equation, Equation 3.5. The field spectrum $\phi(\mathbf{k}, \omega)$ can be computed in a general case provided there is a sufficient knowledge of the propagation channel and of the Tx antennas (note that we have not made, so far, any simplifying assumptions regarding the propagation channel). Knowing the field, which is given by the inverse Fourier transform in Equation 3.6, and Rx antenna properties, one may further compute the signal at the antenna output and, hence, the channel matrix $\mathbf{G}$. The result will, of course, depend on the Rx antenna design details. In order to find a fundamental limit, imposed by the wave equations (Equation 3.5) on the channel capacity (Equation 3.1), we have to avoid any design-specific details. Thus, as we did earlier, we assume that the Rx antennas are ideal field sensors (with no size, no coupling between them, etc.), and consequently, the signal at the antenna output is proportional to the field (any of the six field components may be used). Hence, the channel matrix entries g_{ij} must satisfy the same wave equation as the field itself. In general, different Tx antennas will produce different plane-wave spectra around the Rx antennas, and hence, the wave equation is

$$\left(|\mathbf{k}|^2 - (\omega/c)^2 \right) \mathbf{g}_j(\mathbf{k}, \omega) = 0, \tag{3.12}$$

---

* We should note that mutual coupling between antenna elements is not taken into account in the present study. Based on the results in [30,31], if this effect is accounted for, the capacity may actually decrease for $n > N_{opt}$ since $d < \lambda/2$ in that region.

where $\mathbf{g}_j(\mathbf{k},\omega)$ is the plane-wave spectrum produced by jth Tx antenna. To simplify things further, we employ the narrowband assumption: $\omega = $ const, and hence, $|\mathbf{k}| = \omega/c$ is constant (the case of a frequency-selective channel can be analyzed in a similar way — see below). The channel matrix entries for given locations of the Rx antennas can be found using the inverse Fourier transform in the wave vector domain:

$$\mathbf{g}_j(\mathbf{r},\omega) = \frac{1}{(2\pi)^3} \int \mathbf{g}_j(\mathbf{k},\omega) e^{-j\mathbf{k}\cdot\mathbf{r}} d\mathbf{k}, \ g_{ij} = \mathbf{g}_j(\mathbf{r}_i,\omega), \quad (3.13)$$

where $\mathbf{r}_i$ is the position vector of ith Rx antenna, and $\mathbf{g}_j(\mathbf{r},\omega)$ is the channel "vector," i.e., propagation factor from jth Tx antenna to an Rx antenna located at position $\mathbf{r}$. The integration in Equation 3.13 is performed on a hypersurface $|\mathbf{k}| = \omega/c$. As we show below, this results in a very important consequence. Consider, for simplicity, the 2D case (the 3D case can be considered in a similar way). In this case, the integration in Equation 3.13 is performed along the line given by

$$k_x^2 + k_y^2 = \left(\omega/c\right)^2 \rightarrow k_x = \pm\sqrt{\left(\omega/c\right)^2 - k_y^2} . \quad (3.14)$$

Assume that the Rx antenna is a linear array of elements located on the OX axis, i.e., $r_y = 0$. In this case, Equation 3.13 reduces to

$$\mathbf{g}_j(x,\omega) = \frac{1}{(2\pi)^2} \int_{-k_{max}}^{k_{max}} \mathbf{g}_j(k_x,\omega) e^{-jk_x \cdot r_x} dk_x,$$

$$g_{ij} = \mathbf{g}_j(x_i,\omega), \quad (3.15)$$

where $k_{max} = \omega/c$ due to Equation 3.14. At this point, we ignored the evanescent waves with $|\mathbf{k}| > k_{max}$ because they decay exponentially with distance and can be ignored at distances more than a few λ from the source [15,16]. Note that computing g_{ij} corresponds to sampling $\mathbf{g}_j(x,\omega)$ with sampling points being x_i. Let us now apply the Nyquist sampling theorem to Equation 3.15. According to it, a band-limited signal, $\mathbf{g}_j(k_x,\omega)$ in our case (it is band-limited in k_x-domain), can be exactly recovered from its samples taken at a rate equal at least to twice the maximum signal frequency (Nyquist rate). In our case, the Nyquist rate is $2k_{max}$ and the sampling interval is

$$\Delta x_{min} = 2\pi/(2k_{max}) = \lambda/2, \quad (3.16)$$

where $\lambda = 2\pi c/\omega$ is the wavelength. There is no loss of information associated with the sampling since the original channel "vector" $\mathbf{g}_j(\mathbf{r},\omega)$ (as well as the field itself) can be recovered exactly from its samples at $x = 0$, $\pm\Delta x_{min}$, $\pm 2\Delta x_{min}$, ... This means that by locating the field sensors at sampling points, which are separated by Δx_{min}, we are able to recover all the information transmitted by electromagnetic waves to the receiver. Hence, the channel capacity is not reduced. This implies, in turn, that the minimum spacing between antennas is half a wavelength:

$$d_{min} = \Delta x_{min} = \lambda/2. \qquad (3.17)$$

Locating antennas closer to each other does not provide any additional information and, hence, does not increase the channel capacity. It should be noted that the same half-wavelength limit was established in Section 3.4 using the channel correlation argument, i.e., locating antennas closer will increase correlation and, hence, the capacity will not increase. However, while the channel correlation argument may produce some doubts as to whether the limit is of fundamental nature or not (correlation depends on a scenario considered), the spatial sampling argument demonstrates explicitly that the limit is of fundamental nature because it follows directly from Maxwell equations (i.e., the wave equation), without any simplifying assumptions as, for example, the geometrical optics approximation [18] (when evaluating correlation, we have to use it to make the ray tracing valid). Note that the spatial sampling arguments hold also for a broadband channel (the smallest wavelength, corresponding to the highest frequency, should be used in this case to find Δx_{min}) and for the case of 2D and 3D antenna apertures. However, in the latter two cases, the minimum distance (i.e., the sampling interval) is different [21]. If one uses a 2D antenna aperture (i.e., 2D sampling), the sampling interval is

$$\Delta x_{min,2} = \lambda/\sqrt{3}, \qquad (3.18)$$

and in the case of 3D aperture,

$$\Delta x_{min,3} = \lambda/\sqrt{2}. \qquad (3.19)$$

While the minimum distance in these two cases is different from the 2D case, $\Delta x_{min} < \Delta x_{min,2} < \Delta x_{min,3}$ (i.e., each additional dimension possesses fewer degrees of freedom than the previous one), the numerical values are quite close to each other.

Another interpretation of the minimum distance effect can be made through a concept of the number of degrees of freedom. As the sampling argument shows, for any limited region of space (1D, 2D or 3D), there is a limited number of degrees of freedom possessed by the EM field itself. No antenna design or its specific location can provide more. This is a fundamental limitation imposed by the laws of electromagnetism (Maxwell equations) on the MIMO channel capacity.

An important remark, often overlooked, on using the sampling theorem to find the minimum antenna spacing is worth mentioning. The sampling theorem guarantees that the original band-limited signal can be recovered from its samples, provided that the infinite number of samples are used (band-limited signal cannot be time limited!). Hence, the half-wavelength limit, as derived using the sampling theorem, holds true only asymptotically, when $n \rightarrow \infty$. For finite n, the sampling series does not represent exactly anymore the continuous signal (field) due to the truncation error [19]. This is often overlooked in the array processing area [27] when the minimum antenna (sensor) spacing is derived for $n = \infty$, while the number of antennas is actually finite. In the latter case, the optimal number of antennas may be larger than that given by Equation 3.11, i.e., the minimum spacing may be less than $\lambda/2$ because a slight oversampling is required to reduce the truncation error. The maximum truncation error of the sampling series for a given limited space region (i.e., the antenna aperture in our case) decreases to zero as the number of terms in the sampling series (i.e., the number of antennas in our case) increases and provided that there is a small oversampling [20]. Below we present some truncation error bounds and discuss them in the context of spatial sampling for the MIMO system.

3.5.1   Bounds on Truncation Error in Sampling Series

Consider reconstruction of a band-limited signal $x(t)$ from its samples $x(n\Delta)$:

$$x(t) = \sum_{n=-\infty}^{\infty} x(n\Delta) \sin c \left(f_s t - n \right), \tag{3.20}$$

where $\sin c(t) = \sin(\pi t)/(\pi t)$, $\Delta = 1/f_s$ and $f_s \geq 2 f_{max}$ are the sampling interval and frequency, respectively, f_{max} is the maximum frequency in the spectrum of $x(t)$,

$$x(t) = \int_{-f_{max}}^{f_{max}} S_x(f) e^{j2\pi ft} df , \tag{3.21}$$

where $S_x(f)$ is the spectrum of $x(t)$. When the series in Equation 3.20 is truncated to $|n| \leq N$,

$$x_N(t) = \sum_{n=-N}^{N} x(n\Delta) \sin c \left(f_s t - n \right), \tag{3.22}$$

the truncation error is

$$\varepsilon(t) = x(t) - x_N(t) = \sum_{|n|>N} x(n\Delta)\sin c\left(f_s t - n\right). \tag{3.23}$$

Several bounds to $|\varepsilon(t)|$ are known [19,20], depending on the nature of the signal and the interval of interest. When the recovered signal $x_N(t)$ is considered over a finite interval only (i.e., limited antenna aperture), $|t| \leq T = N\Delta$, $|\varepsilon(t)|$ can be bounded as [20]:

$$\frac{|\varepsilon(t)|}{\sqrt{E}} \leq \frac{\sqrt{2}}{\pi}\left|\sin\frac{\pi t}{\Delta}\right|\sqrt{\frac{T\Delta}{T^2 - t^2}}, \quad |t| \leq T, \tag{3.24}$$

where E is the signal's energy,

$$E = \int_{-f_{max}}^{f_{max}} \left|S_x(f)\right|^2 df. \tag{3.25}$$

As Equation 3.24 indicates, when $\Delta \to 0$ (i.e., increasing oversampling) for fixed T (i.e., more antennas for fixed antenna aperture), or when $T \to \infty$ for fixed Δ, we obtain $|\varepsilon(t)| \to 0$. In practical terms, as the mean squared error (MSE),

$$\overline{\varepsilon^2} = T^{-1}\int_0^T \left|\varepsilon(t)\right|^2 dt,$$

becomes smaller than the noise power, $\overline{\varepsilon^2} < \sigma_0^2$, its impact on the capacity is small, and hence, it can be neglected. A tighter bound can be obtained from Equation 3.24 by using the energy carried out by the truncated samples instead of the total energy E [20]. We also note that Equation 3.24 does not necessarily require oversampling.

Another bound to $|\varepsilon(t)|$, which does involve oversampling, is of the following form [19]:

$$\frac{|\varepsilon(t)|}{\max\left\{|x(t)|\right\}} \leq \frac{4}{\pi^2 N(1-\alpha)}, \quad -\infty < t < \infty, \tag{3.26}$$

where $\alpha = f_{max}/f_s$ is the oversampling ratio. Note that this bound limits the error over the entire range of t. Clearly, as $N \to \infty$, the truncation error $|\varepsilon(t)| \to 0$

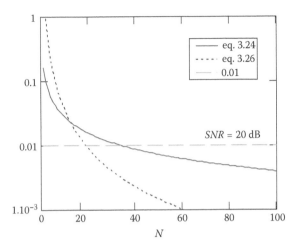

FIGURE 3.4
Normalized mean square truncation error bounds vs. the number of samples; $\alpha = 0.8$.

for any $\alpha < 1$. We note that Equation 3.24 and Equation 3.26 also justify the use of truncated series in the time-domain as any transmission spans a finite number of symbols. The difference between time-domain and spatial domain cases is that in the former case, the number of samples is much larger, and hence, the truncation error is much smaller. On the contrary, since the number of antennas in many practical systems is small, the truncation error may be, generally speaking, significant. The smaller the α (i.e., larger oversampling), the smaller the N is required for the same bound. While the convergence of $|\varepsilon(t)|$ to 0 in Equation 3.24 and Equation 3.26 is slow in N, these bounds, in many cases, overestimate the error, which converges to 0 much faster with N. Additionally, it should be noted that the bound in Equation 3.26 applies to the entire range, $-\infty < t < \infty$, while the function is recovered from the samples in $-T < t < T$, and hence, significant contribution to the error may come from the interval $|t| > T$, i.e., outside of the antenna aperture. This interval does not contribute anything to the capacity for aperture-limited system, and hence, this part of the error is irrelevant. Needless to say, more accurate bounds can be obtained if more details are known about the signal [35].

Figure 3.4 illustrates the normalized MSE $\overline{\varepsilon^2}/c$ using the bounds in Equation 3.24 and Equation 3.26, where $c = E$ and $c = \max|x(t)|^2$, respectively, vs. the number of samples. Clearly, for more than ten samples the error is already small.

3.5.2 Impact of Truncation Error on the Capacity

As the discussion above demonstrates, even for small oversampling, the truncation error goes to zero as the number of antennas (samples) increases.

While the truncation error expressions above are useful on their own (in particular, because they provide confidence that indeed a truncated sampling series can be good enough), they not only overestimate the error in many cases, but also do not indicate explicitly the effect of the truncation on the capacity.

A way to overcome this difficulty is to consider the true mean squared error and to compare it with the noise power. When the squared truncation error averaged over the antenna aperture is less than the noise power, MSE < 1/SNR, it is negligible as one is able to recover almost all the information conveyed by the EM field to the antenna aperture (but, possibly, not outside of the aperture) in given noise. For example, using Figure 3.4, SNR = 20 dB corresponds to MSE < 0.01 and $N > 20$ or $N > 35$ using Equation 3.24 or Equation 3.26, respectively. It should not be surprising that these bounds are different because different normalizations are used in Equation 3.24 and Equation 3.26; also the nature of the bounds themselves is different, i.e., Equation 3.26 implies oversampling but Equation 3.24 does not (it is clear from Figure 3.4 that oversampling results in much smaller truncation error when N is not too small). Note also that larger SNR requires a larger number of samples to make the truncation error small (less than the noise). Using Equation 3.26, the required number of samples, which provides negligible truncation error for given SNR ρ, can be estimated as

$$N > \frac{4\sqrt{\rho}}{\pi^2(1-\alpha)}.$$

Since the truncation error is zero for an infinite number of samples and the required spacing is $d_{min} = \lambda/2$ in this case, one may expect that the actual minimum antenna spacing is quite close to half a wavelength for a finite but large number of antennas. The channel correlation argument, which roughly does not depend on n, also confirms this. Detailed analysis shows that the truncation error effect can be eliminated by approximately a 10% increase in the number of antennas for many practical cases. Figure 3.5 illustrates the effect of oversampling by considering the MIMO capacity vs. the number of antennas for given (fixed) aperture length (linear antenna) $L = 5\lambda$ for different realizations of an independent identically distributed (i.i.d.) Rayleigh fading channel. Clearly, there exists an optimum number of antennas n_{max}; using more antennas does not result in higher capacity for any channel realization. Remarkably, this maximum is only slightly larger than that in Equation 3.11, i.e., spatial sampling and correlation arguments agree well. There is, however, one significant difference between these two arguments: while the latter is valid "on average" (i.e., for the mean capacity), the former is valid for each channel realization (i.e., for the instantaneous capacity) and not only on average. Clearly, the sampling argument is more powerful in this respect.

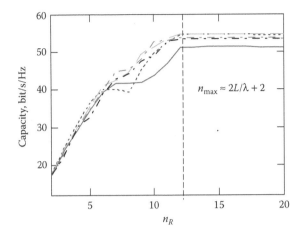

FIGURE 3.5

MIMO channel capacity vs. the number of Rx antennas for $L = 5\lambda$, $n_T = 10$. Capacities of five different realizations of a Rayleigh fading channel are shown. Capacity saturation for each of them is clear.

Keeping this in mind, one may say, based on the sampling theorem, that the optimal number of antennas for a given aperture size is given approximately by Equation 3.11. Due to the reciprocity of Equation 3.1, the same argument holds true for the transmit antennas as well. Hence, using Equation 3.2 and Equation 3.11, the maximum MIMO capacity can be found for a given aperture size.

It should be noted that, as it was indicated above, in some practical cases increasing n over N_{opt} in Equation 3.11 may result in higher SNR due to antenna gain increase (i.e., more power collected by the Rx antenna elements) and, consequently, in logarithmic increase in capacity. However, if this increase does take place, it is very slow (logarithmic) and it does not occur if the SNR is fixed, i.e., when one factors out the effect of the antenna gain. From a physical perspective, the total power collected by the antenna array cannot exceed the power collected by the ideal continuous aperture of the same size, which equals the total power delivered by the electromagnetic wave to the given space region. Consequently, the array antenna gain vs. the number of elements for a fixed aperture is limited by the gain of a continuous antenna (with the same aperture). As an example, Figure 3.6 shows the gain of a uniform linear array of isotropic elements vs. the number of elements, computed using the well-known model [26], and compares it to the gain of continuous linear antenna (aperture) of the same size $L = 5\lambda$. Clearly, the array gain saturates at about the same point as the capacity, $n \approx 2L/\lambda$, which corresponds to $d \approx \lambda/2$, and equals that of the continuous aperture at that point (this is explained, of course, by the convergence of the array antenna pattern to that of the linear aperture as the number of array

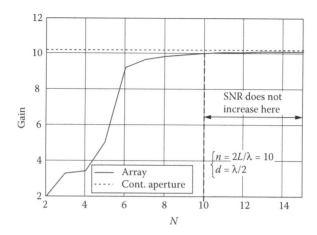

FIGURE 3.6
Linear uniform array gain vs. the number of (isotropic) elements for $L = 5\lambda$. Continuous aperture (1D) gain is also indicated.

elements increases within the fixed aperture). Hence, there is no increase in SNR beyond that point and, consequently, no increase in capacity can be expected due to higher SNR. As a side remark, we note that the similar gain saturation effect can be observed for an array of collinear short dipoles and other elements.

As it was indicated above, in many practical cases the minimum spacing can be substantially larger than $\lambda/2$. For example, when all the multipath components arrive within a narrow angle spread $\Delta \ll 1$, $d_{min} \approx \lambda/(2\Delta) \gg \lambda/2$ [8] (remarkably, the same limit follows from the spatial sampling argument). Hence, fewer antennas can be accommodated within a given aperture, $N_{opt} \approx 2L\Delta/\lambda + 1$, and, consequently, the MIMO capacity is smaller for a given aperture size.

A note is in order on practical value of the minimum spacing results above. Since these results were derived under the assumption of ideal antennas (isotropic field sensors with no mutual coupling), it is expected that practical implementation of real antenna arrays operating in real scattering environments may result in some deviations from the results above. For example, it was observed in the literature that mutual coupling (neglected in the present study) may have a significant effect on the capacity [28–31]. Since both positive (i.e., higher capacity) [28,29] and negative [30,31] effects of the mutual coupling have been reported, one has to conclude that this effect depends significantly on the environment, antenna design and also on the assumptions, i.e., whether the SNR is assumed to be fixed or affected by the coupling, whether the matching network takes the coupling into account, etc. [32]. This demonstrates once more that numerous details tend to hide the fundamental principles behind them. A side remark is that when this effect is positive, smaller element spacing becomes feasible without significant capacity loss [29]. This clearly shows that practical implementation may somewhat deviate from our idealistic theoretical analysis.* There exists, however, a final number of degrees of freedom possessed by the electromagnetic field itself, which ultimately limits any practical system (with or without mutual coupling, etc.). The way these degrees of freedom are used in practice by realistic antennas may of course differ from what the idealistic theory above suggests.

We should also note that applications of the sampling theorem to electromagnetic problems have a long history. Among others, these applications include spatial sampling techniques in optics [33], which is electromagnetics at very high, i.e., optical, frequencies, and also in near-field measurements and numerical analysis of electromagnetic fields [34]. A significant difference, however, with the present analysis is that, while the former deals mostly with free-space propagation, the latter considers a (rich) scattering environment, where the advantages of MIMO systems are most pronounced (as free-space propagation results in rank-deficient channel and, hence, low capacity, unless the Tx-Rx antenna spacing is very small). Additionally, while the earlier applications have used the sampling theorem only as a convenient analysis tool, when the sampling theorem is considered in the context of MIMO systems a fundamental link with information theory becomes clear.

With respect to the minimum antenna spacing, it is interesting to note that the MIMO capacity analysis of waveguide channels, which is based on a rigorous electromagnetic approach and does not involve the usage of the sampling theorem, indicates that the minimum antenna spacing is about $\lambda/2$ as well [12,14]. This is discussed in detail in the next section.

3.6   MIMO Capacity of Waveguide Channels**

The case of an ideal waveguide MIMO channel (i.e., lossless uniform waveguide) is especially interesting because the relationship between information theory and electromagnetics manifests itself in the clearest form in this scenario. We further consider such a waveguide unless otherwise indicated.

Arbitrary electromagnetic fields inside of a waveguide can be presented as a linear combination of the modes [15,16],

---

* While mutual coupling can have a significant effect on antenna array pattern (especially in the sidelobe region) even for $d > \lambda/2$, the MIMO capacity is not significantly affected by it in that case [28,30,31]. A possible explanation for this is that the channel matrix, which includes the effect of mutual coupling, is known to the receiver and taken into account in the processing, and, hence, the mutual coupling is implicitly compensated for. Consequently, most of our results for $d > \lambda/2$ will hold true even if the effect of mutual coupling is taken into account.

** This section is based on [12,14].

$$\mathbf{e}(x,y,z) = \sum_n \alpha_n \mathbf{E}_n(x,y)e^{-jk_{zn}z}$$

$$\mathbf{h}(x,y,z) = \sum_n \beta_n \mathbf{H}_n(x,y)e^{-jk_{zn}z},$$

(3.27)

where $\mathbf{E}_n(x,y)$ and $\mathbf{H}_n(x,y)$ are the normalized modal functions of the electric and magnetic fields, α_n and β_n are the expansion coefficients (mode amplitudes), k_{zn} is the axial component of the wave vector and n is the (composite) mode index. The modal functions $\mathbf{E}_n(x,y)$ and $\mathbf{H}_n(x,y)$ give the field variation in the transverse directions (x,y) and the variation along the axial direction (z) is given by $e^{-jk_{zn}z}$. While the particular form of the modal functions depends on the guide cross-section and may be difficult to find in explicit form (unless some symmetry is present), an important general property of the modal functions of a lossless cylindrical waveguide is their orthogonality in the following sense [16],

$$\iint_S \mathbf{E}_n \mathbf{E}_m dxdy = \delta_{mn}$$

$$\iint_S \mathbf{H}_n \mathbf{H}_m dxdy = \delta_{mn}$$

(3.28)

$$\iint_S \mathbf{E}_n \mathbf{H}_m dxdy = 0,$$

where the integrals are over the guide cross-sectional area S, $\delta_{mn} = 1$ if $m = n$ and 0 otherwise. For a given frequency, there exist a finite number of propagating modes and all the other modes are evanescent, i.e., they decay exponentially with z.

Equation 3.27 and Equation 3.28 immediately suggest the transmission strategy for a waveguide channel, which is to use all the eigenmodes (or simply modes) as independent subchannels since they are orthogonal, and it is well known that the MIMO capacity is maximum for independent subchannels. In this case, the maximum number of independent subchannels equals the number of modes and there is no loss in capacity if *all* the modes are used. For lossy and/or non-uniform waveguides, there exist some coupling between the modes [16], and hence, the capacity is smaller (due to the power loss as well as to the mode coupling). Thus, the capacity of a lossless waveguide will provide an upper bound for a true capacity since some loss and non-uniformity is always inevitable. It should be noted that if the coupling results in the normalized mode correlation being less than approximately 0.5, the capacity decrease is not significant [23]. We further assume that the waveguide is lossless and is matched at both ends. Figure 3.7 shows the system block diagram. At the Tx end, all the possible modes are excited

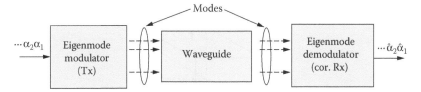

FIGURE 3.7
MIMO system architecture for a wavelength channel. (© 2005 IEEE.)

using any of the well-known techniques (i.e., eigenmode modulator) and at the Rx end the transverse electric field is measured on the waveguide cross-sectional area (proper spatial sampling may be used to reduce the number of field sensors) and is further correlated with the distribution functions of each mode (i.e., spatial correlation receiver). The signals at the correlator outputs are proportional to the corresponding transmitted signals since the modes are orthogonal, and, hence, there is no cross-coupling between different Tx signals. Thus, the channel matrix (i.e., Tx end-Rx end-correlator outputs) for this system can be expressed using the modal functions (for simplicity, we use only the E-field — the H-field can be used in the same way) as follows [14]:

$$G_{ij} = \iint_S \mathbf{E}_i(x,y)\mathbf{E}_j(x,y)dxdy, \tag{3.29}$$

and, for a uniform lossless waveguide, $\mathbf{G} = \mathbf{I}_N$, where $\mathbf{I}_N$ is a $N \times N$ identity matrix, and N is the number of modes. Clearly, the capacity achieves its maximum (Equation 3.2) in this case. Knowing the number of modes N, the maximum MIMO capacity can easily be evaluated. The maximum capacity (we call it further simply "capacity") of the present MIMO architecture described above does not vary along the waveguide length and it increases with the number of modes, as one would intuitively expect. If not all the available modes are used, the capacity decreases accordingly. The capacity may also decrease if the Rx antennas measure the field at some specific points rather than the field distribution along the cross-sectional area (since the mode orthogonality cannot be efficiently used in this case). In order to evaluate the maximum capacity, we further evaluate the number of modes.

3.6.1 Rectangular Waveguide Capacity

Let us consider first a rectangular waveguide located along the OZ axis (see Figure 3.8). The field distribution at the XY plane (cross-section of the waveguide) for E and H modes is given by well-known expressions [16] and the variation along the OZ axis is given by $e^{-jk_z z}$, where j is an imaginary unit, and k_z is the longitudinal component of the wavenumber:

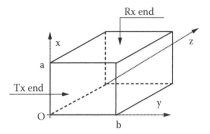

FIGURE 3.8
Rectangular waveguide geometry. (© 2005 IEEE.)

$$k_z = \sqrt{\left(\frac{\omega}{c_0}\right)^2 - \gamma_{mn}^2}, \quad \gamma_{mn}^2 = \left(\frac{\pi m}{a}\right)^2 + \left(\frac{\pi n}{b}\right)^2, \tag{3.30}$$

where ω is the radial frequency, c_0 is the speed of light, and m and n designate the mode (note that E and H modes with the same (m,n) pair have the same γ_{mn}). The sign of k_z is chosen in such a way that the field propagates along the OZ axis (i.e., from the Tx end to the Rx end). The case of $\gamma_{mn} > \omega/c$ corresponds to the evanescent field, which decays exponentially with z and is negligible at a few wavelengths from the source [16]. Assuming that the Rx end is located far enough from the Tx end (i.e., at least a few wavelengths), we neglect the evanescent field. Hence, the maximum value of γ_{mn} is $\gamma_{mn,\max} = \omega/c$. This limits the number of modes that exist in the waveguide at a given frequency ω. All the modes must satisfy the following inequality, which follows from Equation 3.30:

$$\left(\frac{m}{a'}\right)^2 + \left(\frac{n}{b'}\right)^2 \leq 4, \tag{3.31}$$

where $a' = a/\lambda$, $b' = b/\lambda$ and λ is the free-space wavelength; and $m\ n = 1, 2, \ldots$ for E mode and $m, n = 0, 1, \ldots, m + n \neq 0$ for H mode. Using a numerical procedure and Equation 3.31, the number of modes N can be easily evaluated. A closed-form approximate expression can be obtained for large a' and b' by observing that Equation 3.31 is, in fact, an equation of ellipse in terms of (m,n) and all the allowed (m,n) pairs are located within the ellipse. Hence, the number of modes is given approximately by the ratio of areas:

$$N \approx 2\frac{S_e/4}{S_0} = \frac{2\pi ab}{\lambda^2} = \frac{2\pi S_w}{\lambda^2}, \tag{3.32}$$

where $S_e = 4\pi a'b'$ is the ellipse area, $S_0 = 1$ is the area around each (m,n) pair, $S_w = ab$ is the waveguide cross-sectional area, the factor $1/4$ is due to the

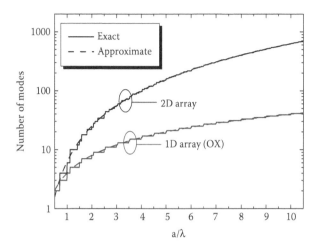

FIGURE 3.9
Number of modes in a rectangular waveguide for *a=b*. (© 2005 IEEE.)

fact that only non-negative m and n are considered, and the factor 2 is due to the contributions of both E and H modes. As Equation 3.32 demonstrates, the number of modes is determined by the ratio of the waveguide cross-section area ab to the wavelength squared. As we will see later on, this is true for a circular waveguide as well. Hence, one may conjecture that this is true for a waveguide of arbitrary cross-section as well. This conjecture seems to be consistent with the spatial sampling argument (2D sampling must be considered in this case). In fact, Equation 3.32 gives the number of degrees of freedom the rectangular waveguide is able to support and which can be used for MIMO communication. Figure 3.9 compares the exact number of modes computed numerically using Equation 3.31 and the approximate number (Equation 3.32). As one may see, Equation 3.32 is quite accurate when a and b are greater than approximately a wavelength. Note that the number of modes has a step-like behavior with a/λ, which is consistent with Equation 3.31. Using Equation 3.2 and Equation 3.32, the maximum capacity of the rectangular waveguide channel can be easily evaluated.

The analysis above assumes that the vector E-field (including both E_x and E_y components) is measured on the entire cross-sectional area (or at a sufficient number of points to recover it using the sampling expansion). However, it may happen in practice that only one of the components is measured, or that the field is measured only along the OX (or OY) axis. Apparently, it should lead to the decrease of the available modes.

To analyze this in detail, let us assume that the E-field (both components) is measured along the OX axis only, which corresponds to a 1D antenna array located along the OX axis. Due to this limitation, one can compute the correlations at the Rx using the integration over the OX axis only, since the field distribution along the OY axis is not known. Hence, we need to find the modes that are orthogonal in the following sense:

$$I = \int_0^a \mathbf{E}_\mu \mathbf{E}_\nu \, dx = c\delta_{\mu\nu},$$ (3.33)

where μ and ν are composite mode indices. In this case, one finds that two different E-modes, $E_{m_1 n_1}$ and $E_{m_2 n_2}$, are orthogonal provided that $m_1 \ne m_2$; if these modes have the same m index, they are not orthogonal. The same is true about two H-modes and about one E-mode and one H-mode. This results in a substantial reduction of the number of orthogonal modes since, in the general case, two E-modes are orthogonal if at least one of the indices is different, i.e., if $m_1 \ne m_2$ or $n_1 \ne n_2$. Surprisingly, if one measures only E_x component in this case, the modes are still orthogonal provided that $m_1 \ne m_2$. Hence, if the receive antenna array is located along the OX axis, there is no need to measure the E_y component — it does not provide any additional degrees of freedom that can be used for MIMO communications (recall that only orthogonal modes can be used). The number of orthogonal modes can be evaluated using Equation 3.31:

$$N_x \approx 4a/\lambda.$$ (3.34)

This corresponds to $2a/\lambda$ degrees of freedom for each (E and H) field. Note that this result is similar to that obtained using the spatial sampling argument, i.e., independent field samples (which are, in fact, the degrees of freedom) are located at $\lambda/2$.

A similar argument holds true when the receive array is located along the OY axis. In this case two modes are orthogonal provided that $n_1 \ne n_2$, and there is also no need to measure the E_x component. The number of orthogonal modes is approximately

$$N_y \approx 4b/\lambda.$$ (3.35)

Figure 3.10 shows the MIMO capacity of a rectangular waveguide of the same geometry as in Figure 3.8 for SNRρ = 20 dB. Note that the capacity saturates as a/λ increases. This is because Equation 3.2 saturates as well as N increases:

$$\lim_{N \to \infty} C = \rho/\ln 2.$$ (3.36)

C in Equation 3.2 can be expanded as

$$C = \frac{\rho}{\ln 2} \sum_{i=0}^{\infty} \frac{(-1)^i}{i+1} \left(\frac{\rho}{N} \right)^i.$$ (3.37)

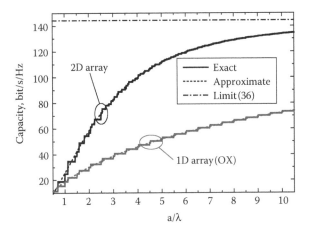

FIGURE 3.10
MIMO capacity in a rectangular waveguide for $a=b$ and SNR=20 dB. (© 2005 IEEE.)

For large N, i.e., for small ρ/N, this series converges very fast, and it can be approximated by the first two terms:

$$C \approx \frac{\rho}{\ln 2}\left(1 - \frac{\rho}{2N}\right). \qquad (3.38)$$

The capacity does not change substantially when the contribution of the second term is small:

$$\frac{\rho}{2N} \ll 1 \Rightarrow N > N_{max} \approx \rho. \qquad (3.39)$$

N_{max} is the maximum "reasonable" number of antennas (modes) for given SNR (or vice versa): if N increases above this number, the capacity does not increase significantly. It may be considered as a practical limit (since further increase in capacity is very small and it requires for very large increase in complexity). Using Equation 3.32 and Equation 3.34, one finds the maximum "reasonable" size of the waveguide for the case of 2D and 1D arrays correspondingly:

$$\frac{a_{max}}{\lambda} \approx \sqrt{\frac{\rho}{2\pi}} \text{ (2D array)}, \quad \frac{a_{max}}{\lambda} \approx \frac{\rho}{4} \text{ (1D OX array)}. \qquad (3.40)$$

Note that Figure 3.10 shows, in fact, the fundamental limit of the waveguide capacity, which is imposed jointly by the laws of information theory and electromagnetics.

3.6.2 Rectangular Cavity Capacity

The analysis of MIMO capacity in cavities is different from that in waveguides in one important aspect. Namely, the modes of a cavity exist only for some finite discrete set of frequencies (recall that, as in the case of waveguide, we consider a lossless cavity). Hence, there may be no modes for an arbitrary frequency. To avoid this problem, we evaluate the number of modes for a given bandwidth, $f \in [f_0, f_0 + \Delta f]$, starting at f_0. For a rectangular cavity, the wave vector must satisfy [16]:

$$k^2 = \left(\frac{\pi m}{a}\right)^2 + \left(\frac{\pi n}{b}\right)^2 + \left(\frac{\pi p}{c}\right)^2 = \left(\frac{\omega}{c_0}\right)^2, \tag{3.41}$$

where c is the waveguide length (along the OZ axis in Figure 3.8), and p is a non-negative integer; $m, n = 1, 2, 3, \ldots, p = 0, 1, 2, \ldots$ for E-modes, and $m, n = 0, 2, 3, \ldots, p = 1, 2, \ldots$ for H-modes ($m = n = 0$ is not allowed). Noting that Equation 3.41 is a equation of a sphere in terms of ($\pi m/a$, $\pi n/b$, $\pi p/c$), the number of modes with $k \in [k_0, k_0 + \Delta k]$ can be found as the number of (m,n,p) points between two spheres with radii of k_0 and $k_0 + \Delta k$, correspondingly. Using the ratio of areas approach described above, the number of modes is approximately:

$$N_c \approx 2\frac{V_e/8}{V_0} = \frac{8\pi V_c}{\lambda^3}\frac{\Delta f}{f_0}, \tag{3.42}$$

where $V_e = 4\pi k^2 \Delta k$ is the volume between the two spheres, $V_0 = \pi^3/V_c$ is the volume around each (m,n,p) point, $V_c = abc$ is the cavity volume; factor 2 is due to two types of modes, and factor $1/8$ is due to the fact that only non-negative values of (m,n,p) are allowed. An important conclusion from Equation 3.42 is that the number of modes is determined by the cavity volume expressed in terms of wavelength and by the normalized bandwidth. Detailed analysis shows that Equation 3.42 is accurate for large a, b, and c, and if $c/\lambda < f_0/4\Delta f$.

It should be noted that the mode orthogonality for cavities is expressed through the volume integral (over the entire waveguide volume),

$$\iiint_{V_c} E_\mu E_\nu dV = c\delta_{\mu\nu}, \tag{3.43}$$

and, hence, all the modes are orthogonal provided that the field is measured along all three dimensions, which, in turn, means that 3D arrays must be used, which may not be feasible in practice. If only 2D arrays are used, then the mode orthogonality is expressed as for a waveguide, i.e., Equation 3.28, and, consequently, only those modes are orthogonal that have different (m,n)

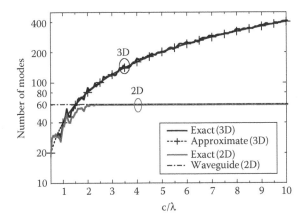

FIGURE 3.11
Number of orthogonal modes in a rectangular cavity for $a = 4\lambda$, $b = 2\lambda$ and $\Delta f/f_0 = 0.2$.

indices. The use of a 2D array results in significant reductions of the number of modes for large c, as Figure 3.11 demonstrates. Note that for small c, there is no loss in the number of orthogonal modes. This is because different p corresponds in this case to different (m,n) pairs (this can also be seen from Equation 3.41). However, as c increases, different p may include the same (m,n) pairs, which results in the number loss if a 2D array is used. In fact, the 2D case with large c is the same as the waveguide case (with the same cross-sectional area), as it should be. The value of c for which the cavity has the same number of orthogonal modes as the corresponding waveguide can be found from the following equality:

$$N_c \approx N_w \Rightarrow \frac{c_t}{\lambda} = \frac{f_0}{4\Delta f}. \tag{3.44}$$

Hence, if 2D antenna arrays are used and $c \geq c_t$, the waveguide model provides approximately the same results as the cavity model does, i.e., the cross-section has the major impact on the capacity, while the effect of cavity length is negligible. The waveguide model should be used to evaluate the number of orthogonal modes (and capacity) in this case because it is simpler to deal with. For example, a long corridor can be modeled as a waveguide rather than a cavity (despite the fact that it is closed and looks like a cavity). Figure 3.12 shows the MIMO capacity in the cavity. While the capacity of a 2D array system saturates like the waveguide capacity, which is limited by a and b, the capacity of a 3D system is larger and saturates at the value given by Equation 3.36. It should be noted that Equation 3.36 is the capacity limit due to the information theory laws, and Equation 3.32, Equation 3.34, Equation 3.35 and Equation 3.44 are the capacity limits due to the laws of electromagnetism (i.e., limited due to the number of degrees of freedom of the EM field).

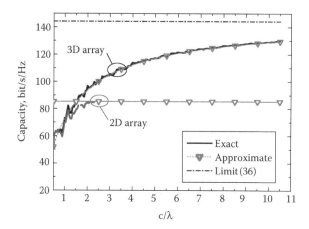

FIGURE 3.12
Capacity in a rectangular cavity for $a = 4\lambda$, $b = 2\lambda$ and $\Delta f/f_0 = 0.2$.

3.7 Spatial Capacity of Waveguide Channels

While the maximum in Equation 3.9 is difficult to find in general since one of the constraints is a partial differential equation with an arbitrary boundary condition, it can be found in an explicit closed form for some waveguide channels. Consider, for example, a lossless uniform waveguide. Using Equation 3.27, we conclude that (i) the optimizations over $p(\mathbf{x})$ and $\mathbf{E}$ can be carried out separately (since they are independent of each other) and (ii) the optimization over $\mathbf{E}$ is equivalent to optimization over α_n (since the expansion coefficients determine the field uniquely). When the Tx does not know the channel, $\mathbf{x}$ is i.i.d. complex Gaussian because (i) the channel is AWGN and Gaussian distribution maximizes the entropy, and (ii) the lack of the channel knowledge at the Tx forces the covariance of $\mathbf{x}$ to be the identity matrix, i.e., no "preferred direction" in the eigenspace (see [2] for more details), $p(\mathbf{x}) = CN(0, P_T / n_T \mathbf{I})$ and the capacity is given by Equation 3.1 [2]. Further optimization of Equation 3.9 over α_n subject to $tr(\mathbf{GG}^+) = n_T$ results in $\alpha_n = 1$ and the channel correlation matrix $\mathbf{R} = \mathbf{GG}^+ = \mathbf{I}$, i.e., all the modes are independent and carry the same power. Hence, all the capacity results above give, in fact, the spatial capacity, i.e., the maximum capacity limited by the laws of electromagnetics. One may say that the spatial capacity concept encompasses the limits in information transmission due to both the information theory and electromagnetics.

When there is mode coupling and, additionally, different modes experience different attenuation (e.g., lossy waveguide with different γ_i), one has to consider generic correlation matrix $\mathbf{R} \neq \mathbf{I}$. The optimum power allocation in the mode eigenspace can be found by applying the water-filling solution [14] to the eigenvalues λ_i of $\mathbf{R}$,

$$\alpha_i = \sqrt{\left[\nu - \frac{\lambda_i^{-1} n_T}{\rho} \right]_+} , \tag{3.45}$$

where $(x)_+ = x$ if $x \geq 0$ and 0 otherwise, and ν is chosen to satisfy the power constraint,

$$\sum_i \left[\nu - \frac{\lambda_i^{-1} n_T}{\rho} \right]_+ = n_T . \tag{3.46}$$

α_i is large for large eigenvalues, and small or even zero (i.e., no transmission on the eigenmode) for small λ_i. Without loss of generality, we further assume that $\lambda_1 \geq \lambda_2 \geq \ldots \geq \lambda_{n_T}$. Two important conclusions follow from Equation 3.45 and Equation 3.46 [14]. In the large SNR mode, $\rho \gg n_T / \lambda_{n_T}$, all the eigenmodes carry approximately the same power ($\alpha_i \approx 1$). We also note that the same solution applies when all the eigenvalues are equal, $\lambda_i = \lambda \rightarrow \alpha_i = 1 \forall i$, regardless of the SNR. In the small SNR regime, $\rho \leq 1/\lambda_2 - 1/\lambda_1$, all the power is allocated to the largest eigenmode, $\alpha_1 = n_T$, $\alpha_i = 0 \forall i \neq 1$. From this, we conclude that the traditional transmission strategy (i.e., using a single dominant mode only) is optimum in the small SNR regime only. For large SNR, the best strategy is to use all the modes. In all the other cases, between these two extremes, several dominant modes should be used, the exact number being determined by the available SNR and by the eigenvalues λ_i.

Remarkably, this insight requires a joint application of electromagnetic and information-theoretic techniques. It is expected that an efficient design of eigenmode modulators and demodulators (see Figure 3.7) will also call for significant unification of these two presently unrelated areas of research.

3.8 Acknowledgments

The authors would like to acknowledge numerous insightful discussions with C. Charalambous, A. Kouki, A. Molisch, I.E. Telatar, D. McNamara and also the colleagues within the European projects COST 273 and 284.

References

1. G.J. Foschini and M.J Gans. 1998. "On limits of wireless communications in a fading environment when using multiple antennas," *Wireless Personal Communications*, Vol. 6, No. 3, March 1998, pp. 311–335.
2. I.E. Telatar. 1995. "Capacity of multi-antenna Gaussian channels," AT&T Bell Lab. Internal Tech. Memo., June 1995 (*European Trans. Telecom.*, Vol. 10, No. 6, Dec. 1999).
3. D. Gesbert et al. 2003. "From theory to practice: an overview of MIMO space-time coded wireless systems," *IEEE Journal on Selected Areas in Communications*, Vol. 21, No. 3, April 2003, pp. 281–302.
4. A. Paulraj, R. Nabar, and D. Gore. 2003. *Introduction to Space-Time Wireless Communications*, Cambridge: Cambridge University Press.
5. A.M. Sayeed. 2002. "Deconstructing multiantenna fading channels," *IEEE Trans. Signal Processing*, Vol. 50, No. 10, Oct. 2002, pp. 2563–2579.
6. H. Bolcskei et al. 2002. "On the capacity of OFDM-based spatial multiplexing systems," *IEEE Trans. Communications*, Vol. 50, No. 2, Feb. 2002, pp. 225–234.
7. D. Chizhik, G.J. Foschini, and R.A. Valenzuela. 2000. "Capacities of multi-element transmit and receive antennas: correlations and keyholes," *Electronics Letters*, Vol. 36, No. 13, June 22, 2000, pp. 1099–1100.
8. S. Loyka and G. Tsoulos, "Estimating MIMO system performance using the correlation matrix approach," *IEEE Communication Letters*, Vol. 6, No. 1, Jan. 2002, pp. 19–21.
9. S.L. Loyka and J.R. Mosig. 2000. Spatial Channel Properties and Spectral Efficiency of BLAST Architecture, AP2000 Millennium Conference on Antennas & Propagation, Davos, Switzerland, April 9–14, 2000.
10. S. Loyka. 2002. MIMO Channel Capacity: Electromagnetic Wave Perspective, 27th General Assembly of the International Union of Radio Science, Maastricht, the Netherlands, Aug. 17–24, 2002.
11. S.L. Loyka. 2003. On MIMO Channel Capacity, Spatial Sampling and the Laws of Electromagnetism, the 3rd IASTED International Conference on Wireless and Optical Communications (WOC 2003), Banff, Alberta, Canada, July 14–16, 2003, pp. 132–137.
12. S.L. Loyka. 2003. "Multi-antenna capacities of waveguide and cavity channels," *IEEE CCECE'03*, Montreal, May 2003.
13. S. Loyka. 2004. Information Theory and Electromagnetism: Are They Related? (Invited paper), the Joint COST 273/284 Workshop on Antennas and Related System Aspects in Wireless Communications, Chalmers University of Technology, Gothenburg, Sweden, June 2004.
14. S.L. Loyka. 2005. "Multi-antenna capacities of waveguide and cavity channels," *IEEE Trans. Vehicular Technology*, Vol. 54, No. 3, May 2005, pp. 863–872.
15. E.D. Rothwell and M.J. Cloud. 2001. *Electromagnetics*, Boca Raton, FL: CRC Press.
16. R.E. Collin. 1991. *Field Theory of Guided Waves*, New York: IEEE Press.
17. J.D. Gibson, ed. 2002. *The Communications Handbook*, Boca Raton, FL: CRC Press.
18. S.R. Saunders. 1999. *Antennas and Propagation for Wireless Communication Systems*, Chichester, U.K.: Wiley.

19. A.J. Jerry. 1977. The Shannon sampling theorem — its various extensions and applications: a tutorial review, *Proc. IEEE,* Vol. 65, No. 11, Nov. 1977, pp. 1565–1596.

20. Y.I. Khurgin and V.P. Yakovlev. 1977. "Progress in the Soviet Union on the theory and applications of bandlimited functions," *Proc. IEEE,* Vol. 65, No. 7, July 1977, pp. 1005–1029.

21. D.P. Petersen and D. Middleton. 1962. "Sampling and reconstruction of wave-number-limited functions in N-dimensional Euclidean spaces," *Information and Control,* Vol. 5, 1962, pp. 279–323.

22. W.C. Jakes, Jr. 1974. *Microwave Mobile Communications,* New York: John Wiley & Sons.

23. S.L. Loyka. 2001. "Channel capacity of MIMO architecture using the exponential correlation matrix," *IEEE Communication Letters,* Vol. 5, No. 9, Sep 2001, pp. 369–371.

24. T.S. Pollock et al. 2003. "Antenna saturation effects on dense array MIMO capacity," *IEEE ICASSP 2003.*

25. J.W. Wallace and M.A. Jensen. 2002. "Intrinsic capacity of the MIMO wireless channel," *IEEE VTC (Fall),* Sept. 24–28, 2002.

26. R.C. Johnson. 1993. *Antenna Engineering Handbook,* New York: McGraw Hill.

27. D.H. Johnson and D.E. Dudgeon. 1993. *Array Signal Processing,* Upper Saddle River, NJ: Prentice Hall.

28. J.W. Wallace and M.A. Jensen. 2004. "Mutual coupling in MIMO wireless systems: a rigorous network theory analysis," *IEEE Trans. Wireless Commun.,* Vol. 3, No. 4, July 2004, pp. 1317–1325.

29. V. Jungnickel, V. Pohl, and C. von Helmolt. 2003. "Capacity of MIMO systems with closely spaced antennas," *IEEE Communications Letters,* Vol. 7, No. 8, Aug. 2003, pp. 361–363.

30. R. Janaswamy. 2002. "Effect of element mutual coupling on the capacity of fixed length linear arrays," *IEEE Antennas and Wireless Propagation Letters,* Vol. 1, 2002, pp. 157–160.

31. P.S. Kildal and K. Rosengren. 2004. "Correlation and capacity of MIMO systems and mutual coupling, radiation efficiency, and diversity gain of their antennas: simulations and measurements in a reverberation chamber," *IEEE Communications Magazine,* Vol. 42, No. 12, Dec. 2004, pp. 104–112.

32. M.K. Ozdemir, H. Arslan and E. Arvas. 2003. "Mutual coupling effect in multi-antenna wireless communication systems," *IEEE Globecom.,* pp. 829–833.

33. F. Gori. 1993. "Sampling in optics," in *Advanced Topics in Shannon Sampling and Interpolation Theory,* R.J. Marks II, ed., New York: Springer-Verlag.

34. T.B. Hansen and A.D. Yaghjian. 1999. *Plane-Wave Theory of Time-Domain Fields,* New York: IEEE Press.

35. A.I. Zayed. 1993. *Advances in Shannon's Sampling Theory,* Boca Raton, FL: CRC Press.

36. D. Slepian. 1976. "On bandwidth," *Proc. IEEE,* Vol. 64, No. 3, pp. 292–300.

4

Introduction to Space–Time Coding

Antonis D. Valkanas and Alexander D. Poularikas

CONTENTS

4.1   Introduction

The explosion of wired and wireless telecommunication systems marked the end of the second millennium. The number of subscribers to the Internet or to a second-generation mobile network has increased exponentially. Hence, the users of both systems have augmented expectations in services and capacity. For wireless communication systems, a novel direction proposed to resolve capacity requirements in the challenging radio environment is the exploitation of Multiple Element Array (MEA) at both transmitter and receiver sides. Wireless systems with antenna elements at both edges are referred to in the literature as Multiple Input Multiple Output (MIMO) systems in contrast to Single Input Single Output (SISO) antenna systems that have one transmit and one receive antenna. Following this classification, a system with one transmit antenna and several receive antennas is denoted as Single Input Multiple Output (SIMO) system while several transmit antennas and one receive antennas designate a Multiple Input Single Output (MISO) system. In order to include the case where the transmitter and/or receiver antenna elements are not located in the same devices, MIMO systems are referred to as Multiple Transmitters Multiple Receivers (MTMR).

In this chapter, we introduce some general concepts about the signal processing used in MIMO systems, which is commonly called Space–Time Coding (STC) or Space–Time Processing (STP). In the first part of this chapter, we describe the MIMO system model and formulate the general expression of the capacity of a MIMO system. We also present exploitation methodologies of this capacity and a classification of STCs. The next subsection describes the simplest class of STCs, i.e., Space–Time Block Codes (STBCs). Space–Time Trellis Codes (STTCs) that provide diversity and coding gain at the same time are analyzed next. Then, spatial multiplexing methods are tackled through the description of the Vertical Bell Laboratories Architecture Space–Time (V-BLAST) algorithm. Exploitation of the channel capacity in the case where channel state information is available at the transmitter side is briefly introduced in the following subsection. We conclude this chapter by mentioning different interesting space–time coding schemes that appear in the literature but which detailed description surpasses the scope of this chapter.

4.2   MIMO System and Space–Time Coding

4.2.1   MIMO System and Capacity

A MIMO system with M_T transmit antennas and M_R receive antenna is shown in Figure 4.1. From the figure, it seems that space–time coding is charged to perform serial-to-parallel and parallel-to-serial conversion of the transmitted information data. The block responsible for the serial-to-parallel transformation at the transmitter is the Space–Time Encoder. The opposite operation, i.e., the parallel-to-serial conversion, is performed in the Space–Time Decoder at the receiver side. However, space–time coding is more than these transformations, since these codes can realize at the same time several communication processes such as channel encoding/decoding, modulation/demodulation, multiplexing/demultiplexing or equalization.

The transmitted symbols are noted by x_i, with the subscript index i taking integers with non-zero values up to M_T. Similarly, the received signals are

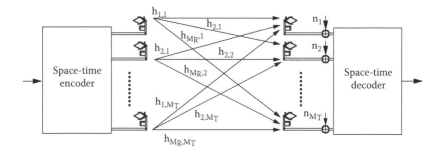

FIGURE 4.1
Block diagram of a MIMO system.

marked by r_j with j taking values from 1 to M_R. Based on this notation, the channel transfer function from the ith transmit antenna to the jth receive antenna are denoted by $h_{j,i}$. In order to have a MIMO system comparable to a SISO system, the sum of the transmitted powers of all antennas M_T must be equal to the transmitted power of a SISO system denoted P. Hence, the transmitted power from each antenna is P/M_T. Additionally, the normalization of the channel coefficient must be such that at each receive antenna j the sum of channel gains of all transmit antennas to receive antenna j is equal to the number of transmit antennas, i.e., M_T. The Additive White Gaussian Noise (AWGN) components are added to each receive antenna and are designated by n_j, with j being an index taking values from 1 to M_R. The AWGN components are assumed identically and independently distributed.

The MIMO system can be described in a matrix form, using the previous notations, as follows:

$$\left.\begin{aligned} r_1 &= \sum_{i=1}^{M_T} h_{1,i} \cdot x_i + n_1 \\ r_2 &= \sum_{i=1}^{M_T} h_{2,i} \cdot x_i + n_2 \\ &\cdots \\ r_{M_R} &= \sum_{i=1}^{M_T} h_{M_R,i} \cdot x_i + n_{M_R} \end{aligned}\right\} \Rightarrow R = H \cdot X + N$$

$$\Rightarrow \begin{bmatrix} r_1 \\ r_2 \\ \vdots \\ r_{M_R} \end{bmatrix} = \begin{bmatrix} h_{1,1} & h_{1,2} & \cdots & h_{1,M_T} \\ h_{2,1} & h_{2,2} & \cdots & h_{2,M_T} \\ \vdots & \vdots & \ddots & \vdots \\ h_{M_R,1} & h_{M_R,2} & \cdots & h_{M_R,M_T} \end{bmatrix} \cdot \begin{bmatrix} x_1 \\ x_2 \\ \vdots \\ x_{M_T} \end{bmatrix} + \begin{bmatrix} n_1 \\ n_2 \\ \vdots \\ n_{M_R} \end{bmatrix}.$$

(4.1)

In Equation 4.1, R is a column vector $M_R \times 1$ composed of the received signals r_j, X is a column vector $M_T \times 1$ composed of the transmit constellation points x_i, N is a column vector $M_R \times 1$ composed of the noise components n_j, and H is a $M_R \times M_T$ matrix with *ji*th component being the channel coefficient $h_{j,i}$.

It has been demonstrated in [1] that MIMO systems provide tremendous capacity. When Channel State Information (CSI) is not available at the transmitter, the capacity of a MIMO system expressed in bits per second per hertz (bps/Hz) can be written as:

$$C = \log_2 \left[\det \left(I_{M_R} + \frac{\rho}{M_T} \cdot H \cdot H^T \right) \right] \text{ bps/Hz.} \qquad (4.2)$$

In Equation 4.2, I_{MR} is the identity matrix of size $M_R \times M_R$, H is the channel matrix of size $M_R \times M_T$ with H^T being its transpose conjugate, and ρ gives the average Signal-to-Noise Ratio (SNR) per receiver branch independent of the number of transmitting antennas M_T.

4.2.2   Methodologies and Diversity

Two methodologies have been developed to exploit the capacity offered by the presence of several transmit and receive antennas in a telecommunication system.

The first method exploits the additional diversity of multiple antennas, namely spatial diversity, to combat channel fading. This method is performance oriented. It targets improving the reliability of the link, which can be achieved by the transmission and reception of several replicas of the same information through independent fading paths and, hence, reduces the probability of simultaneous signal fades. The provision of replicas of the same information at the receiver is referred to as diversity. The number of independent receptions of the same information at the receiver is defined as the "diversity order" or the "diversity gain" of the system. In a MIMO system, the transmit and receiver antennas form $M_T \times M_R$ independent radio links and, hence, can provide a maximum or full diversity gain equal to this product. Diversity techniques are particularly interesting in the case of a severely rich fading environment, but the targeted transmitted rate is equivalent to that of a SISO system. In other words, the additional antennas of the MIMO system are used to support the transmission of a SISO system. Diversity methods are traditionally used in Base Stations (BS). In the downlink, the BS transmits from two or more antennas, while in the uplink the BS receives information via several receive antennas. The diversity approach is particularly important for systems having a relatively small number of transmit antennas that operate at low SNR values. A major drawback of a

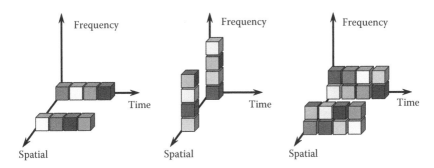

FIGURE 4.2
Diversity exploitation in MIMO systems in three domains.

MIMO system is that the transmitted signals from distinct antennas must be decorellated, and hence, the antenna elements must be sufficiently separated. It has been shown in the literature that the spacing between antenna elements must exceed half the wavelength of the transmitted signals. In practice, the spading exceeds by three and even ten times the signal's wavelength. Therefore, the diversity schemes are popular to mobile/portable devices that have size limitations.

In MIMO systems that target maximizing the transmission and reception diversity, the channel coding is named according to the domains where diversity is applied. Usually, diversity is applied to two or more domains. Hence, we have Space–Time (ST) coding, Space–Frequency (SF) coding and Space–Time–Frequency (STF) coding. Those coding techniques can be visualized in the three-dimensional plane as shown in Figure 4.2.

The methodology described previously is suboptimal for MIMO systems with a large number of transmit antennas when operating at high SNR regimes. Since the capacity of a MIMO system increases according to the number of antennas, this capacity augmentation can be exploited differently for MIMO systems that have no limitations on the number of antennas. There are other methods to exploit the capacity of a MIMO system, which are throughput oriented. In this case, the target of the system is to transfer the maximum possible information data. These techniques are particularly interesting for the case of a Line of Sight (LOS) environment, where the channels have practically no fading. This approach is known as Spatial Multiplexing (SM) or Layer Space–Time (LST). The number of extra degrees of freedom available for communication in a MIMO system is then equal to the minimum between the number of transmit antennas and the number of receive antennas.

Several efforts have been made recently in the literature to combine the two methods described in this section. Some hybrid encoding schemes have been suggested. A tradeoff between diversity and multiplexing has been presented in [2].

4.3 Space–Time Block Codes

Space–Time Block Codes (STBCs) are the simplest type of spatial temporal codes that exploit the diversity offered in systems with several transmit antennas. In 1998, Alamouti designed a simple transmission diversity technique for systems having two transmit antennas [3]. This method provides full diversity and requires simple linear operations at both transmission and reception side. The encoding and decoding processes are performed with blocks of transmission symbols. Alamouti's simple transmit diversity scheme was extended in [4] and [5] thanks to the theory of orthogonal designs for larger numbers of transmit antennas. These codes are referred to in the literature as Orthogonal Space–Time Block Codes (OSTBCs). In this section, we initially describe the simple Alamouti's scheme. A brief overview of STBC based on orthogonal design is given next.

4.3.1 Alamouti's Transmit Technique

Historically, the transmit diversity technique proposed by Alamouti was the first STBC. The encoding and decoding operation is carried out in sets of two modulated symbols. Hence, the information data bits are first modulated and mapped into their corresponding constellation points. Therefore, let us denote by x_1 and x_2 the two modulated symbols that enter the space–time encoder. Usually, in systems with only one transmit antenna, these two symbols are transmitted at two consecutive time instances t_1 and t_2. The times t_1 and t_2 are separated by a constant time duration T. In the Alamouti scheme, during the first time instance, the symbol x_1 and x_2 are transmitted by the first and the second antenna element, respectively. During the second time instance t_2, the negative of the conjugate of the second symbol, i.e., $-x_2^*$, is sent to the first antenna while the conjugate of the first constellation point, i.e., x_1^*, is transmitted from the second antenna. The encoding operation is described in the Table 4.1. The transmission rate is equal to the transmission rate of a SISO system.

The space–time encoding mapping of Alamouti's two-branch transmit diversity technique can be represented by the coding matrix:

$$X_1 = \begin{bmatrix} x_1 & -x_2^* \\ x_2 & x_1^* \end{bmatrix}. \tag{4.3}$$

In the coding matrix X_1, the subscript index gives the transmit rate compared to a SISO system. For Alamouti's scheme, the transmission rate is 1. The rows of the coding matrix represent the transmit antennas while its columns correspond to different time instances.

TABLE 4.1

Encoding of the Alamouti's
Transmit Diversity Scheme.

	Antenna 1	Antenna 2
Time t_1	x_1	x_2
Time t_2	$-x_2^*$	x_1^*

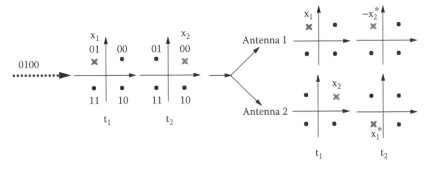

FIGURE 4.3
An example of the Alamouti's encoding process for QPSK.

An example of the space–time encoding process is schematically shown in Figure 4.3. The two modulated symbols are selected from a Quadrature Phase Shift Keying (QPSK) constellation. Their representation in the complex plane is depicted in the figure. The transmitted symbols from the two antennas are sketched to the right side of the figure.

The decoding operation assumes that the fading channel coefficients during the two consecutive transmission time periods, t_1 and t_2, are to remain constant. In other words, the channel coefficients from the first antenna to the jth receiver antenna $h_{j,1}$ and those from the second antenna to the jth receiver antenna $h_{j,2}$ must satisfy the following set of equations:

$$h_{j,1} = h_{j,1}(t) = h_{j,1}(t+T) = h_{j,1}(t_1) = h_{j,1}(t_2)$$
$$h_{j,2} = h_{j,2}(t) = h_{j,2}(t+T) = h_{j,2}(t_1) = h_{j,2}(t_2).$$

(4.4)

The received signals at receiver antenna j during the two time instances are r_j^1 and r_j^2. The received signals satisfy the equations:

$$r_j^1 = h_{j,1} \cdot x_1 + h_{j,2} \cdot x_2 + n_j^1$$
$$r_j^2 = -h_{j,1} \cdot x_2^* + h_{j,2} \cdot x_1^* + n_j^2.$$

(4.5)

In Equation 4.5, the Additive White Gaussian Noise (AWGN) components added at each receiver antenna element j during the transmission time instances t_1 and t_2, are denoted n_j^1 and n_j^2, respectively.

The decoding process is simple and requires signal combining and a maximum likelihood decoding. The combiner block performs elementary linear operations such as additions, multiplications and conjugate transformations of the received signals. However, the fading channel coefficients are considered perfectly known at the receiver. The fading components can be recovered at the receiver side through commonly known procedures, such as pilot training sequences. The processing of the linear combiner and the signals fed to the maximum likelihood decoder are given in the following equations:

$$\tilde{x}_1 = \sum_{j=1}^{M_R} \left\{ h_{j,1}^* \cdot r_j^1 + h_{j,2} \cdot \left(r_j^2\right)^* \right\} = \sum_{i=1}^{2} \sum_{j=1}^{M_R} \left|h_{j,i}\right|^2 \cdot x_1 + \sum_{j=1}^{M_R} \left\{ h_{j,1}^* \cdot n_j^1 + h_{j,2} \cdot \left(n_j^2\right)^* \right\}$$

$$\tilde{x}_2 = \sum_{j=1}^{M_R} \left\{ h_{j,2}^* \cdot r_j^1 - h_{j,1} \cdot \left(r_j^2\right)^* \right\} = \sum_{i=1}^{2} \sum_{j=1}^{M_R} \left|h_{j,i}\right|^2 \cdot x_2 + \sum_{j=1}^{M_R} \left\{ h_{j,2}^* \cdot n_j^1 - h_{j,1} \cdot \left(n_j^2\right)^* \right\}. \tag{4.6}$$

We notice that the decision statistics are composed by an amplification of the transmitted signals and a noise component. The signal amplification is equal to the sum of the amplitudes of all channel coefficients. The noise component is a sum of the receiver antenna noises multiplied by channel fading components.

4.3.2 Orthogonal Space–Time Block Codes

The transmit diversity scheme designed by Alamouti can be used only in a system with two transmit antennas. It turns out that this technique belongs to a general class of codes named Space–Time Block Codes or, more precisely, Orthogonal STBCs, since they are based on the theory of orthogonal designs. The authors of [4] introduced the theory of generalized orthogonal designs in order to create codes for an arbitrary number of transmit antennas.

The general idea behind STBCs construction is based on finding coding matrices X that can satisfy the following condition:

$$X \cdot X^H = p \cdot \left(\sum_{i=1}^{n} |x_i|^2\right) \cdot I_{M_T}. \tag{4.7}$$

In this equation, X^H is the Hermitian of X, p is a constant, I_{MT} is the identity matrix of size $M_T \times M_T$, M_T represents the number of transmit antennas, and n is the number of symbols x_i transmitted per transmission block in X. The generalized theory of orthogonal design is exploited to provide codes that satisfy Equation 4.7.

The orthgonality property of STBCs is reflected in the fact that all rows of X are orthogonal to each other. In other words, the sequences transmitted from two different antenna elements are orthogonal to each other for each transmission block.

For real signal, it is possible to reach full rate. However, it has been proven in [4] that this statement is false for two-dimensional constellations, i.e., complex signals. The encoding and decoding approaches follow the pattern described in Alamouti's scheme.

For complex signals, the theory of orthogonal designs can be used to generate coding matrices that achieve a transmission rate of $1/2$ for the cases of 3 and 4 transmission antennas:

$$X_{1/2} = \begin{bmatrix} x_1 & -x_2 & -x_3 & -x_4 & x_1^* & -x_2^* & -x_3^* & -x_4^* \\ x_2 & x_1 & x_4 & -x_3 & x_2^* & x_1^* & x_4^* & -x_3^* \\ x_3 & -x_4 & x_1 & x_2 & x_3^* & -x_4^* & x_1^* & x_2^* \end{bmatrix}, \tag{4.8}$$

$$X_{1/2} = \begin{bmatrix} x_1 & -x_2 & -x_3 & -x_4 & x_1^* & -x_2^* & -x_3^* & -x_4^* \\ x_2 & x_1 & x_4 & -x_3 & x_2^* & x_1^* & x_4^* & -x_3^* \\ x_3 & -x_4 & x_1 & x_2 & x_3^* & -x_4^* & x_1^* & x_2^* \\ x_4 & x_3 & -x_2 & x_1 & x_4^* & x_3^* & -x_2^* & x_1^* \end{bmatrix}. \tag{4.9}$$

Using the theory of orthogonal design to construct STBCs is not necessarily the optimal approach. There exist some sporadic STBCs mentioned in the literature, [6–8], that can provide a transmission rate of $3/4$ for schemes of either 3 or 4 transmit antennas.

$$X_{3/4} = \begin{bmatrix} x_1 & -x_2^* & x_3^* & 0 \\ x_2 & x_1^* & 0 & -x_3^* \\ x_3 & 0 & -x_1^* & x_2^* \end{bmatrix}, \tag{4.10}$$

$$X_{3/4} = \begin{bmatrix} x_1 & 0 & x_2 & -x_3 \\ 0 & x_1 & x_3^* & x_2^* \\ -x_2^* & -x_3 & x_1^* & 0 \\ x_3^* & -x_2 & 0 & x_1^* \end{bmatrix}. \tag{4.11}$$

It is important to notice that the channel coefficients must remain constant during the transmission of a block of coded symbols X.

The decoding of the STBCs described above can be easily deduced from the encoding matrix. Let us assume that we wish to estimate symbols x_p and that we have defined by r_j^k the received signal from antenna j at time instance k. The values to be added at the linear combiner are:

- $+(h_{j,i})^* \cdot r_j^k$ if we have x_p at column k and line (transmit antenna) i of X
- $-(h_{j,i})^* \cdot r_j^k$ if we have $-x_p$ at column k and line (transmit antenna) i of X
- $+h_{j,i} \cdot (r_j^k)^*$ if we have $(x_p)^*$ at column k and line (transmit antenna) i of X
- $-h_{j,i} \cdot (r_j^k)^*$ if we have $-(x_p)^*$ at column k and line (transmit antenna) i of X

The linear combiner sum is realized for all receive antennas j.

It is important to remember that STBCs based on orthogonal design do not achieve a rate of 1 for complex signal constellations. In [8], it has been shown for 3 and 4 transmit antennas the maximum possible rate is 3/4 with 4 delays. For 5 to 8 transmit antennas, the achievable rate is 1/2 with 8 delays, and for the 9 to 16 case, the rate becomes 5/16 in 16 time instances. In order to achieve the rate of a SISO system, the orthogonal property of STBCs must be broken as described in [9].

4.4 Space–Time Trellis Codes

As mentioned in the previous section, STBCs cannot achieve the transmission rate of a SISO system when complex signals and more than two transmit antennas are available. Furthermore, even though STBCs provide diversity, the capacity of the MIMO system is not fully exploited. It is possible to design codes that provide not only diversity but also some coding gain. The side effect of this additional provision is a serious increase in complexity. More precisely, the code's complexity also increases according to the number of transmission bits of the modulation used. The codes proposed are denoted in the literature as STTCs. These codes are based on the well-known convolutional encoding technique. Initial STTCs were presented in [10].

4.4.1 Encoding and Decoding

The STTC encoder is composed of a convolutional encoder as described in [11]. The data bits are inserted into the convolutional encoder composed by a fixed number of feed forward shift registers. The values of these registers at any instance represent the current state of the encoder. The complexity of

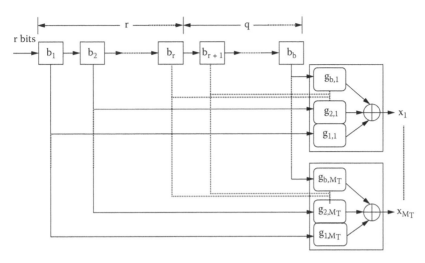

FIGURE 4.4
Space–Time Trellis encoder diagram.

the codes but also their coding gains augment with the number of shift registers. A block diagram of an encoder is shown in Figure 4.4.

The shift registers are connected to multipliers and adders. All multiplications and additions are performed with a modulo equal to the constellation size. The coefficients of the multipliers give the coefficients of the generator polynomial that describes the convolutional encoder.

The encoding process is described by Equation 4.12 where a set $\{b\}$ composed of r input bits and q bits describing the shift registers, i.e., the state of the encoder, is multiplied by the generator polynomial. All operations and values are performed with a modulo equal to the constellation size.

$$
\begin{bmatrix} x_1 & x_2 & \cdots & x_{M_T} \end{bmatrix} = \begin{bmatrix} b_1 & \cdots & b_r & \cdots & b_b \end{bmatrix} \cdot \begin{bmatrix} g_{1,1} & g_{1,2} & \cdots & g_{1,M_T} \\ g_{2,1} & g_{2,2} & \cdots & g_{2,M_T} \\ \vdots & \vdots & \ddots & \vdots \\ g_{b,1} & g_{b,2} & \cdots & g_{b,M_T} \end{bmatrix}. \quad (4.12)
$$

The encoder state of the shift registers is initialized with zeros at the beginning of transmission and is required to reach the zero state at the end of transmission.

Convolutional codes can be represented with a trellis diagram. The trellis code is a tree graph giving the transitions between the states of the registers according to the input bits. The trellis of a STTC composed of four states, designed for QPSK modulation, and applied to a system with two transmit antennas is given as an example in Figure 4.5.

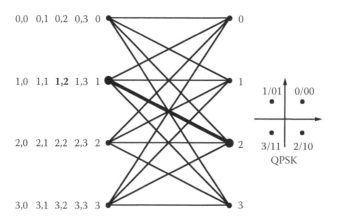

FIGURE 4.5
Trellis of a four-state, STTC-coded QPSK with two transmit antennas.

In this figure, the branches emerging from each state from top to bottom are labeled by x_1,x_2, which are the symbols transmitted from the first and second antenna, respectively. An example is marked in bold in the figure. We assume that we are already at state 1 and that the inputs bits are 1 and 0, which correspond to the QPSK symbol labeled by 2 in the right side of the figure. The next state is then 2, and the symbols transmitted from the two antennas are the QPSK symbols labeled by the numbers 1 and 2.

During the decoding of the STC code, each branch connecting two different states can be assigned with a branch metric as a function of the received signals r_j of the M_R receive antennas, the channel coefficients $h_{j,i}$ and the constellation points marked by the labels of the transitions $x_1 \dots x_{MT}$. This branch metric $BM(a \rightarrow b)$, identifying the transition from state a to state b, is given by:

$$BM(a \rightarrow b) = \sum_{j=1}^{M_R} \left| r_j - \sum_{i=1}^{M_T} h_{j,i} \cdot x_i \right|^2. \tag{4.13}$$

The Viterbi algorithm can be used to compute the path metric with the lowest accumulated metric. The path metric calculation can be simplified by selecting the minimum branch metric arriving at each node or state. The selected branch metric is called the survivor. In Figure 4.5, the bold branch is the survivor for state 2 when $BM(1 \rightarrow 2)$ is inferior to $BM(0 \rightarrow 2)$, $BM(2 \rightarrow 2)$ and $BM(3 \rightarrow 2)$. The Viterbi algorithm is composed of a forward and a backward phase. During the forward phase, the Viterbi algorithm saves at each stage, the sum of the branch metric of the survivor and the previously saved path metric of the state from which the survivor emerges. In other words and following the previous example, the path metric saved at state 2 in

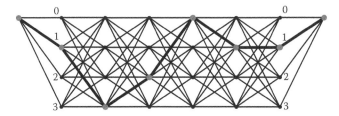

FIGURE 4.6
Example of a trellis path Viterbi selection for a four-state, STTC-coded QPSK with two transmit antennas.

Figure 4.5 is equal to $BM(1{\rightarrow}2)$ plus the previously calculated metric of state 1. After the end of this forward phase, a back trace operation starts by selecting the minimum of the path metrics and tracing back the states of the path selected and, hence, the transmitted sequence.

In Figure 4.6, an example is drawn based on the trellis described in Figure 4.5. The selected path shown in bold format follows the state transitions 0, 1, 3, 2, 0, 1, 1, 0. Based on the four-state STTC of Figure 4.5, the labels of the symbols transmitted are 0, 1, 3, 2, 0, 1, 1 from antenna 1 and 1, 3, 1, 0, 1, 1, 0 from antenna 2.

4.4.2 Performance Analysis and Code Design

In order to conduct a performance analysis of space–time trellis codes, we assume that a sequence of L consecutive symbols is transmitted from each transmit antennas during L consecutive time instances. We can then define the space–time codeword matrix X as:

$$X = \begin{bmatrix} x_1^1 & \cdots & x_L^1 \\ \vdots & \ddots & \vdots \\ x_1^{M_T} & \cdots & x_L^{M_T} \end{bmatrix}. \tag{4.14}$$

The superscript indexes of x_t^i represent the antenna element while the subscript indexes gives the time instance. We will refer by x^i the ith row of X and by x_t the tth column of X.

The pairwise error probability is defined as the probability that the decoder estimates erroneously the transmitted sequence. This case occurs during the maximum likelihood (ML) decoding process of the Viterbi algorithm when:

$$\sum_{t=1}^{L}\sum_{j=1}^{M_R}\left| r_t^j - \sum_{i=1}^{M_T} h_{j,i}^t \cdot x_t^i \right| \geq \sum_{t=1}^{L}\sum_{j=1}^{M_R}\left| r_t^j - \sum_{i=1}^{M_T} h_{j,i}^t \cdot \hat{x}_t^i \right|. \tag{4.15}$$

The modified Euclidian distance between two space–time codeword matrices is identified as:

$$d_h^2\left(X,\hat{X}\right)=\left\|H\cdot\left(\hat{X}-X\right)\right\|^2=\sum_{t=1}^{L}\sum_{j=1}^{M_R}\left|\sum_{i=1}^{M_T}h_{j,i}^t\cdot\left(\hat{x}_t^i-x_t^i\right)\right|^2. \qquad (4.16)$$

The pairwise error probability conditioned by the matrix H of channel fading coefficients is given by the following equation where $erfc(.)$ is the complementary error function:

$$P\left(X,\hat{X}|H\right)=erfc\left(\sqrt{\frac{E_s}{2\cdot N_0}\cdot d_h^2\left(X,\hat{X}\right)}\right). \qquad (4.17)$$

The upper bound of this conditional pairwise error probability is given by the following equation:

$$P\left(X,\hat{X}|H\right)\le\frac{1}{2}\cdot e^{-d_h^2\left(X,\hat{X}\right)\frac{E_s}{4\cdot N_0}}. \qquad (4.18)$$

Tarok et al. in [10] conducted a performance analysis for both Rician and Rayleigh types of channels. The Rayleigh distribution is more representative of the Non-Line-of-Sight (NLOS) environment and is considered in this chapter. Nevertheless, in both channel situations, two different cases must be dissociated. The first case involves slow fading channels, which implies that the channel coefficients remain constant for the time durations equal to L symbols. In the second case, which involves fast or rapid fading channels, the channel coefficients change at each time instance.

In the case of slow Rayleigh fading channels, Tarok et al. in [10] define the codeword difference matrix B between the codeword transmitted and the erroneous codeword as:

$$B=X-\hat{X}=\begin{bmatrix} x_1^1-\hat{x}_1^1 & \cdots & x_L^1-\hat{x}_L^1 \\ \vdots & \ddots & \vdots \\ x_1^{M_T}-\hat{x}_1^{M_T} & \cdots & x_L^{M_T}-\hat{x}_L^{M_T} \end{bmatrix}. \qquad (4.19)$$

Then, the codeword distance matrix A is defined to be the product of matrix B with its Hermitian transpose, i.e., its transpose conjugate. The modified Euclidian distance necessary to define the pairwise error probability of Equation 4.17 becomes:

$$d_h^2\left(X,\hat{X}\right)=\sum_{j=1}^{M_T}\sum_{j=1}^{M_R}\lambda_i\cdot\left|h_j\cdot v_i\right|. \qquad (4.20)$$

In the previous equation, λ_i are the eigenvalues of A, h_j are the jth rows of H and v_i represent the ith eigenvector of A.

The case of fast fading channels is more complex, and we invite the reader to refer to Tarok et al. in [10] for the complete performance analysis. Distinctive treatment is conducted according to the value of the space–time symbolwise Hamming distance. This distance represents the number of differences in the space–time symbol difference vector F at a specific time t:

$$F = \begin{bmatrix} x_t^1 - \hat{x}_t^1 \\ \vdots \\ x_t^{M_T} - \hat{x}_t^{M_T} \end{bmatrix}. \tag{4.21}$$

Tarok et al. defined design criteria for both slow and fast fading Rayleigh channel conditions. The space–time code design rules for slow fading Rayleigh channels are known as the rank and the determinant criteria. The optimum designed code must:

- Maximize the minimum rank of matrix A over all possible pairs of distinct transmission codewords
- Maximize the minimum product of all non-zero eigenvalues of matrix row along the pairs of distinct codewords with the minimum rank

Similarly, for the case of rapid fading Rayleigh channel coefficients design criteria are defined. They are named distance and product criteria. In this case, the optimum codes must:

- Maximize the minimum space–time symbol-wise Hamming distance between all pairs of distinct codewords
- Maximize the minimum product distance along the path with the minimum symbol-wise Hamming distance

The design criteria presented in this section were the first to be presented. Several investigations have been conducted in this area and other more refined criteria have been defined, [11] and [12].

4.5 Spatial Multiplexing

Spatial Multiplexing (SM) techniques have a different orientation than the coding methods presented in the previous sections. The general operation behind SM processing is to break the sequence of information bits into a certain set of sub-streams that will be treated differently. In SM, M_T independent symbols are transmitted from M_T transmitting antennas. Therefore, there are two main advantages of SM techniques. First, these techniques can reach a closer bound to the available capacity of MIMO systems, since the spatial rate is equal to the number of transmit antennas M_T. On the other hand, the complexity requirement, while heavy, remains constant and independent of the modulation in use, which is not the case of STTCs. In general, in order to perform SM, the number of receive antennas must be equal to or greater than the number of transmit antennas.

The main idea behind SM encoding techniques relies on the use of powerful decoding techniques on the receiver side. At the transmitter side, the information sequence is subdivided into several sub-streams through a demultiplexer. The sub-stream where the signal processing is conducted is identified as "Layer." Hence, SM is also called the Layer Space–Time (LST). The demultiplexing operation can be applied to bits or symbols. Consequently, the demultiplexing of the data information stream can be realized in three different ways according to the demultiplexer position in the transmitter chain and the directions of the layers. The encoding processes are known as horizontal, vertical and diagonal SM and are schematically drawn from top to bottom in Figure 4.7.

In the horizontal encoding case, the data bits are demultiplexed into M_T sub-streams that are independently encoded, interleaved and modulated. In vertical encoding SM, the data stream is at first coded, interleaved and

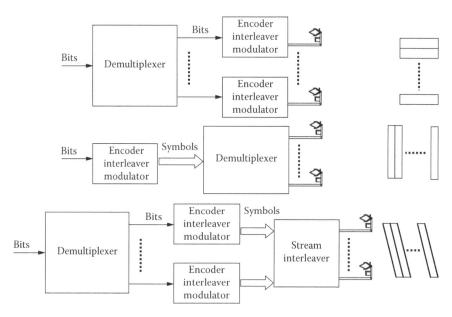

FIGURE 4.7
Spatial Multiplexing Encoding Schemes.

modulated and the resulting symbols are then demultiplexed into M_T sub-streams. Diagonal SM encoding is similar to horizontal encoding with the only difference that, after the final stage, the frames of symbols undergo a stream interleaver, which rotates the transmitted frames by padding with zero some antennas. The forms of the layers for each case are shown in the right side of Figure 4.7.

4.5.1 V-BLAST Algorithm

The vertical SM encoding is referred in [13] as V-BLAST. The detection in vertical architecture is simpler than the detection in diagonal architecture and is preferable to the sub-optimal horizontal case. The information data are at first coded, interleaved and mapped into their corresponding symbols. Then, the symbols are demultiplexed to the M_T transmit antennas. Hence, the symbols transmitted are independent of each other. Each set of M_T symbols composes a transmission vector as shown in the second scheme of Figure 4.7.

Assuming that the receiver is aware of the modulation used, the decoding at the receiver becomes possible through the V-BLAST algorithm. In vertical SM, detection and estimation of the transmitted symbols is performed in a vector-by-vector basis, since each vector represents the layer of the SM coding. The algorithm is initialized and operates on each vector in a symbol-per-symbol basis by iteratively detecting and estimating the transmitted symbols. We assume that the inter-symbol interference is negligible and that the system is quasi-stationary, which implies that the matrix of the channel coefficients remains constant. We also assume that the channel matrix is perfectly known at the receiver side.

The algorithm is based on linear interference suppression in combination with interference cancellation, methods commonly used in multi-user environments. The algorithm also performs a rearrangement of the symbol detection in an optimal sense. In general, V-BLAST is characterized as an interference avoidance technique. The received signals are perceived as a set of signals issued from different users. Each signal is an interferer to the rest of the signals. The best signal stream in terms of SNR is selected for detection and is also removed from the remaining signals. This way, the remaining signals have one interferer less. Optimal combining is repeated until all signals are detected.

Three consecutive phases take place:

- Linear interference suppression through Minimum Mean Square Estimation (MMSE) or Zero Forcing (ZF)
- Interference cancellation of the symbols detected
- Reordering of the detection process through SNR post-detection

The initial form of V-BLAST described in the literature uses ZF for the suppression of the linear interference. This operation is realized in three steps:

$$w_{k_i} = \left(G_i\right)_{k_i}$$

$$y_{k_i} = w_{k_i}^T \cdot R_i \qquad\qquad (4.22)$$

$$\hat{x}_{k_i} = Q\left(y_{k_i}\right).$$

The first step consist on finding the zero-forcing vector w_{ki} that satisfies:

$$w_i^T \cdot \left(H\right)_j = \begin{cases} 0 & j > i \\ 1 & j = i \end{cases}. \qquad\qquad (4.23)$$

In Equation 4.23, $(H)_j$ represents the jth column of H which is a vector composed of all the channel coefficients produced from transmit antenna j. It turns out that the condition of Equation 4.23 is realized by taking w_i to be the equivalent jth column from the Moore-Penrose pseudo-inverse. Hence, the algorithm finds the zero-forcing vector w_{ki} through consecutively calculating the Moore-Penrose pseudo-inverse. In the first line of Equation 4.22, i is the iteration number, k_i is the index of the transmit antenna which symbol is detected during iteration i and G_i is the Moore-Penrose pseudo-inverse at iteration stage i. With w_{ki}, the statistic decision y_{ki} is obtained from the received vector R_i at stage i, and is quantized (represented by the operator $Q(\cdot)$) in order to obtain the k_ith detected symbol.

The interference of the detected symbol is cancelled in the second step of the algorithm by removing its interference from the received signal:

$$R_{i+1} = R_i - \hat{x}_{k_i} \cdot \left(H\right)_{k_i}$$

$$\qquad\qquad (4.24)$$

$$G_{i+1} = H_{k_i}^{\pm}.$$

A new received vector R_{i+1} is formed by removing from R_i the estimated transmit signal of the detected symbol. The Moore-Penrose pseudo-inverse (represented by the + sign as superscript) is recalculated from the previous stage transfer matrix after annulling its k_ith row. The operations realized in Equation 4.24 can be seen as redefining a new system where the antenna that used to transmit the previous detected symbol is extinguished. This is achieved by removing the signal produced by the detected symbol and by annulling the channel coefficients related to the estimated symbol's antenna.

The last step of the process is charged to select the optimum order for decoding. In other words, the layers are rearranged in the process in order to minimize the probability of error. This means finding an index k_i of stage *i*. The method proposed in the algorithm is to select the antenna with the best post-detection SNR. The post-detection SNR of each transmit antenna *j* is the absolute value of the *j*th column of the Moore-Penrose pseudo-inverse G_i at stage *i*. If *S* is the optimum ordering and $S = \{k_1, k_2, \ldots k_{MT}\}$, the search for the best SNR post-detection at stage *i* is performed for the set $\{k_i, \ldots, k_{MT}\}$, since the symbols transmitted from antennas k_1 to k_i already have been detected in previous stages. The best post-detection SNR is the minimum absolute value calculated:

$$k_i = \underset{j \in S, j \notin \{k_1 \ldots k_{i-1}\}}{\operatorname{argmin}} \left\| \left(G_i \right)_j \right\|^2 . \tag{4.25}$$

The generalized algorithm is composed of two phases, an initialization and a recursive phase. The initialization phase consists of calculating the Moore-Penrose pseudo-inverse:

$$G_1 = H^+$$
$$i = 1. \tag{4.26}$$

The recursive phase is composed of the previously described steps:

$$k_i = \underset{j \in S, j \notin \{k_1 \ldots k_{i-1}\}}{\operatorname{argmin}} \left\| \left(G_i \right)_j \right\|^2$$

$$w_{k_i} = \left(G_i \right)_{k_i}$$

$$y_{k_i} = w_{k_i}^T \cdot r_i$$

$$\hat{x}_{k_i} = Q \left(y_{k_i} \right) \tag{4.27}$$

$$R_{i+1} = R_i - \hat{x}_{k_i} \cdot \left(H \right)_{k_i}$$

$$G_{i+1} = H_{\overline{k_i}}^{\pm}$$

$$i = i + 1.$$

4.6 Space–Time Coding with CSI Knowledge at the Transmitter

The coding schemes described in the previous sections involve space–time processing when no channel state information (CSI) is available at the transmitter side. However, in a duplex system, it is sometimes possible to provide

CSI at the transmitter. The optimum exploitation of the CSI at the transmitter is linear pre-filter or pre-equalization realized with the Singular Value Decomposition (SVD) theorem, [14]. The channel matrix can be written as:

$$H = U \cdot D \cdot V^H. \tag{4.28}$$

In this equation, matrix U and V are composed of the eigenvectors of the channel and D is a diagonal matrix composed of the eigenvalues of H. It becomes possible to left multiply the transmit matrix X with V and to left multiply the received matrix R with U^H. In this case, the received signals become:

$$
\begin{aligned}
U^H \cdot R &= U^H \cdot \left(H \cdot (V \cdot X) + N \right) \\
&= U^H \cdot \left((U \cdot D \cdot V^H) \cdot (V \cdot X) + N \right) \\
&= D \cdot X + U^H \cdot N.
\end{aligned}
\tag{4.29}
$$

The MIMO system becomes equivalent to a set of parallel SISO systems as shown in Figure 4.8. Each SISO is amplified by an eigenvalue of D. The capacity of the MIMO system is then the sum of distinct parallel SISO channel capacities. The number of parallel connections is equal to the number of non-zero eigenvalues or, in other words to the rank of H. Each SISO sub-channel forms an eigenmode and optimal energy allocation can be applied through the water-pouring algorithm.

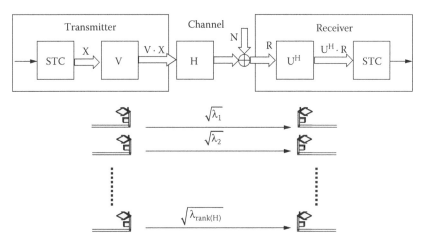

FIGURE 4.8
Exploitation of CSI knowledge at the transmitter.

The SVD method requires at least the transfer of the matrix V back to the transmitter, and complex calculation tasks, such as diagonalization, must be performed. There are several other methods that can sub-optimally use CSI at the transmitter [15]. Such a method is the antenna selection method described more analytically in [16].

Antenna selection is a simple and efficient technique to exploit CSI at the transmitter. The technique is referred to as Transmit Selection Diversity (TSD) in [17]. The purpose of the TSD algorithm is to select the "best" subset of transmit antennas that optimizes reception according to some sort of criteria. Performance results of different schemes including TSD for Orthogonal Frequency Division Multiplexing (OFDM) systems are provided in Figure 4.9. In OFDM, diversity is generally applied in the spatial and frequency domain. In the figure, Symbol Error Rate (SER) is shown in terms of SNR when Quadrature Amplitude Modulation with 16 constellation points (QAM-16) is used. The lines marked with stars are Space–Frequency Block Codes while the circles are used for the performances of Space–Frequency Trellis Coding schemes. Plain lines represent performances when CSI is unknown at the transmitter side for a MIMO system of 2 transmit and 2 receive antennas, i.e., (2, 2). The dotted curves shows the SER when TSD is used and 2 transmit antennas are selected out of 3 transmit antennas. Similarly, the dashed lines are representing the case where 2 transmit antennas are selected out of 4 transmit antennas.

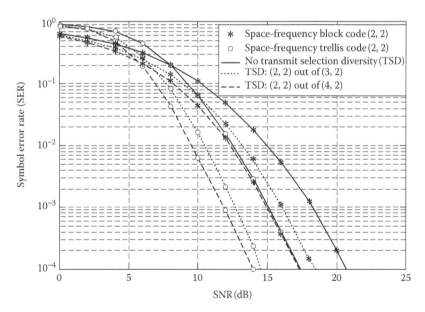

FIGURE 4.9
Performance results for different SF schemes in an OFDM system for QAM-16.

4.7 Other Space–Time Coding Schemes

The space–time coding schemes presented in the previous sections are just some of the best-known schemes for MIMO systems. It is obvious that there are several other possible schemes. The description of all these STC techniques goes beyond the scope of this chapter. In this section, we provide some general but not exhaustive directions of STC found in the literature.

A major open research direction is the design of STC in the case where the channel matrix cannot be determined at the receiver. In this case, differential schemes and techniques can be used in combination with classic ST coding methods such as STBCs.

Another class of STCs not mentioned in the previous subsections is the extension of STTCs with serial or parallel concatenate convolutional schemes. In particular, recursive codes that perform iterative decoding, such as turbo codes, can be applied to MIMO systems forming a new category of STC. Some research has also been carried out for Low Density Parity Check (LDPC) codes.

The most common form of STC is the combination of two distinct codes, namely an inner and an outer code. The outer codes are usually well-known bit-encoding schemes, such as reed-solomon or bit interleaved convolutional codes, and they are placed before the modulator in the transmission chain. Their task is to provide coding gain to the system. The inner codes are positioned after the modulator and operate on a symbol basis. The inner codes such as STBCs are mainly charged to provide diversity to the system and to perform the serial-to-parallel signal conversion. It is evident that the names "inner" and "outer" comes from the position of the codes in the transmission chain. This combined scheme is considered in several emerging standards and technologies such as the IEEE 802.16 standard, which is known to the general public by the certification mark WiMAX.

References

1. G.J. Foschini and M.J. Gans. 1998."On limits of wireless communications in a fading environment when using multiple antennas," *IEEE Wireless Personal Communications*, Vol. 6, March 1998, pp. 311–335.
2. L. Zheng and D.N.C. Tse. 2003. "Diversity and multiplexing: a fundamental tradeoff in mulitple antenna channels," *IEEE Transactions on Information Theory*, Vol. 49, No. 5, May 2003, pp. 1073–1096.
3. S.M. Alamouti. 1998. "A simple transmit diversity technique for wireless communications," *IEEE Journal on Selected Areas in Communications*, Vol. 16, No. 8, October 1998, pp. 1451–1458.
4. V. Tarok, H. Jafarkhami, and A.R. Calderbank. 1999. "Space-time block codes from orthogonal designs," *IEEE Transactions on Information Theory*, Vol. 45, No. 5, July 1999, pp. 1456–1467.

5. V. Tarok, H. Jafarkhami, and A.R. Calderbank. 1999. "Space-time block codes for wireless communications: performance results," *IEEE Journal on Selected Areas in Communications*, Vol. 17, No. 3, March 1999, pp. 451–460.

6. G. Ganesan and P. Stoica. 2001. "Space-time block codes: a maximum SNR approach," *IEEE Transactions on Information Theory*, Vol. 47, No. 4, May 2001, pp. 1650–1656.

7. X.B. Liang. 2003. "Orthogonal designs with maximal rates," *IEEE Transactions on Information Theory*, Vol. 49, No. 10, October 2003, pp. 2468–2503.

8. O. Tirkkonen and A. Hottinen. 2002. "Square-matrix embeddable space-time block codes for complex signal constellations," *IEEE Transactions on Information Theory*, Vol. 48, No. 2, February 2002, pp. 384–395.

9. A. Boariu and D.M. Ionescu. 2003. "A class of nonorthogonal rate-one space-time block codes with controlled interference," *IEEE Transactions on Wireless Communications*, Vol. 2, No. 2, March 2003, pp. 270–395.

10. V. Tarok, N. Seshadri, and A.R. Calderbank. 1998. "Space-time codes for high data rate wireless communication: performance criterion and code construction," *IEEE Transaction on Information Theory*, Vol. 44, No. 2, March 1998, pp. 744–765.

11. R.S. Blum. 2002. "Some analytical tools for the design of space-time convolutional codes," *IEEE Transactions on Communications*, Vol. 50, No. 10, October 2002, pp. 1593–1599.

12. Y. Liu, M.P. Fitz, and O.Y. Takeshita. 2002. "A rank criterion for QAM space-time codes," *IEEE Transaction on Information Theory*, Vol. 48, No. 12, December 2002, pp. 3062–3079.

13. G.D. Golden, G.J. Foschini, R.A. Valenzuela, and P.W. Wolniansky. 1999. "Detection algorithm and initial laboratory results using the V-BLAST space-time communication architecture," *Electronics Letters*, Vol. 35, No. 1, January 7, 1999, pp. 14–15.

14. R. Horn and C. Johnson. 1985. *Matrix Analysis*, Cambridge, U.K.: Cambridge University Press.

15. A.D. Valkanas and P.I. Dallas. 2003. "Adaptive modulation and coding in space-time-frequency block and trellis for (2, 2) MIMO OFDM systems," *International OFDM Workshop (InOWo)*, 2003.

16. I. Bahceci, T.M. Duman, and Y. Altunbasak. 2003. "Antenna selection for multiple-antenna transmission systems: performance analysis and code construction," *IEEE Transaction on Information Theory*, Vol. 49, No. 10, October 2003, pp. 2669–2681.

17. A.D. Valkanas. 2004. "Adaptive Space-Frequency Coding for Multiple-Input Multiple-Output Orthogonal Frequency Division Multiplexing Systems," Ph.D. Dissertation, University of Alabama in Huntsville.

Acronyms

AWGN   Additive White Gaussian Noise
BS   Base Station
CSI   Channel State Information

LDPC	Low Density Parity Check
LOS	Line of Sight
LST	Layer Space Time
MEA	Multiple Element Array
MIMO	Multiple Input Multiple Output
MISO	Multiple Input Single Output
ML	Maximum Likelihood
MMSE	Minimum Mean Square Estimation
MTMR	Multiple Transmitter Multiple Receiver
NLOS	Non Line of Sight
OFDM	Orthogonal Frequency Division Multiplexing
OSTBC	Orthogonal Space–Time Block Code
QAM	Quadrature Amplitude Modulation
QPSK	Quadrature Phase Shift Keying
SER	Symbol Error Rate
SF	Space–Frequency
SIMO	Single Input Multiple Output
SISO	Single Input Single Output
SM	Spatial Multiplexing
SNR	Signal to Noise Ratio
ST	Space–Time
STBC	Space–Time Block Code
STC	Space–Time Code/Coding
STF	Space–Time-Frequency
STP	Space–Time Process/Processing
STTC	Space–Time Trellis Code
SVD	Singular Value Decomposition
TSD	Transmit Selection Diversity
V-BLAST	Vertical Bell Laboratories Architecture Space Time
WiMAX	Worldwide Interoperability for Microwave Access
ZF	Zero Forcing

5

Feedback Techniques for MIMO Channels

David J. Love and Robert W. Heath, Jr.

CONTENTS

5.1 Introduction

Recent wireless innovations are improving bandwidth efficiency and reducing the deleterious effects of fading in wireless communication channels. Multiple antennas offer high capacity, enhanced resistance to interference, and reductions in fading thanks to diversity when arrays are used at both transmitter and receiver in a MIMO (multiple-input multiple-output) configuration. Broadband signaling, such as orthogonal frequency division multiplexing

(OFDM), simplifies equalization in channels with significant multipath interference thus enabling larger bandwidth channels. Multiuser processing
allows space, time, and frequency resources to be distributed among multiple
users to improve network performance and system capacity.

The capacity, quality, and complexity obtained from these new technologies are substantially improved when information about the channel state is
available at the transmitter. For example, equipped with channel state information, the transmitter can customize the transmitted waveforms to provide
higher link capacity and throughput, improve system capacity by more
efficiently sharing the channel with multiple users, increase range by exploiting diversity due to spatial and frequency selectivity, and simplify multiuser
receivers through known interference cancellation.

Research often assumes that channel state information is known perfectly
at the transmitter. In most systems, however, the only opportunity for the
transmitter to learn about the channel is through a low rate feedback control
channel. This is difficult because

1. the feedback rate is fixed and is generally small to reduce system
   overhead,

2. there are several channel parameters (proportional to the product of
   the number of channel taps, number of antennas, and number of
   users) to send each feedback interval, and

3. frequent feedback is needed since the wireless channel is changing
   over time due to mobility of the transmitter, receiver, or scatterers
   (characterized by the coherence time).

The implication of these three points is that channel state information must
be highly compressed. This motivates the study of feedback techniques as
a means for practically informing the transmitter about the channel state.

This chapter provides an overview of feedback techniques for MIMO communication channels. We emphasize instantaneous feedback, i.e., those methods that attempt to inform the transmitter about the instantaneous channel
state. Our approach and presentation are based on the framework illustrated
in Figure 5.1. The key idea is to treat the (estimated) channel coefficients as
a source and then to apply specially developed source coding algorithms to
quantize and compress the channel coefficients. This is not traditional source
coding, wherein the object is to reconstruct the source, because the performance metrics in a communication system are system parameters such as
SNR, bit-error-rate, throughput, and capacity. The objective of the quantizer
in Figure 5.1 is to compress (typically with a lossy algorithm) channel state
information such that performance is maximized under some assumptions
about the spatial formatting, e.g., space–time block coding or spatial multiplexing. We will present a vector quantization framework for quantizing
wireless channels that includes identifying the characteristics of the channel
that must be quantized and deriving source codes based on communication-
theoretic measures of distortion.

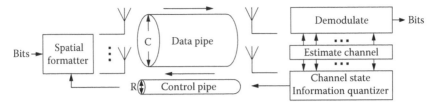

FIGURE 5.1
This figure illustrates a communication link with a feedback control channel. The receiver uses an estimate of the channel both to demodulate the data and to generate quantized channel state information. This information is sent to the transmitter using the feedback control channel. The transmitter, in turn, uses this information to customize the transmitted signal for the channel. The rate of the feedback channel R is typically much smaller than the rate of the data pipe C since a large control channel penalizes the overall efficiency of the system.

This chapter is organized as follows. First we motivate the problem of channel quantization and then review related work in Section 5.2. We introduce the notion of limited feedback MIMO communication in Section 5.3. We discuss two approaches to limiting feedback: direct channel quantization and quantized signal adaptation. The latter only quantizes features of the channel that are necessary to adjust the transmitted signal. In Section 5.4 we describe various algorithms for quantized signal adaptation for narrowband channels. Under some assumptions about the channel statistics we develop designs for codebooks suitable for beamforming and precoding in MIMO systems. We examine their performance in some practical correlated channels in Section 5.5. In Section 5.6 we study the extension to broadband MIMO-OFDM channels. We discuss per-tone quantization and interpolation for exploiting the correlation of adjacent subcarriers. The emphasis is on beamforming, though we discuss extensions to more general precoding. A detailed bibliography is included in the References section.

We use the following notation throughout this chapter. We use $\mathbf{I}_k$ to denote the $k \times k$ identity matrix, $E_y[\cdot]$ to denote expectation with respect to y, $*$ to denote matrix conjugate transposition, argmax to denote a function that returns a single global maximizer, $\lambda_j\{\mathbf{A}\}$ to denote the jth largest singular value of $\mathbf{A}$, $\|\cdot\|_2$ to denote the matrix two-norm, $\|\cdot\|_F$ to denote the matrix Frobenius norm, $(\mathbf{A})_k$ to denote the kth column of $\mathbf{A}$, $\mathbf{0}$ to denote a matrix of all zeros, $\mathcal{L}(M_t, M)$ to denote the set of $M_t \times M$ matrices with maximum singular value less than or equal to one (i.e., $\mathcal{L}(M_t, M) = \{\mathbf{F} \in \mathbb{C}^{M_t \times M} | \lambda_1\{\mathbf{F}\} \leq 1\}$), and $\mathcal{U}(M_t, M)$ to denote the set of $M_t \times M$ matrices with orthonormal columns.

5.2   Prior Work and Motivation

Single-user wireless links can be classified as narrowband or broadband channels on the basis of channel memory [88]. In this section, we review relevant literature in both these areas, with an emphasis on MIMO systems. Feedback issues have been considered in IEEE 802.16 fixed wireless systems [1], third generation cellular [33], and the next generation wireless local area network standard IEEE 802.11n [84].

To motivate the problem we consider a narrowband M_t transmit and M_r receive antenna MIMO system (we consider broadband extensions in Section 5.6). At baseband the channel can be represented by an $M_r \times M_t$ random matrix $\mathbf{H}$. This yields a channel input/output relationship at the kth transmission of

$$\mathbf{Y}_k = \mathbf{H}\mathbf{X}_k + \mathbf{N}_k \qquad (5.1)$$

where $\mathbf{X}_k$ is an $M_t \times T$ transmit matrix, $\mathbf{Y}_k$ is an $M_r \times T$ receive matrix, and $\mathbf{N}_k$ is an $M_r \times T$ additive white Gaussian noise matrix. The variable T simply denotes the number of temporal transmissions in the space–time signal. Space–time signal design can be partitioned into two cases: open-loop and closed-loop MIMO. In an open-loop MIMO system, $\mathbf{X}_k$ is designed *independently* of the channel conditions. Open-loop MIMO includes the popular spatial multiplexing or Bell Labs Space–Time (BLAST) system [20], orthogonal space–time block codes [4,90], space–time trellis codes [36,45,50,91], more recent space–time codes [11,14,15,35], as well as codes proposed by one of the authors [24,27,28]. Alternatively, closed-loop MIMO signals are designed *as a function* of the channel conditions. Closed-loop algorithms benefit from their channel adaptive nature by providing improved error rate, better spectral efficiency, and simplified decoding [6,70].

A transmit technique of interest in closed-loop MIMO is linear precoding of an open-loop signal. Linear precoding works by constructing $\mathbf{X}_k = \mathbf{F}\mathbf{S}_k$ where $\mathbf{S}_k$ is formed from one of the aforementioned open-loop algorithms and $\mathbf{F}$ is a linear precoding matrix determined based on $\mathbf{H}$. A common application of precoding for MIMO systems is to reduce the error rate, improve the throughput, or decrease the complexity of a space–time code designed without channel information. Initial work on precoding for spatial multiplexing was done by Scaglione et al. [81], Sampath et al. [77], Zhou et al. [106], and Heath et al. [29]. Jöngren et al. pioneered precoding for space–time block codes in [40] before more recent work on this subject [3,23,46,101,103,106,107]. The precoding matrix can be derived from the channel (see, e.g., [39,75,81]) or the channel statistics (see, e.g., [2,21,34,68,77,97, 104,106,107]). In general the precoding method depends on the type of receiver used. A special case of linear precoding is transmit beamforming, where $\mathbf{F} = \mathbf{f}$ is an M_t dimensional beamforming vector, and $\mathbf{S}_k = s_k$ is a complex number taken from a digital constellation. Various forms of $\mathbf{f}$ have been explored over the years, beginning with antenna selection in [86,100] to more general forms in [13,14,51,95], where $\mathbf{f}$ is restricted to be a unit vector and our work in [53], where $\mathbf{f}$ is restricted to have unit gain entries.

One problem with most of the current work in closed-loop MIMO communications is the assumption that complete channel state information is available at the transmitter. Sometimes this can be inferred from channel reciprocity (in time division duplexing (TDD) systems), but in frequency division duplexing (FDD) systems reciprocity will not be present. Besides, TDD systems require careful calibration of the RF devices for exploiting channel reciprocity; this calibration may require additional RF units and typically must be repeated frequently during operation. Thus closed-loop communication in FDD systems, and sometimes in TDD systems as well, requires investigating channel quantization and limited feedback.

The challenge of direct channel quantization [37–39, 46] is the large number of parameters. To illustrate, Table 5.1 provides the number of parameters for different kinds of transmit processing. Notice that more sophisticated forms of precoding require more feedback. Quantization of instantaneous channel state information is the alternative to direct channel quantization. Initial work in this area stems from the channel quantization work of Narula et al. [69] and the work by Heath and Paulraj on beamforming phase quantization [26] and antenna selection [22,29]. This kind of technique was then extended by Love et al. [53,63], Mukkavilli et al. [66,67] Santipach et al. [79,80], and Zhou et al. in [108]. Love et al. and Mukkavilli et al. independently found that the problem of feedback design for beamforming relates to the applied mathematics problem of Grassmannian line packing [10,87]. The codebook vectors are thought of as lines and are designed using subspace coding techniques. An overview of work in limited feedback MIMO can be found in [63b].

In broadband wireless links the feedback problem becomes more acute because of the additional channel parameters. The baseband input-output relationship of a broadband MIMO wireless link can be written as

$$\mathbf{y}(n) = \sum_{l=0}^{L} \mathbf{H}(l)\mathbf{s}(n-l) + \mathbf{v}(n) \tag{5.2}$$

where $\mathbf{y}(n)$ is the received vector at time n, $\mathbf{s}(n)$ is the transmitted signal vector at time n, $\mathbf{v}(n)$ is noise, and $\{\mathbf{H}(l)\}_{l=0}^{L}$ is a multitap channel impulse. In

TABLE 5.1

Example of the Gross Feedback Requirements per User with M_t Transmit Antennas, M_r Receive Antennas, for a Narrowband and N Tone MIMO System

Scheme	Feedback	# Params	# Params for OFDM
Simple power control	Total received power	1	N
Beamforming	Beamforming vector	$2M_t$	$2NM_t$
Eigenmode	Eigenvectors & values	$2M_tM_r$	$2N\,M_tM_r$
Precoding	Precoding matrix	$2M_tM_r$	$2N\,M_tM_r$
Full channel	Channel matrix	$2M_tM_r$	$2N\,M_tM_r$

Note: Feedback requirements for precoding are reduced when there is a special structure such as orthogonal columns.

general the impulse response is time-varying though often assumed time-invariant over a transmission burst. The convolution in Equation 5.2 implies that successive symbols interfere with one another creating intersymbol interference. To mitigate this self-interference, two different approaches are to use receive/transmit equalization or orthogonal frequency division multiplexing (OFDM).

Receive equalization techniques are well known in the communications literature [71] and are used in some form in most wireless systems. The transmit counterpart of receive equalization is conventionally known as Tomlinson-Harashima (TH) precoding [64,92] (see also [16] and the references therein). TH precoding has received less attention in wireless applications due to the requirement of precise knowledge of the impulse response at the transmitter. Fischer et al. have examined TH precoding for MIMO systems with low rate feedback channels [17] using the statistics of the channel. A general investigation, however, has not been made of the effect of feedback on TH precoding.

OFDM is a digital modulation technique where a wideband channel is divided into multiple narrowband channels [31]. Due to interest in OFDM for IEEE 802.11n [84], IEEE 802.16 [1], and fourth generation applications [78], we will address only limited feedback communication in broadband channels with OFDM equalization.

MIMO-OFDM systems make use of the benefits of OFDM modulation to simplify equalization in frequency selective channels. Thanks to OFDM, the equivalent system model is converted from the convolution in Equation 5.2 to a set of N (the number of subcarriers) flat fading channels. We comment more about the system model in Section 5.6. In MIMO-OFDM systems, which combine the capacity benefits of MIMO communication with the simple equalization of OFDM, feedback can be used to improve capacity, as demonstrated by Raleigh and Cioffi in early work on the subject [72]. Performance can also be improved with partial channel information, e.g., the work by Xia et al. who derived an adaptive beamformer for MIMO-OFDM based on imperfect channel state estimates [102]. Since MIMO-OFDM results in a set of narrowband matrix channels, our quantized beamforming and precoding strategies show great promise for MIMO-OFDM.

A simple application of narrowband quantized precoding to MIMO-OFDM can result in substantial feedback, in proportion to the number of subcarriers. Recently Choi and Heath recognized that the correlation between adjacent subcarriers can be used to reduce the amount of feedback [8,9]. Specifically, they proposed a technique where the precoders for only a fraction of the subcarriers were quantized and a special subspace interpolator was used for the other subcarriers. Work by Mondal and Heath bypassed the interpolation stage and instead used a clustering approach to find the best precoder for a cluster of subcarriers [65]. Quantized precoding for MIMO-OFDM has recently been adopted in the IEEE 802.16 standard; further developments are ongoing.

5.3   Limited Feedback MIMO

Due to the substantial performance gains that can be obtained when a MIMO signal is adapted to the channel, the design of limited feedback MIMO systems is an important theoretical and practical problem. The following subsections give an overview of the general ideas behind limited feedback MIMO systems.

5.3.1   System Description

A limited feedback system leverages cooperation between the transmitter and receiver. Consider for illustration purposes the narrowband MIMO communication system in Equation 5.1. In a coherent MIMO system, the receiver has an estimate of the forwardlink channel matrix $\mathbf{H}$ that is used for decoding. The receiver uses its knowledge of $\mathbf{H}$ to design feedback information to send to the transmitter. The transmitter then uses this feedback to adapt the transmitted signal to the channel.

The design of feedback can be partitioned into two main techniques: direct quantization of the channel and indirect quantization of the channel through signal adaptation. In general, either method can be successfully employed. We will show, however, that direct channel quantization lacks the performance of more specialized feedback methods. The reason is that it is often better to design limited feedback adaptation algorithms than to adapt to the quantized channel.

5.3.2   Channel Quantization

Motivated by the large body of VQ work over the last forty years, one approach to limited feedback is to employ channel quantization. The matrix channel quantization problem is reformulated as a VQ problem by stacking the columns of the channel matrix $\mathbf{H}$ into an $M_r \times M_t$ dimensional complex vector. The problem is then to quantize this high dimensional vector, which can be successfully done using a VQ algorithm.

In a vector quantizer, a real or complex valued vector is mapped into one of a finite number of vector realizations, known as a codebook. The codebook is chosen to minimize the average distortion with respect to the source distribution. The Lloyd algorithm or other numerical techniques are typically used to solve for the codebook. The mapping function is a measure of distortion between the unquantized input and the quantized output. The most common example is the mean squared error (MSE), i.e., choose $Q(\mathbf{x})$ such that the expected value of $\left\| \mathbf{x} - Q(\mathbf{x}) \right\|^2$ is minimized. There is, however, one key difference between VQ discussed in the compression literature and channel VQ. In the channel VQ case, the distortion function (or mapping) is typically a non-Euclidean distance measure. With VQ, the distortion is typically a 2-norm like the aforementioned MSE. In channel VQ, however, other distortion functions might be used, such as $2 - 2 \ |\langle \mathbf{x}, Q(\mathbf{x}) \rangle|$. The reason is that communication systems are often invariant to certain aspects of the channel — a property that can be used to reduce the number of degrees of freedom and, thus, the amount of feedback required. One simple example of invariance is closed-loop beamforming. The performance of beamforming is invariant to the channel being multiplied by $e^{j\vartheta}$ for any phase offset ϑ. Narula et al. first used this invariance in [69] for application to closed-loop beamforming. Note that the MSE is not invariant to phase distortion while the second distance mentioned above is, in fact, phase invariant. This invariance was used to reduce the number of feedback parameters.

Direct channel quantization technique does not bind the transmitter to any specific space–time signaling technique. In theory, this gives the transmitter flexibility to choose among different space–time signaling techniques such as beamforming, spatial multiplexing, or space–time coding. Previous work has shown that channel quantization can be employed for multiple-input single-output (MISO) beamforming [12,69] and MIMO precoded orthogonal space–time block coding [39].

5.3.3   Quantized Signal Adaptation

Narula et al.'s work motivated a new approach to limited feedback design. The algorithm in [69] was technically not only quantizing a MISO vector channel, but also quantizing *the optimal beamforming vector*. This different way of looking at the problem motivates a different approach to limited feedback. *The receiver only needs to send back the portion of the channel structure needed for signal design.*

For a fixed transmission technique, performance gains can be achieved by concentrating on improving the transmitter's knowledge of the quantized information needed to adapt the transmitted signal to current channel conditions. In particular, research has concentrated on modifying the precoded space–time block coding model described by Equation 5.1 to allow for quantized signal adaptation. The general approach is to use the fact that precoders are only a function of the channel's right singular vectors, thus yielding a dramatic reduction in the dimensionality of the quantization problem.

Recall that a precoded system transmits $\mathbf{X}_k = \mathbf{F}\mathbf{S}_k$ where $\mathbf{F}$ adapts the space–time block code matrix $\mathbf{S}_k$ to the current channel conditions. Thus, the transmitter does not need to know the complete channel matrix $\mathbf{H}$ but only $\mathbf{F}$. The lack of transmit channel knowledge can then be overcome by designing $\mathbf{F}$ at the receiver and sending $\mathbf{F}$ back to the transmitter using a small number of feedback bits (denoted by B). Practically, the receiver will use a selection function f (where $\mathbf{F} = f(\mathbf{H})$) mapping to a codebook

$$\mathcal{F} = \{\mathbf{F}_1, \mathbf{F}_2, \ldots, \mathbf{F}_N\} \tag{5.3}$$

of possible precoding matrices where $N = 2^B$. Because the codebook has 2^B elements, the chosen matrix $\mathbf{F}$ can be conveyed from the receiver to transmitter using a B bit pattern. This codebook-based approach has been used to study limited feedback beamforming [48,63,67], precoded orthogonal space–time block codes [46,47,57], precoded spatial multiplexing [58,62,73,74,76], and transmit covariance optimization [7,49].

The problem of quantized signal adaptation is similar to channel VQ. Solving the problem requires identifying an objective function and a codebook. The major difference is that the channel is not quantized directly, rather some function of the channel is quantized. Second, the objective is typically not to minimize distortion, rather it is to maximize performance. For example, we may choose the $\mathbf{F}$ such that the SNR (as a function of the channel) is maximized, rather than choosing $\mathbf{F}$ to best approximate the unquantized solution. This is possible since system performance is a function of the chosen precoder; the distortion does not directly play a role but indirectly affects system performance.

5.4   Quantized Signal Adaptation Algorithms

The analysis and design of limited feedback precoders is highly dependent on the type of underlying open-loop system. We will address the design for beamforming, precoded orthogonal space–time block coding, and precoded spatial multiplexing.

5.4.1   Beamforming Example and Summary of Approach

To illustrate the use of quantized signal adaptation consider an M_t transmit antenna by M_r receive antenna narrowband MIMO wireless system performing transmit beamforming with an optimal maximum likelihood (ML) receiver. With beamforming a single transmit stream is weighted by a beamforming vector determined by the channel $\mathbf{H}$. The baseband input-output relationship for this system after combining can be written as

$$y = \left\|\mathbf{H}\mathbf{f}\right\|_2 s + n \tag{5.4}$$

where $\mathbf{H}$ is the $M_r \times M_t$ narrowband channel response, $\mathbf{f}$ is an M_t dimensional beamforming vector, s is a complex number taken from a transmit constellation, n is a noise term, and $\|\cdot\|_2$ is the vector two-norm. Note that the temporal dependence of the symbol has been abstracted out. Transmit beamforming achieves a strong resilience to fading by adapting the vector $\mathbf{f}$ to the current channel conditions. Given a symbol s to be transmitted, the ith entry of the vector $\mathbf{f}s$ is then transmitted on the ith transmit antenna. We will assume that $E\left[|s|^2\right] = \varepsilon_s$ and that n is distributed according to $\mathcal{CN}(0, N_0)$. Note that the average transmit power conditioned on a beamformer is given by $\|\mathbf{f}\|_2^2 \varepsilon_s$. Therefore, we will restrict $\mathbf{f}$ to have unit norm.

Because the vector $\mathbf{f}$ is a function of current channel conditions, some knowledge of $\mathbf{H}$ is necessary at the transmitter. We propose to take a codebook-based approach to use a limited number of feedback bits to convey channel state information. The solution proposed in [61,63] is to restrict $\mathbf{f}$ to be a member of a codebook $\mathcal{F} = \{\mathbf{f}_1, \mathbf{f}_2, ..., \mathbf{f}_N\}$. This codebook is designed off-line, known to both the transmitter and receiver, and fixed for all time. There are two main challenges, given this kind of feedback architecture. First, a method must be determined for choosing a beamforming vector from the codebook $\mathcal{F}$ to send back. The chosen vector is conveyed to the transmitter using $\lceil \log_2 N \rceil$ bits of feedback. Second, we must determine how to construct the codebook $\mathcal{F}$.

In a beamforming system, maximizing the average receive SNR corresponds to simultaneously minimizing the probability of error and maximizing the channel capacity. Therefore, the receiver, which has knowledge of $\mathbf{H}$ from training, chooses one of the vectors in $\mathcal{F}$ to maximize the receive SNR. Assuming optimal maximum ratio combining, the post-combining average SNR is given by

$$\mathrm{SNR}_{out} = \|\mathbf{Hf}\|_2^2 \frac{\varepsilon_s}{N_0}$$

Therefore, the receiver can choose the transmit beamforming vector by performing a brute force search over the N vectors in the codebook $\mathcal{F}$ to solve

$$\mathbf{f} = \underset{\tilde{\mathbf{f}} \in \mathcal{F}}{\mathrm{argmax}} \left\|\mathbf{H}\tilde{\mathbf{f}}\right\|_2^2$$

We can address the codebook design issue by defining a distortion and solving for a codebook that minimizes the distortion. Consider the distortion

$$D(\mathcal{F}) = E\left[\left\|\mathbf{Hf}_{opt}\right\|_2^2 - \underset{\mathbf{f} \in \mathcal{F}}{\max}\|\mathbf{Hf}\|^2\right] \qquad (5.5)$$

defined by taking an expectation of the beamforming gain loss (i.e., the difference between the beamforming gain obtained by using the optimal unquantized vector $\mathbf{f}_{opt}$ and the beamforming gain obtained using the codebook $\mathcal{F}$) with respect to the channel. This distortion can be bounded as [63]

$$D(\mathcal{F}) \leq E\left[\|\mathbf{H}\|_2^2\right] E\left[\left(1 - \underset{\mathbf{f} \in \mathcal{F}}{\max}|\mathbf{v}^*\mathbf{f}|^2\right)\right] \qquad (5.6)$$

$$\leq E\left[\|\mathbf{H}\|_2^2\right]\left(\frac{\delta^2}{4}N\left(\frac{\delta}{2}\right)^{2(M_t-1)} + \left(1 - N\left(\frac{\delta}{2}\right)^{2(M_t-1)}\right)\right) \qquad (5.7)$$

where $\mathbf{v}$ is the dominant right singular vector of $\mathbf{H}$ and

$$\delta = \min_{1 \leq k < l \leq N} \sqrt{1 - \left|\mathbf{f}_k^* \mathbf{f}_l\right|^2} \qquad (5.8)$$

Differentiating the bound in Equation 5.7 shows that δ should be maximized. Note that

$$d(\mathbf{f}_1, \mathbf{f}_2) = \sqrt{1 - \left|\mathbf{f}_1^* \mathbf{f}_2\right|^2}$$

is a *subspace distance*. Thus we should think of the codebook as a finite set of *subspaces* rather than a finite set of vectors. The set of M-dimensional subspaces in $\mathbb{C}^{M_t}$ is known as the Grassmann manifold $\mathcal{G}(M_t, M)$.

From the theory of Grassmannian manifolds [5], the design of the codebook $\mathcal{F}$ was found to relate to *Grassmannian line packing*. Grassmannian line packing is an exciting area of coding theory where finite sets of subspaces are designed such that the minimum distance between any pair of subspaces is as large as possible. These kinds of codes can be designed using techniques from the theory of non-coherent code design [32], numerical techniques [10], alternating projection algorithms [93], and algebraic techniques [87].

Figure 5.2 shows the symbol error rate performance of a four transmit and five receive antenna beamforming system transmitting 16-QAM. Optimal maximum ratio combining is used at the receiver. Signal adaptive beamforming using a 6-bit codebook designed with the criterion in [63] outperforms 40-bit (2 bits per complex entry) channel quantization by approximately 1 dB. Limited feedback signal adaptive beamforming also performs within 0.7 dB of full transmit channel knowledge, unquantized beamforming.

The reason for the dramatic performance gains with the limited feedback signal adaptive approach over channel quantization is because the quantization problem focuses *strictly on the singular vector structure of the channel*. The 40-bit channel quantization has such large quantization error that the fragile eigen-structure of the channel is often mangled at the transmitter. The lack of reliable eigen-structure information at the transmitter causes a loss in performance for the beamformer.

5.4.2 Precoded Orthogonal Space–Time Block Codes

As other chapters have shown, orthogonal space–time block coding (OST-BCing) is an effective technique to combat the effects of fading with multiple

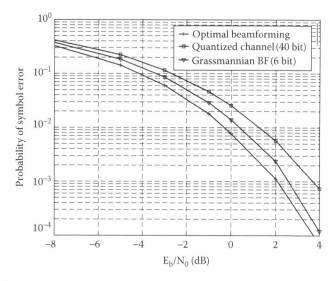

FIGURE 5.2
Performance of limited feedback beamformer for a four-transmit and five-receive antenna system using 16-QAM. This plot demonstrates the benefits of using quantized signal adaptation rather than direct channel quantization.

transmit antennas. The performance of OSTBCs, however, can be improved by leveraging channel knowledge at the transmitter [47].

At the kth transmission, an OSTBC encoder maps a block of n_s symbols s_1, s_2, ..., s_{n_s} taken from a constellation S to an $M \times T$ (with $M < M_t$) space–time code matrix $\mathbf{C}(k) = [\mathbf{c}_1(k)\,\mathbf{c}_2(k)\dots\mathbf{c}_T(k)]$. The power is constrained so that [40]

$$\mathbf{C}(k)\mathbf{C}(k)^* = \left(\sum_{l=1}^{n_s}|s_l|^2\right)\mathbf{I}_M \tag{5.9}$$

for all k. In addition, we will assume that the constellation has been normalized so that $E_{s_l}\left[|s_l|^2\right] = 1$.

The space–time block codeword $\mathbf{C}(k)$ is precoded by an $M_t \times M$ matrix $\mathbf{F}$. As in the limited feedback beamforming case, this matrix will be restricted to lie in a codebook $\mathcal{F} = \{\mathbf{F}_1, \mathbf{F}_2, \dots, \mathbf{F}_N\}$. The matrix $\mathbf{F}$ is restricted according to a peak power constraint for all $\mathbf{F} \in \mathcal{F}$ such that $\lambda_1\{\mathbf{F}\} \leq 1$. The matrix $\mathbf{F}$ will be further restricted, but this restriction will be developed later.

We will assume a narrowband MIMO system with a spatially uncorrelated, memoryless linear channel that is constant over several codeword transmissions before independently changing. These assumptions allow the system input-output expression to be written as

$$\mathbf{Y} = \sqrt{\frac{\rho}{M}}\mathbf{HFC} + \mathbf{N} \tag{5.10}$$

where $\mathbf{H}$ is the $M_r \times M_t$ channel matrix with independent entries distributed as $\mathcal{CN}(0,1)$, ρ is the signal-to-noise ratio (SNR), and $\mathbf{N}$ is an $M_r \times T$ noise matrix with independent entries distributed according to $\mathcal{CN}(0,1)$. The receiver has perfect knowledge of $\mathbf{H}$ and $\mathbf{F}$, and it decodes $\mathbf{Y}$ using optimal ML decoding.

To implement a limited feedback precoded OSTBCing system, two main problems must be addressed. First, we must develop a selection criterion to choose $\mathbf{F}$ from $\mathcal{F}$. Second, we must determine how to design the codebook $\mathcal{F}$ to maximize some performance criterion.

We will define performance with respect to the symbol error rate (SER) given $\mathbf{H}$ denoted by $Pr(ERROR|\mathbf{H})$. Using the orthogonality properties of OSTBCs, it can be shown that

$$Pr\left(ERROR|\mathbf{H}\right) \le exp\left(-\gamma \left\|\mathbf{HF}\right\|_F^2\right) \tag{5.11}$$

where γ is a function that depends on ρ, M, and S [46,47]. Because the SNR, constellation, and signaling architectures are fixed during transmission, maximizing $\left\|\mathbf{HF}\right\|_F$ corresponds to minimizing the SER bound. This yields a natural precoder selection criterion. The receiver chooses the linear precoder $\mathbf{F}$ according to

$$\mathbf{F} = \underset{\tilde{\mathbf{F}} \in \mathcal{F}}{\mathrm{argmax}} \left\|\mathbf{H}\tilde{\mathbf{F}}\right\|_F \tag{5.12}$$

Consider the optimal precoder $\mathbf{F}_{opt}$ chosen to maximize $\left\|\mathbf{HF}_{opt}\right\|_F^2$ over the set of maximum singular value constrained matrices $\mathcal{L}(M_t, M)$. This matrix $\mathbf{F}_{opt}$ will not be unique over $\mathcal{L}(M_t, M)$ because for arbitrary $\mathbf{U} \in \mathcal{U}(M,M)$, $\left\|\mathbf{HF}_{opt}\right\|_F = \left\|\mathbf{HF}_{opt}\mathbf{U}\right\|_F$. Denote singular value decomposition (SVD) of $\mathbf{H}$ by

$$\mathbf{H} = \mathbf{V}_L \Sigma \mathbf{V}_R^* \tag{5.13}$$

where $\mathbf{V}_R \in \mathcal{U}(M_t, M_t)$, $\mathbf{V}_L \in \mathcal{U}(M_r, M_r)$, and Σ is an $M_r \times M_t$ diagonal matrix with $\lambda_j\{\mathbf{H}\}$ at entry (j,j). Let $\bar{\mathbf{V}}_R$ denote the matrix taken by selecting the first M columns of $\mathbf{V}_R$. The following lemma summarizes $\mathbf{F}_{opt}$.

LEMMA 5.1
[57] The precoder matrix $\mathbf{F}_{opt} = \bar{\mathbf{V}}_R$ maximizes $\left\|\mathbf{HF}_{opt}\right\|_F$ over $\mathcal{L}(M_t, M)$.

Lemma 5.1 tells us that not only should $\lambda_1\{\mathbf{F}_{opt}\} = 1$ but $\lambda_j\{\mathbf{F}_{opt}\} = 1$ for $1 \le j \le M$. Intuitively, this means that we should perform waterfilling to maximize $\left\|\mathbf{HF}\right\|_F^2$ compound channel when transmitting with a peak power constraint. When the transmitter has no *a priori* knowledge of $\mathbf{H}$, the transmitter will have no knowledge of $\bar{\mathbf{V}}_R$. Therefore, we will use a limited feedback path to convey a suboptimal precoder. The following lemma shows that *any* suboptimal linear precoder should always have unit singular values under a maximum singular value constraint in order to minimize the SER.

LEMMA 5.2

[57] *Let* $\mathbf{F}' \in \mathcal{L}(M_t, M)$ *have a singular value decomposition* $\mathbf{F}' = \mathbf{U}_L \mathbf{\Gamma} \mathbf{U}_R^*$ *such that* $\lambda_M\{\mathbf{F}'\} < 1$. *Given this,* $\|\mathbf{HF}'\|_F \leq \|\mathbf{H\tilde{F}}\|_F$ *when* $\tilde{\mathbf{F}} = \mathbf{U}_L [\mathbf{I}_M \ 0]^T$.

Proof: Let $\mathbf{F}'$ be defined as in the lemma. Using (i) the invariance of the Frobenius norm to unitary transformation, (ii) the singular value bound for $\mathbf{F}'$, and (iii) the lemma's definition of $\tilde{\mathbf{F}}$,

$$\|\mathbf{HF}'\|_F = \|\mathbf{HU}_L \mathbf{\Gamma}\|_F \leq \|\mathbf{HU}_L [\mathbf{I}_M \ 0]^T\|_F = \|\mathbf{H\tilde{F}}\|_F$$

Lemma 5.2 is important because it tells us that the precoder should *always* be designed to have singular values that are as large as possible. Because of the maximum singular value restriction, this singular value maximization result corresponds to designing precoders in the set $\mathcal{U}(M_t, M)$ rather than $\mathcal{L}(M_t, M)$. For this reason, we will restrict our codebook precoders to lie in $\mathcal{U}(M_t, M)$ (i.e., $\mathbf{F}^*\mathbf{F} = \mathbf{I}_M$).

We would like the quantized equivalent channel $\mathbf{HF}$ to provide SER performance close to that provided by the optimal precoded equivalent channel $\mathbf{HF}_{opt}$ when $\mathbf{F}$ is restricted to $\mathcal{F}$. One method for doing this was proposed in [57] by using the average of the loss in received channel power as a measure of distortion. Using this definition of performance loss, the *average* distortion of the codebook is given by

$$E_{\mathbf{H}}\left[\min_{\mathbf{F}\in\mathcal{F}}\left(\|\mathbf{HF}_{opt}\|_F^2 - \|\mathbf{HF}\|_F^2\right)\right] \tag{5.14}$$

It was shown in [54,55,57] that the average distortion can be bounded as

$$E_{\mathbf{H}}\left[\min_{\mathbf{F}\in\mathcal{F}}\left(\|\mathbf{HF}_{opt}\|_F^2 - \|\mathbf{HF}\|_F^2\right)\right] \leq E_{\mathbf{H}}\left[\lambda_1^2\{\mathbf{H}\}\right] E_{\mathbf{H}}\left[\min_{\mathbf{F}\in\mathcal{F}}\frac{1}{2}\|\mathbf{\tilde{V}}_R \mathbf{\tilde{V}}_R^* - \mathbf{FF}^*\|_F^2\right] \tag{5.15}$$

This bound on the average distortion can be thought of as the product of two different terms. The first term relates to the distribution of the maximum channel singular value, which is not under the control of the system designer. The second term represents the "quality" of the codebook $\mathcal{F}$ and can therefore be optimized.

Before dealing further with the codebook design, we must review some properties and notations dealing with finite subsets of $\mathcal{U}(M_t, M)$. The column space of each matrix in $\mathcal{U}(M_t, M)$ corresponds to an M-dimensional subspace of $\mathbb{C}^{M_t}$. The set of all subspaces in $\mathbb{C}^{M_t}$ is known as the complex Grassmann manifold $\mathcal{G}(M_t, M)$ [105]. Just as in the traditional Euclidean space, the Grassmann manifold can have distances defined on it. The *chordal distance* between subspaces $\mathcal{P}_{F_1}$ and $\mathcal{P}_{F_2}$, where $\mathcal{P}_F$ denotes the subspace generated by $\mathbf{F}$, is given by

$$d(\mathbf{F}_1, \mathbf{F}_2) = \frac{1}{\sqrt{2}} \left\| \mathbf{F}_1 \mathbf{F}_1^* - \mathbf{F}_2 \mathbf{F}_2^* \right\|_F \qquad (5.16)$$

Consider the set of subspaces $V = \{\mathcal{P}_{F_1}, \mathcal{P}_{F_2}, \ldots, \mathcal{P}_{F_N}\}$ generated by the codebook matrices $F = \{\mathbf{F}_1, \mathbf{F}_2, \ldots, \mathbf{F}_N\}$. This finite set of subspaces can be thought of as a subspace code or subspace packing in $\mathcal{G}(M_t, M)$. Just as in binary coding theory, we can define the minimum distance of a packing:

$$\delta = \min_{1 \le k < l \le N} d(\mathbf{F}_k, \mathbf{F}_l) \qquad (5.17)$$

In applied mathematics, the Grassmannian subspace packing problem is the problem of choosing N subspaces (i.e., designing an N element subspace code) in order to maximize the minimum distance δ.

Using a metric ball approach, the "codebook quality" term in Equation 5.15 can be approximately bounded as [57]

$$E_H \left[\min_{\mathbf{F} \in F} \frac{1}{2} \left\| \bar{\mathbf{V}}_R \bar{\mathbf{V}}_R^* - \mathbf{F} \mathbf{F}^* \right\|_F^2 \right] \lesssim M + N \left(\frac{\delta}{2\sqrt{M}} \right)^{2 M_t M + o(M_t)} \left(\frac{1}{4} \delta^2 - M \right) \qquad (5.18)$$

This bound is a decreasing function of δ when $2M_t M + o(M_t) > 2/3$, which can be easily proven by differentiating the bound and noting that $\delta < \sqrt{M}$. Thus maximizing δ approximately minimizes Equation 5.18. This establishes that a low-distortion, high-performance codebook can be designed by packing subspaces in the Grassmann manifold using the chordal distance metric. Note that when $M = 1$, this bound corresponds to the beamforming scenario discussed in the previous section.

The experiment shown in Figure 5.3 compares the performance of antenna subset selection [23] with 8-bit chordal distance precoding on an 8×1 wireless system transmitting an $M = 2$ Alamouti code constructed from 16-quadrature amplitude modulation (QAM). For comparison, the SER performance for an Alamouti code transmitted on a 2×1 system is also shown. For example, at an error rate of 10^{-3}, antenna subset selection gives an 8 dB gain over the two antenna open-loop system. An additional gain of approximately 1.4 dB results from using 8-bit feedback chordal distance precoding.

5.4.3 Precoded Spatial Multiplexing

In spatial multiplexing, a bit stream is demultiplexed into M different bit streams that are independently modulated using the same constellation $\mathcal{S}$.

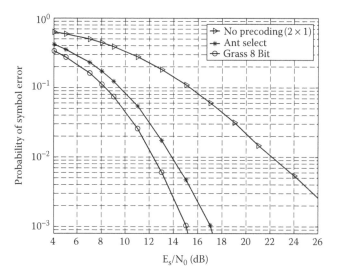

FIGURE 5.3
Symbol error rate comparison of two-antenna open-loop Alamouti space–time coding, eight transmit antenna subset selection combined with an Alamouti code, and eight antenna 8-bit feedback precoding combined with an Alamouti code. The constellation was assumed to be 16-QAM, and the receiver was assumed to have a single antenna.

At time instance K, the M symbols can be written as a vector $\mathbf{s}_k = [s_{k,1}\, s_{k,2} \cdots s_{k,M}]^T$ with $E_{\mathbf{s}_k}\,[\mathbf{s}_k\mathbf{s}_k^*] = \mathbf{I}_M$. In precoded spatial multiplexing, the symbol vector $\mathbf{s}_k$ is multiplied by an $M_t \times M$ matrix $\mathbf{F}$ before transmission producing a length M_t dimensional vector

$$\mathbf{x}_k = \sqrt{\frac{\varepsilon_s}{M}}\,\mathbf{F}\mathbf{s}_k$$

where ε_s is the total transmit energy, M_t is the number of transmit antennas, and $M_t \geq M$.

This formulation allows the baseband, discrete-time input-output relationship to be written as

$$\mathbf{y}_k = \sqrt{\frac{\varepsilon_s}{M}}\,\mathbf{H}\mathbf{F}\mathbf{s}_k + \mathbf{n}_k \qquad (5.19)$$

where $\mathbf{H}$ is an $M_t \times M_r$ channel matrix and $\mathbf{n}_k$ is a length M_r noise vector. The receiver, which is assumed to have perfect knowledge of $\mathbf{H}$ and $\mathbf{F}$, decodes the received vector $\mathbf{y}_k$ using a vector decoder to produce a hard decoded symbol vector $\hat{\mathbf{s}}_k$. We assume the channel matrix $\mathbf{H}$ to have independent $\mathcal{CN}(0,1)$ entries, and the noise vector $\mathbf{n}_k$ to have independent $\mathcal{CN}(0, N_0)$ entries.

The precoding matrix **F** is chosen at the receiver from a finite set of possible precoding matrices $\mathcal{F} = \{\mathbf{F}_1, \mathbf{F}_2, \ldots, \mathbf{F}_N\}$ known to both the transmitter and receiver. The receiver sends the binary index of the chosen precoding matrix back to the transmitter over a limited capacity, zero-delay feedback link. As in the precoded OSTBC scenario, each $\mathbf{F} \in \mathcal{F}$ has unit column vectors that are orthogonal. This orthogonality assumption is not especially restrictive since it follows from the form of the optimal spatial multiplexing precoders derived in [81] that consist of a matrix with orthogonal columns pre-multiplied by a diagonal matrix.

As with beamforming and precoded OSTBCing, we need to (i) determine how to choose the precoded **F** given a codebook $\mathcal{F}$ and (ii) design the codebook $\mathcal{F}$. Unlike the other algorithms, spatial multiplexing is often implemented with a variety of different decoder algorithms. We will briefly discuss the different decoders and their accompanying precoder selection criterion.

5.4.3.1 Precoder Selection for Spatial Multiplexing

Maximum Likelihood Receiver Inspired Selection

The ML receiver decodes to $\hat{\mathbf{s}}_k$ by solving

$$\hat{\mathbf{s}}_k = \underset{\mathbf{s} \in \mathcal{S}^M}{\operatorname{argmin}} \left\| \mathbf{y}_k - \sqrt{\frac{\varepsilon_s}{M}} \mathbf{HFs} \right\|_2^2 \tag{5.20}$$

To improve probability of error performance, it is of interest to find a closed-form expression for the probability of symbol vector error. Unfortunately, a closed-form expression for the probability of symbol vector error has yet to be derived in the literature. One solution is to upper bound the probability of symbol vector error for high signal-to-noise ratios (SNR) using the vector Union Bound [30]. Since we assume that the SNR ε_s/N_0 is held constant, the Union Bound is solely a function of the receive minimum distance of the multidimensional constellation $\mathcal{S}^M$ [30]. This motivates the following criterion for ML receivers.

SC-ML: Pick **F** such that

$$\mathbf{F} = \underset{\mathbf{F} \in \mathcal{F}}{\operatorname{argmax}} \ \underset{\mathbf{s}_1, \mathbf{s}_2 \in \mathcal{S}^M : \mathbf{s}_1 \neq \mathbf{s}_2}{\min} \sqrt{\frac{\varepsilon_s}{M}} \left\| \mathbf{H}\tilde{\mathbf{F}}\left(\mathbf{s}_1 - \mathbf{s}_2\right) \right\|_2 \tag{5.21}$$

Linear Receiver Inspired Selection

Unlike ML decoding, linear receivers can be designed in different ways depending on the desired decoding matrix **G**. A linear decoder decodes to the vector $\hat{\mathbf{s}}_k = \mathbf{Q}(\mathbf{Gy}_k)$ where **G** is an $M \times M_r$ matrix and $\mathbf{Q}(\cdot)$ is a function that performs single-dimension maximum likelihood decoding for each entry of a vector.

Minimum Singular Value

It was shown in [29] that the performance of a spatial multiplexing system using a linear receiver is characterized by the minimum substream SNR.

Maximizing the minimum singular value provides a close approximation to maximizing the minimum substream SNR [29]. The reason for this is that as *card*(W) grows large, the probability of an error vector lying collinear to the minimum singular value direction goes to one. This motivates the following criterion

SC-MSV: Pick **F** such that

$$F = \underset{\tilde{F} \in \mathcal{F}}{\arg\max} \ \lambda_{\min} \{H\tilde{F}\} \tag{5.22}$$

Minimum Mean Squared Error Receiver: Minimizing some function of the mean squared error matrix has been studied in [81] to improve the overall system performance. The average mean squared error when decoding with a minimum mean squared error (MMSE) linear decoder is

$$\overline{MSE}(F) = \frac{\varepsilon_s}{M} \left(I_M + \frac{\varepsilon_s}{MN_0} F^* H^* H F \right)^{-1} \tag{5.23}$$

We can use Equation 5.23 to derive a selection criterion for choosing **F** from $\mathcal{F}$.

SC-MSE: Choose **F** such that

$$F = \underset{\tilde{F} \in \mathcal{F}}{\arg\min} \ m\left(\overline{MSE}(\tilde{F}) \right) \tag{5.24}$$

where $m(\cdot)$ is either $tr(\cdot)$ (the trace of a matrix) or $det(\cdot)$ (the determinant of a matrix).

Capacity Selection

The final criterion can be used for both ML and linear receivers. When the transmitter precodes with **F** before transmission, we can view the system as experiencing an effective channel of **HF**. Thus the mutual information assuming independent and identically distributed Gaussian signaling for a fixed realization of **H** and a fixed **F** is given by

$$I(F) = \log_2 det\left(I_M + \frac{\varepsilon_s}{MN_0} F^* H^* H F \right) \tag{5.25}$$

Therefore, a capacity selection can be derived from Equation 5.25.

SC-Capacity: Choose **F** such that

$$\mathbf{F} = \arg\max_{\mathbf{F}\in\mathcal{F}} I(\tilde{\mathbf{F}}) \tag{5.26}$$

5.4.3.2　Codebook Design for Spatial Multiplexing

As with precoding for OSTBCing we will think of the codebook as a set of subspaces in $\mathbb{C}^{M_t}$. Thus, the limited feedback precoded spatial multiplexing codebook $\mathcal{F}$ represents a set, or packing, of subspaces in the Grassmann manifold. We will use the subspace coding interpretation to simplify notation and analysis.

The measure on $\mathcal{U}(M_t,M)$ induces a normalized invariant measure μ on $\mathcal{G}(M_t,M)$. Volumes can be computed within $\mathcal{G}(M_t,M)$ using this induced measure. In addition to the chordal distance, a number of other distances can be defined on $\mathcal{G}(M_t,M)$ [5]. The *projection two-norm,* and *Fubini-Study* distances between the two subspaces $\mathcal{P}_{\mathbf{F}_1}$ and $\mathcal{P}_{\mathbf{F}_2}$ are given by

$$d_{proj}(\mathbf{F}_1,\mathbf{F}_2) = \left\|\mathbf{F}_1\mathbf{F}_1^* - \mathbf{F}_2\mathbf{F}_2^*\right\|_2$$

$$d_{FS}(\mathbf{F}_1,\mathbf{F}_2) = \arccos\left|det\left(\mathbf{F}_1^*\mathbf{F}_2\right)\right|$$

Let $\mathcal{S} = \{\mathcal{P}_{\mathbf{F}_1}, \mathcal{P}_{\mathbf{F}_2}, ..., \mathcal{P}_{\mathbf{F}_N}\}$ denote the set of subspaces obtained from the column space of the matrices in $\mathcal{F}$. The minimum distance of this packing is

$$\delta = \min_{1\le i<j\le N} d(\mathbf{F}_i,\mathbf{F}_j) \tag{5.27}$$

where $d(\cdot,\cdot)$ is a distance function on $\mathcal{G}(M_t,M)$.

Understanding the optimal unquantized matrix will allow us to understand how to design $\mathcal{F}$. The first important point to notice is contained in the following lemma.

LEMMA 5.3

[54,55] *Let* $\mathbf{H}^*\mathbf{H} = \mathbf{V}_R\Sigma\mathbf{V}_R^*$ *where* $\mathbf{V}_R \in \mathcal{U}(M_t,M_t)$ *and* Σ *has diagonal entries* $\lambda_1^2 \ge \lambda_2^2 \ge ... \ge \lambda_{M_t}^2 \ge 0$. *The optimal unquantized precoding matrix (i.e.,* $\mathcal{F}\subseteq\mathcal{U}(M_t,$ *M)) for each of the criteria in Section 5.4 is* $\overline{\mathbf{V}}_R$, *the first M columns of* $\mathbf{V}_R$.

For *SC-ML* it was shown in [30] that the receive minimum distance is approximately maximized by maximizing the minimum singular value of **HF**. *SC-MSE* using the trace cost function chooses $\mathbf{F}\in\mathcal{F}$ that minimizes Equation 5.23. For high SNR with N_0 scaled to unity, the trace of Equation 5.23 can be approximated by $tr\left[(\mathbf{F}^*\mathbf{H}^*\mathbf{H}\mathbf{F})^{-1}\right]$. However,

$$\min_{\mathbf{F}\in\mathcal{F}} tr\left(\left(\mathbf{F}^*\mathbf{H}^*\mathbf{H}\mathbf{F}\right)^{-1}\right) \le \min_{\mathbf{F}\in\mathcal{F}} \frac{M}{\lambda_{\min}^2\{\mathbf{H}\mathbf{F}\}} \qquad (5.28)$$

Therefore this bound requires the maximization of $\lambda_{\min}\{\mathbf{H}\mathbf{F}\}$.

Thus the *SC-MSV* and the bounds on *SC-ML* and *SC-MSE* using the trace cost function all lead to maximizing $\lambda_{\min}\{\mathbf{H}\mathbf{F}\}$. We will therefore use our codebook to minimize

$$E_{\mathbf{H}}\left[\lambda_{\min}^2\{\mathbf{H}\mathbf{F}_{opt}\} - \max_{\mathbf{F}\in\mathcal{F}}\lambda_{\min}^2\{\mathbf{H}\mathbf{F}\}\right] = E_{\mathbf{H}}\left[\lambda_M^2\{\mathbf{H}\} - \max_{\mathbf{F}\in\mathcal{F}}\lambda_{\min}^2\{\mathbf{H}\mathbf{F}\}\right] \qquad (5.29)$$

for *SC-ML*, *SC-MSE* using the trace cost function, and *SC-MSV* where once again $\lambda_M\{\mathbf{H}\}$ is the Mth largest singular value of $\mathbf{H}$. Using the subspace packing properties of $\mathcal{F}$, we will show that Equation 5.29 can be bounded by a decreasing function of the minimum projection two-norm distance of $\mathcal{F}$, denoted by δ_{proj}.

> **Codebook Design Criterion 1** [58]: Codebooks for systems using *SC-ML*, *SC-MSV*, and *SC-MSE* using the trace cost function to select $\mathbf{F}$ from $\mathcal{F}$, should be designed by using an N subspace packing in the Grassmann manifold that maximizes the minimum projection two-norm distance.

It was shown in [54,55] that selecting $\mathbf{F}\in\mathcal{F}$ using *SC-MSE* with the determinant cost function is equivalent to maximizing $\tilde{I}(\mathbf{F})$ where

$$\tilde{I}(\mathbf{F}) = det\left(\mathbf{I}_M + \frac{\varepsilon_s}{MN_0}\mathbf{F}^*\mathbf{H}^*\mathbf{H}\mathbf{F}\right)$$

Thus the codebook design criterion for *SC-Capacity* and *SC-MSE* with the determinant cost function must be identical.

Note that

$$\tilde{I}(\mathbf{F}) \ge \left|det\left(\bar{\mathbf{V}}_R^*\mathbf{F}\right)\right|^2 det\left(\mathbf{I}_M + \frac{\varepsilon_s}{MN_0}\bar{\Sigma}^T\bar{\Sigma}\right) \qquad (5.30)$$

Using the above analysis and observing that

$$\tilde{I}(\mathbf{F}_{opt}) = det\left(\mathbf{I}_M + \frac{\varepsilon_s}{MN_0}\bar{\Sigma}^T\bar{\Sigma}\right)$$

we will use the cost function

$$E_{\mathrm{H}}\left[det\left(\mathbf{I}_M + \frac{\varepsilon_s}{MN_0}\overline{\Sigma}^T\overline{\Sigma}\right) - \max_{\mathbf{F}\in\mathcal{F}}\left|det\left(\overline{\mathbf{V}}_R^*\mathbf{F}\right)\right|^2 det\left(\mathbf{I}_M + \frac{\varepsilon_s}{MN_0}\overline{\Sigma}^T\overline{\Sigma}\right)\right]$$

$$= E_{\mathrm{H}}\left[det\left(\mathbf{I}_M + \frac{\varepsilon_s}{MN_0}\overline{\Sigma}^T\overline{\Sigma}\right)\right]\left(1 - E_{\mathrm{H}}\left[\max_{\mathbf{F}\in\mathcal{F}}\left|det\left(\overline{\mathbf{V}}_R^*\mathbf{F}\right)\right|^2\right]\right)$$

$$(5.31)$$

where Equation 5.31 follows from the independence of Σ and $\mathbf{V}_R$.

Using the subspace packing interpretation of $\mathcal{F}$, this distortion cost function can be bounded by a decreasing function of the minimum Fubini-Study distance of $\mathcal{F}$, denoted by δ_{FS}. This yields the following design criterion.

> **Codebook Design Criterion 2** [58]: Codebooks for systems using *SC-MSE* with the determinant cost function or *SC-Capacity* to select $\mathbf{F}$ from $\mathcal{F}$ should be designed by using an N subspace packing in the Grassmann manifold that maximizes the minimum Fubini-Study distance.

Finding good Grassmannian packings for arbitrary M_t, M, and N is actually quite difficult [10]. The codebook design difficulty is only increased for the non-standard projective two-norm and Fubini-Study distances [5]. One practical technique for designing good packings is to use the non-coherent constellation designs from [32]. This design algorithm usually yields codebooks with large minimum distances and can be easily modified to work with any distance function on the Grassmann manifold.

Figure 5.4 demonstrates the problems associated with directly quantizing the matrix channel $\mathbf{H}$ and the benefits of using a precoder codebook. A 4×2 wireless system using $M = 2$ and 16QAM was simulated. For a probability of symbol vector error of 10^{-2}, directly quantizing the channel with 16 bits of feedback performs approximately 4.7 dB worse than a 6-bit limited feedback precoder. With only 6 feedback bits, the limited feedback precoder obtains performance almost identical to the unquantized MMSE precoder with the maximum singular value power constraint.

5.5 Effect of Spatial Correlation

In the previous sections, the channel was modeled as a spatially uncorrelated Rayleigh fading channel. This assumption was used to derive codebook design criteria. Real channels will experience spatial correlation. In this section, we will present some results that show that diversity performance is not compromised for our codebook designs.

There are several models for spatial correlation. One of the most popular models was studied in the Information Society Technologies (IST) Multi-Element Transmit and Receive Antennas (METRA) Project [18] funded by

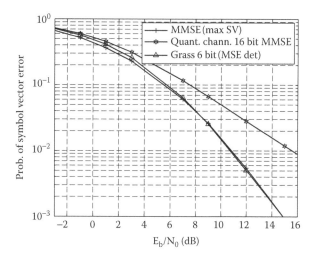

FIGURE 5.4
Comparison of the probability of symbol vector errors for 16-bit direct channel quantization, unquantized MMSE, and 6-bit limited feedback precoding on a 2 substream 4×2 system using 16QAM.

the European Union. The IST METRA model generates the channel **H** according to [96]

$$\mathbf{H} = \mathbf{R}_R^{1/2} \mathbf{G} \mathbf{R}_T^{1/2} \tag{5.32}$$

where **G** is a random matrix with i.i.d. $\mathcal{CN}(0,1)$ entries, $\mathbf{R}_R$ is the receive antenna correlation matrix, and $\mathbf{R}_T$ is the transmit antenna correlation matrix. The correlation matrices $\mathbf{R}_R$ and $\mathbf{R}_T$ can be measured or computed. Measured correlation matrices were discussed in [44], while computational methods for obtaining the correlations were studied in [19,83,96].

Note that the IST METRA model is not a fully correlated model [47] and can be thought of as a transmit-receive correlated model. The correlation is isolated to only transmit and receive correlation. This kind of correlation simplification was experimentally verified in [42–44,82,89,98]. Recently, this model has been adopted by the IEEE 802.11n wireless local area network working group as a MIMO channel model [96] and has been studied for use in IEEE 802.20 working group on mobile broadband wireless access [85]. Because of its wide use, this model is one of the cornerstones by which most MIMO algorithms are tested before implementation.

System performance on transmit-receive correlated models is highly dependent on the correlation matrices. It is therefore imperative to understand the requirements that $\mathbf{R}_R$ and $\mathbf{R}_T$ must maintain to guarantee full diversity order. We will characterize the diversity order of beamforming and precoded OSTBCing in transmit-receive correlated fading assuming optimal combining.

THEOREM 5.1

[60] Beamforming and combining over memoryless, correlated Rayleigh fading channels provide a diversity order of $M_t M_r$ if and only if (i) the vectors in the beamformer codebook $\mathcal{F}$ span $\mathbb{C}^{M_t}$, (ii) the rank of $\mathbf{R}_T$ is M_t, and (iii) the rank of $\mathbf{R}_R$ is M_r.

Theorem 5.1 provides the intuition that the limited feedback codebook *must always* form a basis for the vector space of possible dominant singular value directions. Thus, it is impossible for the optimal direction to be orthogonal to all of the codebook vectors. The correlation matrices change the distributions of left and right singular vectors. It is important, however, that the correlations allow for the full number of degrees-of-freedom.

THEOREM 5.2

[52] Precoded OSTBCing provides a diversity of order $M_t M_r$ if and only if (i) the matrix $\mathbf{E} = [\mathbf{F}_1\ \mathbf{F}_2\ \cdots\ \mathbf{F}_N]$ has a rank of M_t, (ii) the rank of $\mathbf{R}_T$ is M_t, and (iii) the rank of $\mathbf{R}_R$ is M_r.

If given a sufficient rate of feedback, limited feedback algorithms can obtain *identical* diversity orders to the diversity order from perfect channel knowledge closed-loop schemes. To obtain full diversity order, Theorem 5.2 tells us that precoded OSTBCing requires $\log_2(M_t/M)$ bits of feedback because the dimension of $\mathbf{F}_i$ for all i is $M_t \times M$. This differs from the result in Theorem 5.1, which says that limited feedback beamforming requires $\log_2(M_t)$ bits of feedback. Thus, precoded OSTBCing can provide full diversity order with lower rate feedback. This benefit, however, comes at the expense of a performance loss for higher feedback amounts.

5.6 Feedback in Broadband Channels with MIMO-OFDM

Previous sections addressed the theory of limited feedback MIMO communication based on precoder quantization and studied their practical performance in narrowband channels. Most developing wireless standards though are broadband, typically employing some form of OFDM modulation. In this section we describe how the aforementioned concept of limited feedback can be extended to the MIMO-OFDM setting. The key idea is to reduce feedback by exploiting the correlation of adjacent subcarriers. We describe the application to beamforming in detail. Extensions to precoding are more briefly described. Limited feedback for MIMO-OFDM has recently been adopted as an option in the IEEE 802.16 standard.

5.6.1 MIMO-OFDM System Model and Limited Feedback

OFDM modulation uses a cyclic prefix and discrete Fourier transform (DFT) operations to convert the linear convolution in Equation 5.2 to a multiplication.

By sending information in the frequency domain, OFDM converts a linear convolution into a set of parallel narrowband channels. Precoding can thus be applied independently per subcarrier.

Consider a MIMO-OFDM system with an M-point DFT, and cyclic prefix that is longer than L, the memory of the discrete-time channel. For illustration in this section we will discuss precoded spatial multiplexing systems. Extensions to other forms of space–time coding follow in a similar fashion. Let

$$\mathsf{H}(k) = \sum_{l=0}^{L} \mathbf{H}(l)e^{j2\pi kl/M}$$

be the multivariate response of the channel in the frequency domain. Mathematically the input-output relationship of a MIMO-OFDM system with precoding is

$$\mathbf{y}(k) = \mathsf{H}(k)\mathbf{F}(k)\mathbf{s}(k) + \mathbf{n}(k) \tag{5.33}$$

where $\mathbf{F}(k)$ is the precoding matrix for subcarrier k and $k = 0, 1, \ldots, M - 1$.

A straightforward application of precoded spatial multiplexing entails that the precoder for each subcarrier $\mathbf{F}(k)$ is quantized and the appropriate index is sent back to the transmitter. For a codebook $\mathcal{F}$ of cardinality $N = 2^b$, this requires feedback of Mb bits to relay the precoders for the whole OFDM symbol. The feedback requirements grow in proportion to M, the number of subcarriers. The channels for adjacent subcarriers, however, are correlated. Based on this, it can also be shown that the adjacent precoders, in some sense, are also correlated [9]. Feedback can thus be reduced by exploiting the correlation and sending back less information per subcarrier.

5.6.2   Interpolated Beamforming in MIMO-OFDM

First we consider the special case of transmit beamforming. This allows us to present the essence of the extension to MIMO-OFDM before getting into the full complexity of precoding. With transmit beamforming, the receiver processing consists of matched filtering and scalar detection for each subcarrier. The combined signal at subcarrier k can be expressed as

$$r(k) = \mathbf{z}^{H}(k)\{\mathsf{H}(k)\mathbf{f}(k)s(k) + \mathbf{n}(k)\}, \quad 1 \leq k \leq N \tag{5.34}$$

where $\mathbf{z}(k)$ is the receive combining vector, chosen to be $\mathsf{H}(k)\mathbf{f}(k)$, which in this case corresponds to maximum ratio combining at the receiver.

A simple beamforming method for using the correlation of the beamforming vectors is to combine the neighboring subcarriers into a cluster. The idea is to choose the cluster size to be less than the coherence bandwidth of the channel, thus all subcarriers see the same effective channel. Then a representative beamforming vector is chosen for the whole cluster. A simple approach, which we call clustering, is to use the beamforming vector for the middle subcarrier of the cluster for all the subcarriers in the cluster [8]. A better approach, which we call modified clustering, chooses the best vector according to some criteria [65] motivated by error rate or squared error.

An alternative to clustering is to use interpolation to reconstruct the transmit beamforming vectors given a fraction of the beamformers $\mathbf{f}(lK + 1)$ where K is the cluster size (which divides N) and $l = 0, 1, ..., N/K - 1$. Conventional interpolation, however, does not preserve the structure in the beamforming problem. First, all beamforming vectors are unit norm due to the transmit power constraint. This means that the optimal beamforming vector is a point on a sphere. The interpolator should give outputs that are also on the sphere. Second the optimal beamforming vector is not unique. The reason is that if one examines the effective channel gain $\Gamma(k) = \|H(k)\mathbf{f}(k)\|^2$, both $\mathbf{f}(k)$ and $e^{j\varphi} \mathbf{f}(k)$ have exactly the same value. Thus the best beamforming vector is not unique.

To solve this problem, we proposed a modification of the spherical interpolator in [99] that includes a phase rotation parameter [8]. This interpolator performs a weighted average of the beamforming vectors and renormalizes the result to place it on the unit sphere. Given $\{\mathbf{f}(lK + 1), 0 \leq l \leq N/K - 1\}$, the proposed interpolator is expressed as

$$\hat{\mathbf{f}}(lK + k; \vartheta_l) = \frac{(1 - c_k)\mathbf{f}_Q(lK + 1) + c_k\{e^{j\vartheta_l}\mathbf{f}_Q((l + 1)K + 1)\}}{\left\|(1 - c_k)\mathbf{f}_Q(lK + 1) + c_k\{e^{j\vartheta_l}\mathbf{f}_Q((l + 1)K + 1)\}\right\|} \qquad (5.35)$$

where $c_k = (k - 1)/K$ is the linear weight value, $\mathbf{f}(N + 1) = \mathbf{f}(1)$ are the quantized transmit beamforming vectors, $1 \leq k \leq K$, and ϑ_l is a parameter for phase rotation with $0 \leq l \leq N/K - 1$. The role of ϑ_l is to remove the distortion caused by the arbitrary phase rotation of the optimal beamforming vectors.

To maximize the performance of the interpolator, the receiver evaluates the optimal phase $\{\vartheta_\lambda, 0 \leq l \leq N/K - 1\}$ and sends this information back to the transmitter along with the quantized beamforming vectors. In practice, the set of possible phase values is quantized to $\Theta = \left\{0, \frac{2\pi}{P}, \frac{4\pi}{P}, ..., \frac{2(P-1)\pi}{P}\right\}$, and P is a desired number of quantization levels (we have found that $P = 4$ is sufficient). The feedback of $\{\vartheta_l\}$ slightly increases the amount of feedback information compared to clustering but results in better performance than the simple clustering approach.

The phase can be optimized in different ways. We find that maximizing the minimum effective channel gain where $0 \leq l \leq N/K - 1$ and

$$\vartheta_l = \arg\max_{\vartheta \in \Theta} \min\{\|H(lK + k)\mathbf{f}(lK + k; \vartheta)\|^2, 1 \leq k \leq K\} \qquad (5.36)$$

works well.

5.6.3   Interpolated Precoding in MIMO-OFDM

As with beamforming, a variety of methods can also be used for the more general case of precoding in MIMO-OFDM including clustering, modified clustering, and interpolation. Because the extension of the interpolator is not trivial, we discuss it here.

The challenge with precoder interpolation is that the unitary property $\mathbf{F}^H(k)\mathbf{F}(k) = \frac{1}{M}\mathbf{I}_M$ and the invariance to right multiplication by a unitary matrix, i.e., $\mathbf{F}(k)$ and $\mathbf{F}(k)\mathbf{Q}$ where $\mathbf{Q}$ is a square unitary matrix give the same performance, must be respected. The latter property implies that the optimal precoder is not unique. The matrix $\mathbf{Q}$, however, has a substantial influence on the performance of an interpolator. Based on this observation, we propose the following interpolation algorithm, which utilizes a unitary matrix $\mathbf{Q}_l$ and the quantized precoders $\mathbf{F}_Q(lK + 1)$ for $l = 0, 1, \ldots, N/K - 1$

$$\mathbf{Z}(lK + m) = (1 - c_m)\mathbf{F}_Q(lK + 1) + c_m\mathbf{Q}_l\mathbf{F}_Q\big((l+1)K + 1\big) \tag{5.37}$$

$$\hat{\mathbf{F}}(lK + m; \mathbf{Q}_l) = \mathbf{Z}(lK + m)\left\{\mathbf{Z}^H(lK + m)\mathbf{Z}(lK + m)\right\}^{-\frac{1}{2}} \tag{5.38}$$

The proposed algorithm in Equation 5.37 and Equation 5.38 is a non-obvious extension of the beamformer interpolator in Equation 5.35. Note that $\hat{\mathbf{F}}(k; \vartheta)$ is the projection of $\mathbf{Z}(k)$ into $\mathcal{U}(M_t, M_s)$ with respect to the Frobenius norm (see Section III-F of [94]). Notice that $e^{j\vartheta}$ of Equation 5.35 is replaced by $\mathbf{Q}_l$ in Equation 5.37. In beamformer interpolation, $e^{j\vartheta}$ removes the distortion caused by the arbitrary phase rotation of beamforming vectors. Similarly, the role of $\mathbf{Q}_l$ is to get rid of the distortion caused by the non-uniqueness of the precoding matrices.

Given the subsampling rate K and the codebook $\mathcal{F}$, the quantized precoders $\{\mathbf{F}(lK + 1), 0 \le l \le N/K - 1\}$ are obtained by using the same criterion as precoding design, e.g., *SC-MSE* or *SC-Capacity*. Since we are quantizing $\mathbf{Q}_l$ for limited feedback, we can numerically find the best $\mathbf{Q}$ by selecting a matrix from a codebook $\mathcal{Q}$ of unitary matrices. Given a codebook $\mathcal{Q}$, we can find the best $\mathbf{Q}$ in a similar way as before by optimizing

$$\mathbf{Q}_l = \arg\min_{\bar{\mathbf{Q}} \in \mathcal{Q}} \max MSE(\hat{\mathbf{F}}(lK + m; \bar{\mathbf{Q}})). \tag{5.39}$$

The performance and complexity are dependent on $|\mathcal{Q}|$, the cardinality of $\mathcal{Q}$.

A reasonable way to design $\mathcal{Q}$ is to employ the Frobenius matrix norm and maximize the minimum norm between the entries of $\mathcal{Q}$. A variety of other approaches for codebook design are also possible; optimal quantization of $\mathbf{Q}$ is still under investigation.

5.6.4 Performance Example

To illustrate the potential performance of limited feedback precoding for MIMO-OFDM we provide a performance example in this section. In the simulation, we consider a MIMO-OFDM system with parameters $M_t = 4$, $M_r = 2$, $M_s = 2$, $N = 64$, and $K = 8$. We assumed that the discrete-time channel impulse responses had 6 taps with uniform profile and i.i.d. complex Gaussian distribution; the channels between different transmit and receive antenna pairs were independent; $\mathbf{n}(k)$ was i.i.d. zero mean complex Gaussian; the feedback channel had no delay and no transmission error; quadrature phase shift keying (QPSK) modulation and MMSE detection were used for BER simulations; and the transmit power was identically distributed to all subcarriers.

We compared antenna subset selection [29] where the best set of 2 out of 4 antennas is chosen for each subcarrier, simple clustering, the proposed interpolator, and MMSE detection with the optimal precoder (with and without quantization). To quantize the precoding matrices, we used the codebook with $|\mathcal{F}| = 64$ requiring 6 bits of feedback for each precoding matrix which was designed according to [54]. Also, we considered the codebook given in [9]. Consequently, we used *64 feedback bits* for the proposed method. In antenna subset selection, the transmit antenna was selected independently for each subcarrier, so 166 feedback bits were used. We used 48 feedback bits for clustering and 384 bits for the optimal precoder with feedback for all subcarriers. For comparison, we also considered the optimal precoder without quantization.

Figure 5.5 shows the coded BER performance of the proposed and existing precoding methods. The maximum minimum singular value criterion [29] was used for antenna subset selection and $\mathbf{Q}_l$ for the proposed scheme was determined by Equation 5.39. For channel coding, we used a convolutional code with generator polynomials $g_0 = 133_8$ and $g_1 = 171_8$ with coding rate $1/2$, the interleaver defined in the IEEE 802.11a standard extended for 64 subcarriers, and soft Viterbi decoding. The frame length was 30 OFDM symbols and each OFDM symbol transmitted 128 bits. The proposed precoder outperformed clustering and antenna subset selection, and had only 0.1 dB loss compared to the optimal MMSE with quantization. The quantization loss was about 1 dB even with per-subcarrier quantization. Clustering obtains nearly as good performance with less feedback required.

5.7 Conclusions

In this chapter we presented an overview of closed-loop MIMO communication systems. We discussed both direct channel quantization and quantized signal design, choosing to focus on the latter because it typically requires less feedback. As concrete examples we discussed strategies for quantized

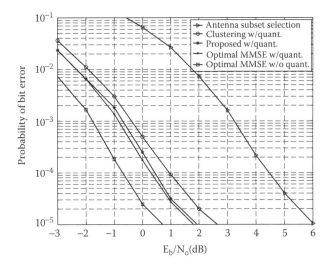

FIGURE 5.5

A bit error rate comparison of different limited feedback precoding schemes for MIMO-OFDM with $M_t = 4$, $M = 2$, and $M_r = 2$. MMSE precoding with the trace constraint is considered and compared with antenna subset selection per tone, simple clustering and optimal approaches with and without quantization.

beamforming and precoding for both spatial multiplexing and space–time coding. We also summarized some extensions of quantized beamforming and precoding for spatial multiplexing with MIMO-OFDM.

Other important aspects of limited feedback MIMO communication were not considered in this chapter due to a lack of space. Perhaps the most important omission is multi-mode precoding, where the number of spatial streams M is also varied in the channel. Results along these lines are available in [25,56,59]. We also overlooked the combination of adaptive modulation and limited feedback, though naturally the two will be combined in a real system. The effect of channel estimation error, delay in the feedback channel, and, more generally, the effect of time variation of feedback performance remain to be studied.

References

1. IEEE-SA Standards Board. 2001. "IEEE Standard for local and metropolitan area networks part 16: air interface for fixed broadband wireless access systems," IEEE Std 802.16-2001, Dec. 2001, 16.-2001.
2. J. Akhtar and D. Gesbert. 2003. "A closed-form precoder for spatial multiplexing over correlated MIMO channels." *Proc. IEEE Glob. Telecom. Conf.*, 4, Dec. 2003, pp. 1847–1851.
3. J. Akhtar and D. Gesbert. 2004. "Extending orthogonal block codes with partial feedback." *IEEE Trans. Wireless Comm.*, 3, Nov. 2004, pp.1959–1962.

4. S.M. Alamouti. 1998. "A simple transmit diversity technique for wireless communications." *IEEE Jour. Select. Areas in Commun.* 16, Oct. 1998, pp. 1451–1458.

5. A. Barg and D. Yu. Nogin. 2002. "Bounds on packings of spheres in the Grassmann manifold." *IEEE Trans. Info. Th.*, 48, Sept. 2002, 2450–2454.

6. G. Bauch and J. Hagenauer. 2002. "Smart versus dumb antennas-capacities and FEC performance." *IEEE Commun. Lett.*, 6, Feb. 2002, 55–57.

7. R.S. Blum. 2003. "MIMO with limited feedback of channel state information." *Proc. IEEE Int. Conf. Acoust., Speech and Sig. Proc.*, 4, April 2003, 89–92.

8. J. Choi and R.W. Heath, Jr. 2004. "Interpolation based transmit beamforming for MIMO-OFDM with limited feedback." *Proc. Int. Conf. on Commun.*, Paris, France, 1, June 2004, 249–253.

9. J. Choi and R.W. Heath, Jr. 2004. "Interpolation based transmit precoding for MIMO-OFDM with limited feedback." *Proc. Glob. Telecom. Conf.*, Dallas, Texas, 1, Nov.-Dec. 2004, 214–218.

10. J.H. Conway, R.H. Hardin, and N.J.A. Sloane. 1996. "Packing lines, planes, etc.: packings in Grassmannian spaces." *Experimental Mathematics*, 5, 139–159.

11. M.O. Damen, K. Abed-Meraim, and J.C. Belfiore. 2002. "Diagonal algebraic space-time block codes." *IEEE Trans. Inf. Th.*, 48, March 2002, 628–636.

12. R.T. Derryberry, S.D. Gray, D.M. Ionescu, G. Mandyam, and B. Raghothaman. "Transmit diversity in 3G CDMA systems." *IEEE Comm. Mag.*, 40, April 2002, 68–75.

13. P.A. Dighe, R.K. Mallik, and S.S. Jamuar. 2003. "Analysis of transmit-receive diversity in Rayleigh fading." *IEEE Trans. Commun.*, 51, April 2003, 694–703.

14. H. El Gamal and M.O. Damen. 2003. "Universal space-time coding." *IEEE Trans. Inf. Th.*, 49, May 2003, 1097–1119.

15. H. El Gamal and A.R. Hammons, Jr. 2003. "On the design of algebraic space-time codes for MIMO block-fading channels." *IEEE Trans. Inf. Th.*, 49, Jan. 2003, 151–163.

16. R.F.H. Fischer. 2002. *Precoding and Signal Shaping for Digital Transmission*, IEEE, New York: Wiley-Interscience.

17. R.F.H. Fischer, C. Windpassinger, A. Lampe, and J.B. Huber. 2002. "Tomlinson-Harashima precoding in space-time transmission for low-rate backward channel." *Proc. of Int. Zur. Sem. on Broadband Commun.*, pp. 7–1 to 7–6, Feb. 2002.

18. J.R. Fonollosa, R. Gaspa, X. Mestre, A. Pages, M. Heikkila, J.P. Kermoal, L. Schumacher, A. Pollard, and J. Ylitalo. 2002. "The IST METRA project." *IEEE Comm. Mag.*, 40, July 2002, 78–86.

19. A. Forenza, D.J. Love, and R.W. Heath, Jr. 2004. "A low complexity algorithm to simulate the spatial covariance matrix for clustered MIMO channel models." *Proc. IEEE Veh. Technol. Conf.*, May 2004.

20. G.J. Foschini. 1996. "Layered space-time architecture for wireless communication in a fading environment when using multiple antennas." *The Bell Sys. Tech. Jour.*, 1, 41–59.

21. A. Goldsmith, S.A. Jafar, N. Jindal, and S. Vishwanath. 2003. "Capacity limits of MIMO channels." *IEEE Jour. Select. Areas in Commun.*, 21, June 2003, 684–702.

22. D. Gore, R.W. Heath, Jr., and A. Paulraj. 2002. "Transmit selection in spatial multiplexing systems." *IEEE Comm. Letters*, 6, Nov. 2002, 491–493.

23. D.A. Gore and A.J. Paulraj. 2002. "MIMO antenna subset selection with space-time coding." *IEEE Trans. Sig. Proc.*, 50, Oct. 2002, 2580–2588.

24. R.W. Heath, Jr. 2001. Space-time Signaling in Multi-Antenna Systems, Ph.D. thesis, Stanford University, Nov. 2001.

25. R.W. Heath, Jr. and D.J. Love. 2003. "Multi-mode antenna selection for spatial multiplexing with linear receivers." *Proc. of Allerton Conf. on Comm. Cont. and Comp.*, Oct. 2003.
26. R.W. Heath, Jr. and A. Paulraj. 1998. "A simple scheme for transmit diversity using partial channel feedback." *Proc. of the Asil. Conf. on Sig. Sys. and Comp.*, 2, Nov. 1998, 1073–1078.
27. R.W. Heath, Jr. and A. Paulraj. 1999. "Transmit diversity using decision-directed antenna hopping." *ICC Comm. Theory Mini-Conf.*, June 1999, 141–145.
28. R.W. Heath, Jr. and A. Paulraj. 2002. "Linear dispersion codes for MIMO systems based on frame theory." *IEEE Trans. on Sig. Proc.*, 50, Oct. 2002, 2429–2441.
29. R.W. Heath, Jr., S. Sandhu, and A. Paulraj. 2001. "Antenna selection for spatial multiplexing with linear receivers." *IEEE Comm. Letters*, 5, April 2001, 142–144.
30. R.W. Heath, Jr. and A. Paulraj. 2001. "Antenna selection for spatial multiplexing systems based on minimum error rate." *Proc. IEEE Int. Conf. on Commun.*, Vol. 7, June 2001, 2276–2280.
31. J. Heiskala and J. Terry. 2002. *OFDM Wireless LANs: A Theoretical and Practical Guide*, Indianapolis: SAMS.
32. B.M. Hochwald, T.L. Marzetta, T.J. Richardson, W. Sweldens, and R. Urbanke. 2000. "Systematic design of unitary space-time constellations." *IEEE Trans. Info. Th.*, 46, Sept. 2000, 1962–1973.
33. H. Holma and A. Toskala, eds. 2001. *WCDMA for UMTS: Radio Access for Third Generation Mobile Communications, Revised Edition*, New York: John Wiley & Sons.
34. Z. Hong, K. Liu, R.W. Heath, Jr., and A.M. Sayeed. 2003. "Spatial multiplexing in correlated fading via the virtual channel representation." *IEEE Jour. Select. Areas in Commun.*, 21, June 2003, 856–866.
35. H. Jafarkhani. 2001. "A quasi-orthogonal space-time block code." *IEEE Trans. Commun.*, 49, Jan. 2001, 1–4.
36. H. Jafarkhani and N. Seshardi. 2003. "Super-orthogonal space-time trellis codes." *IEEE Trans. Inf. Th.*, 49, April 2003, 937–950.
37. G. Jöngren and M. Skoglund. 2000. "Utilizing quantized feedback in orthogonal space-time block coding." *Proc. Glob. Telecom. Conf.*, 2, Nov. 2000, 995–999.
38. G. Jöngren and M. Skoglund. 2001. "Improving orthogonal space-time block codes by utilizing quantized feedback information." *Proc. Int. Symp. on Info. Theo.*, June 2001, 220.
39. G. Jöngren and M. Skoglund. 2004. "Quantized feedback information in orthogonal space-time block coding." *IEEE Trans. Info. Th.*, 50, Oct. 2004, 2473–2486.
40. G. Jöngren, M. Skoglund, and B. Ottersten. 2002. "Combining beamforming and orthogonal space-time block coding." *IEEE Trans. Info. Th.*, 48, March 2002, 611–627.
41. M. Kang and M.-S. Alouini. 2003. "Largest eigenvalue of complex wishart matrices and performance analysis of MIMO MRC systems." *IEEE Jour. Select. Areas in Commun.*, 21, April 2003, 418–426.
42. J.P. Kermoal, L. Schumacher, F. Frederiksen, and P.E. Mogensen. 2001. "Polarization diversity in MIMO radio channels: experimental validation of a stochastic model and performance assessment." *Proc. IEEE Veh. Technol. Conf.*, 1, Fall 2001, 22–26.
43. J.P. Kermoal, L. Schumacher, P.E. Mogensen, and K.I. Pedersen. 2001. "Experimental investigation of correlation properties of MIMO radio channels for indoor picocell scenarios." *Proc. IEEE Veh. Technol. Conf.*, 1, Sept. 2000, 14–21.

44. J.P. Kermoal, L. Schumacher, K.I. Pedersen, P.E. Mogensen, and F. Frederiksen. 2002. "A stochastic MIMO radio channel model with experimental validation." *IEEE Jour. Select. Areas in Commun.*, 20, Aug. 2002, 1211–1226.

45. C. Kose and R.D. Wesel. 2002. "Universal space-time trellis codes." *Proc. Glob. Telecom. Conf.*, Nov. 2002, 1108–1112.

46. E.G. Larsson, G. Ganesan, P. Stoica, and W.-H. Wong. 2002. "On the performance of orthogonal space-time block coding with quantized feedback." *IEEE Commun. Lett.*, 6, Nov. 2002, 487–489.

47. E.G. Larsson and P. Stoica. 2003. *Space-Time Block Coding for Wireless Communications*, New York: Cambridge University Press.

48. K.N. Lau, Y. Liu, and T.-A. Chen. 2003. "Optimal partial feedback design for MIMO block fading channels with causal noiseless feedback." *Proc. IEEE Int. Conf. on Commun.*, 4, May 2003, 1663–2667.

49. V. Lau, Y. Liu, and T.-A. Chen. 2004. "On the design of MIMO block-fading channels with feedback-link capacity constraint." *IEEE Trans. Commun.*, 52, Jan. 2004, 62–70.

49b. V.K.N. Lau, Y. Liu, and T.-A. Chen. 2004. "Capacity of memoryless channels and block-fading channels with designable cardinality-constrained channel state feedback." *IEEE Trans. Info. Th.*, 50(9), Sept. 2004, 2038–2049.

50. X. Lin and R.S. Blum. 2002. "Systematic design of space-time codes employing multiple trellis coded modulation." *IEEE Trans. Commun.*, 50, April 2002, 608–615.

51. T.K.Y. Lo. 1999. Maximum ratio transmission. *IEEE Trans. Commun.*, 47, Oct. 1999, 1458–1461.

51b. D.J. Love. 2004. Feedback Methods for Multiple-Input Multiple-Output Wireless Systems, Ph.D. thesis, The University of Texas at Austin, May 2004.

52. D.J. Love and R.W. Heath, Jr. "Diversity performance of precoded orthogonal space-time block codes using limited feedback." *IEEE Commun. Lett.*, 8, May 2004, 305–307.

53. D.J. Love and R.W. Heath, Jr. 2003. "Equal gain transmission in multiple-input multiple-output wireless systems." *IEEE Trans. Commun.*, 51, July 2003, 1102–1110.

54. D.J. Love and R.W. Heath, Jr. 2003. "Limited feedback precoding for spatial multiplexing systems." *Proc. IEEE Glob. Telecom. Conf.*, 4, Dec. 2003, 1857–1861.

55. D.J. Love and R.W. Heath, Jr. 2003. "Limited feedback precoding for spatial multiplexing systems using linear receivers." *Proc. IEEE Mil. Comm. Conf.*, 1, Oct. 2003, 627–632.

56. D.J. Love and R.W. Heath, Jr. 2004. "Multi-mode precoding using linear receivers for limited feedback MIMO systems." *Proc. IEEE Int. Conf. on Commun.*, 1, June 2004, 448–452.

57. D.J. Love and R.W. Heath, Jr. 2005. "Limited feedback unitary precoding for orthogonal space-time block codes." *IEEE Trans. Sig. Proc.*, 53, Jan. 2005, 64–73.

58. D.J. Love and R.W. Heath, Jr. 2005. "Limited feedback unitary precoding for spatial multiplexing systems." *IEEE Trans. Info. Th.*, 51, Aug. 2005, 2967–2976.

59. D.J. Love and R.W. Heath, Jr. 2005. "Multi-mode precoding for MIMO wireless systems." *IEEE Trans. Sig. Proc.*, 53, Oct. 2005, 3674–3687.

60. D.J. Love and R.W. Heath, Jr. 2005. "Necessary and sufficient conditions for full diversity order in correlated Rayleigh fading beamforming and combining systems." *IEEE Trans. Wireless Comm.*, 4(1), Jan. 2005, 20–23.

61. D.J. Love, R.W. Heath, Jr., and T. Strohmer. 2003. "Grassmannian beam-forming for multiple-input multiple-output wireless systems." *Proc. Int. Conf. on Commun.*, 4, May 2003, 2618–2622.

62. D.J. Love and R.W. Heath, Jr. 2003. "Grassmannian precoding for spatial multiplexing systems." *Proc. of Allerton Conf. on Comm. Cont. and Comp.*, Oct. 2003.

63. D.J. Love, R.W. Heath, Jr., and T. Strohmer. 2003. "Grassmannian beam-forming for multiple-input multiple-output wireless systems." *IEEE Trans. Info. Th.*, 49, Oct. 2003, 2735–2747.

63b. D.J. Love, R.W. Heath, Jr., W. Santipach, and M.L. Honig. 2004. "What is the value of limited feedback for MIMO channels?" *IEEE Commun. Mag.*, 42, Oct. 2004, 54–59.

64. H. Miyakawa and H. Harashima. 1969. "A method of code conversion for digital communication channels with intersymbol interference." *Transactions of the Institute of Electronic Communication Engineering, Japan*, 52-A, June 1969, 272–273.

65. B. Mondal and R.W. Heath, Jr. 2005. "Algorithms for quantized precoding for mimo ofdm." *Proc. of Third SPIE Int. Symp. on Fluctuations and Noise*, Austin, TX, May 2005.

66. K.K. Mukkavilli, A. Sabharwal, and B. Aazhang. 2001. "Design of multiple antenna coding schemes with channel feedback." *Proc. of the 35th Annual Asil. Conf. on Sig. Sys. and Comp.*, Nov. 2001, 1009–1013.

67. K.K. Mukkavilli, A. Sabharwal, E. Erkip, and B. Aazhang. 2003. "On beamforming with finite rate feedback in multiple antenna systems." *IEEE Trans. Info. Th.*, 49, Oct. 2003, 2562–2579.

68. R.U. Nabar, H. Bolcskei, V. Erceg, D. Gesbert, and A.J. Paulraj. 2002. "Performance of multiantenna signaling techniques in the presence of polarization diversity." *IEEE Trans. on Sig. Proc.*, 50, Oct. 2002, 2553–2562.

69. A. Narula, M.J. Lopez, M.D. Trott, and G.W. Wornell. 1998. "Efficient use of side information in multiple-antenna data transmission over fading channels." *IEEE Jour. Select. Areas in Commun.*, 16, Oct. 1998, 1423–1436.

70. A. Paulraj, R. Nabar, and D. Gore. 2003. *Introduction to Space-Time Wireless Communications*, New York: Cambridge University Press.

71. J.G. Proakis. 2001. *Digital Communications*, 4th edition, Boston: McGraw Hill.

72. G.G. Raleigh and J.M. Cioffi.1998. "Spatio-temporal coding for wireless communication." *IEEE Jour. Select. Areas in Commun.*, 46, March 1998, 357–366.

73. J.C. Roh and B.D. Rao. 2004. "Channel feedback quantization methods for MISO and MIMO systems." *Proc. IEEE Int. Symp. Personal, Indoor, and Mobile Radio Commun.*, 2, Sept. 2004, 805–809

74. J.C. Roh and B.D. Rao. 2004. "An efficient feedback method for MIMO systems with slowly time-varying channels." *Proc. IEEE Wireless Comm. and Net. Conf.*, 2, March 2004, 760–764.

75. J.C. Roh and B.D. Rao. 2004. "Multiple antenna channels with partial channel state information at the transmitter." *IEEE Trans. Wireless Comm.*, 3, March 2004, 677–688.

76. M.A. Sadrabadi, A.K. Khandani, and F. Lahouti. 2004. A new method of channel feedback quantization for high data rate MIMO systems." *Proc. IEEE Glob. Telecom. Conf.*, 1, Nov.–Dec. 2004, 91–95.

77. H. Sampath and A. Paulraj. 2002. "Linear precoding for space-time coded systems with known fading correlations." *IEEE Commun. Lett.*, 6, June 2002, 239–241.

78. H. Sampath, S. Talwar, J. Tellado, V. Erceg, and A. Paulraj. 2002. "A fourth-generation MIMO-OFDM broadband wireless system: design, performance, and field trial results." *IEEE Comm. Mag.*, 40, Sept. 2002, 143–149.

79. W. Santipach and M.L. Honig. 2003. "Asymptotic performance of MIMO wireless channels with limited feedback." *Proc. IEEE Mil. Comm. Conf.*, 1, Oct. 2003, 141–146.

80. W. Santipach, Y. Sun, and M.L. Honig. 2003. "Benefits of limited feedback for wireless channels." *Proc. of Allerton Conf. on Comm. Cont. and Comp.*, Oct. 2003.

81. A. Scaglione, P. Stoica, S. Barbarossa, G.B. Giannakis, and H. Sampath. 2002. "Optimal designs for space-time linear precoders and decoders." *IEEE Trans. Sig. Proc.*, 50, May 2002, 1051–1064.

82. L. Schumacher, J.P. Kermoal, F. Frederiksen, K.I. Pedersen, A. Algans, and P. E. Mogensen. 2001. MIMO channel characterisation. Technical report, European IST-1999-11729 Project METRA, Feb. 2001.

83. L. Schumacher, K.I. Pedersen, and P.E. Mogensen. 2002. "From antenna spacings to theoretical capacities — guidelines for simulating MIMO systems." *Proc. IEEE Int. Symp. Personal, Indoor, and Mobile Radio Commun.*, 2, Sept. 2002, 587–592.

84. M.B. Shoemake. 2004. Status of project IEEE 802.11n. http://grouper.ieee.org/groups/802/11/Reports/tgn update.htm. Accessed on March 3, 2004.

85. I. Sohn, G.D. Golden, H. Lee, and J.W. Ahn. 2003. "Overview of METRA model for MBWA MIMO channel." IEEE C802.20-03-50, May 2003.

86. N.R. Sollenberger. 1993. "Diversity and automatic link transfer for a TDMA wireless access link." *Proc. Glob. Telecom. Conf.*, 1, Nov.–Dec. 1993, 532–536.

87. T. Strohmer and R.W. Heath, Jr. 2003. "Grassmannian frames with applications to coding and communications." *Applied and Computational Harmonic Analysis*, 14, May 2003, 257–275.

88. G.L. Stuber, ed. 1996. *Principles of Mobile Communications*, Norwell, MA: Kluwer.

89. Y. Tan, M. Pereira, M. Mewburn, and M. Faulkner. Investigation of singular value distributions of MIMO channels in indoor environment. Available at *http://www.telecommunications.crc.org.au/content/ConfPapers/Tan.pdf*.

90. V. Tarokh, H. Jafarkhani, and A.R. Calderbank. 1999. "Space-time block codes from orthogonal designs." *IEEE Trans. Inf. Th.*, 45, Jul. 1999, 1456–1467.

91. V. Tarokh, N. Seshadri, and A.R. Calderbank. 1998. "Space-time codes for high data rate wireless communication: performance criterion and code construction." *IEEE Trans. Inf. Th.*, 44, March 1998, 744–765.

92. M. Tomlinson. 1971. "New automatic equalizer employing modulo arithmetic." *Electronics Letters*, 7, March 1971, 138–139.

93. J.A. Tropp, I. Dhillon, R.W. Heath, Jr., and T. Strohmer. 2005. "Designing structured tight frames via an alternating projection method." *IEEE Trans. Inf. Th.*, 51, Jan. 2005, 188–209.

94. J.A. Tropp, R.W. Heath, Jr., and T. Strohmer. 2003. "An alternating projection algorithm for constructing quasi-orthogonal CDMA signature sequences." *Proc. Int. Symp. on Info. Theo.*, 407.

95. C.-H. Tse, K.-W. Yip, and T.-S. Ng. 2000. "Performance tradeoffs between maximum ratio transmission and switched-transmit diversity." *Proc. Int. Symp. on Pers., Ind. and Mobile Radio Comm.*, 2, Sept. 2000, 1485–1489.

96. V. Erceg et al. 2003. "Indoor MIMO WLAN channel models." IEEE 802.11-03/161r2, Sept. 2003.

97. E. Visotsky and U. Madhow. 2001. "Space-time transmit precoding with imperfect feedback." *IEEE Trans. Info. Th.*, 47, Sept. 2001, 2632–2639.

98. J.W. Wallace and M.A. Jensen. 2002. "Modeling the indoor MIMO wireless channel." *IEEE Trans. Antennas Propagat.*, 50, May 2002, 591–599.

99. G.S. Watson. 1983. *Statistics on Spheres*, New York: Wiley.

100. A. Wittneben. 1995. "Analysis and comparison of optimal predictive transmitter selection and combining diversity for DECT." *Proc. IEEE Glob. Telecom. Conf.*, 2, Nov. 1995, 1527–1531.

101. W.-H. Wong and E.G. Larsson. 2003. "Orthogonal space-time block coding with antenna selection and power allocation." *Elect. Lett.*, 39, Feb. 2003, 379–381.

102. P. Xia, S. Zhou, and G.B. Giannakis. 2004. "Adaptive MIMO-OFDM based on partial channel state information." *IEEE Trans. on Sig. Proc.*, 52, Jan. 2004, 202–213.

103. T. Xiaofeng, H. Harald, Y. Zhuizhuan, Q. Haiyan, and Z. Ping. 2001. "Closed loop space-time block code." *Proc. Veh. Technol. Conf.*, 2, Oct. 2001, 1093–1096.

104. Y. Zhao, R.S. Adve, and T.J. Lim. 2004. "Optimal STBC precoding with channel covariance feedback for minimum error probability." *EURASIP J. Applied Signal Processing*, Sept. 2004, 1257–1265.

105. L. Zheng and D.N.C. Tse. 2002. "Communication on the Grassmann manifold: a geometric approach to the noncoherent multiple-antenna channel." *IEEE Trans. Info. Th.*, 48, Feb. 2002, 359–383.

106. S. Zhou and G.B. Giannakis. 2002. "Optimal transmitter eigen-beamforming and space-time block coding based on channel mean feedback." *IEEE Trans. on Sig. Proc.*, 50, Oct. 2002, 2599–2613.

107. S. Zhou and G.B. Giannakis. 2003. "Optimal transmitter eigen-beamforming and space-time block coding based on channel correlations." *IEEE Trans. Inf. Th.*, 49, Jul. 2003, 1673–1690.

108. S. Zhou, Z. Wang, and G.B. Giannakis. 2004. "Performance analysis of transmit-beamforming with finite-rate feedback." *Proc. of 38th Conf. on Info. Sciences and Systems*, Princeton University, March 2004.

6

Antenna Selection in MIMO Systems

Neelesh B. Mehta and Andreas F. Molisch

6.1 Introduction

MIMO systems, which employ multiple transmit and receive antenna elements, substantially improve the data rates that can be transmitted over the channel and the reliability with which they can be received without any additional bandwidth. Higher data rates are achieved by transmitting multiple data streams simultaneously using spatial multiplexing techniques. For spatially uncorrelated channels, the data rates even increase linearly with the minimum of the number of transmit and receive antenna elements [1]. Increased reliability is achieved by exploiting spatial diversity to significantly reduce the probability that the channel is in a deep fade. Orthogonal space–time block codes and space–time trellis codes are examples of diversity techniques tailored to MIMO systems. A single input multiple output (SIMO) system, which combines the many received copies of the transmitted signal to improve reliability, is another example of a spatial diversity system.

While MIMO systems perform impressively, they also increase the hardware and signal processing complexity, power consumption, and component size in the transmitter and the receiver [2]. One of the main culprits behind this increase in complexity is that each receive antenna element requires a dedicated radio frequency (RF) chain that comprises a low noise amplifier, a frequency down-converter, and an analog-to-digital converter, and each transmit antenna element requires an RF chain that comprises a digital-to-analog converter, a frequency up-converter, and a power amplifier. Moreover, processing the signals received in spatial multiplexing schemes or with STTCs calls for sophisticated receivers whose complexity increases, sometimes exponentially, with the number of transmit and receive antenna elements.

This increase in complexity has inhibited the widespread adoption of MIMO systems. For example, the third-generation cellular system specification (3GPP) currently supports only an optional two antenna space–time transmit diversity scheme and does not require the handsets to have more than one antenna element [3]. Sophisticated techniques that employ spatial multiplexing or support more antenna elements have met with considerable opposition in 3GPP. While the adoption of MIMO has made headway in the next-generation wireless local area network (WLAN) standard IEEE 802.11n, which aims to transfer raw information at rates greater than 100 Mbps over a 20-MHz bandwidth, complexity considerations are likely to make the adopted MIMO scheme an elementary one that supports a small number of antenna elements.

Antenna selection is a solution that addresses some of the complexity drawbacks associated with MIMO systems. It reduces the hardware complexity of transmitters and receivers by using fewer RF chains than the number of antenna elements. While the antenna elements are typically cheap, and in some cases are just a patch of copper, the RF chains are considerably more expensive. In antenna selection, a subset of the available antenna elements is adaptively chosen by a switch, and only signals from the chosen subset are processed further by the available RF chains.

Given its promise as a low-complexity solution, antenna selection has received considerable attention recently. It has been considered at the transmitter (transmit antenna selection [TAS]), at the receiver (receive antenna selection [RAS]), and at both the transmitter and the receiver (transmit and receive antenna selection [T-RAS]). Its performance has been explored from various angles such as capacity and outage for spatial multiplexing systems, and diversity order and array gain for space–time coded systems. The *diversity order* quantifies the effectiveness in avoiding deep fades and is defined as the slope of average symbol error probability vs. input signal to noise ratio (SNR) curve for high SNRs. The *array gain* quantifies the improvement in average SNR seen at the combiner output when signals received by the multiple antenna elements are combined.

As we shall see, antenna selection — for a variety of MIMO techniques — achieves the full diversity inherent in the system at the expense of a small loss in array gain compared to a full complexity system, i.e., a system that can always allocate RF chains to each and every antenna element. Another way of stating this is that for the same number of RF chains, using additional antenna elements with antenna selection outperforms a system that lacks additional antenna elements. Antenna selection has been found to modify, on a fundamental level, the optimum Gaussian signaling required to transmit information at the maximum possible rate. Considerable effort has also been spent to develop various criteria, both optimal (but complicated) and suboptimal (but simple), to implement antenna selection algorithms.

This chapter studies various aspects of antenna selection in some depth, and assumes basic knowledge about the operation of MIMO systems. The reader is referred to the tutorial papers in [2,4] for a more basic introduction. We first analyze the performance of antenna selection for different MIMO systems. We then consider the algorithms and criteria to implement antenna selection. We also consider antenna selection performance in the presence of non-idealities, such as imperfect channel estimates, multipath dispersion, fast fading, etc. Finally, we discuss a class of antenna selection algorithms that are based on knowledge of the statistical variations of the channel. Using statistical information instead of the instantaneous channel state information often significantly reduces the demands on the transmitter and receiver. We show that even a moderate spatial correlation, when properly exploited by antenna selection, helps recover, to a great extent, the array gain loss that is incurred by antenna selection.

In this chapter, we will focus most of our attention on the simplest setup in which the receiver has full channel state information (CSI) while the transmitter has no CSI, and knows only which antenna elements to use in case of TAS and T-RAS. However, we do briefly discuss how other scenarios, such as the availability of CSI at the transmitter and the presence of co-channel interference in addition to noise, can be handled.

6.2 MIMO System Model

Let N_t and N_r denote the number of antenna elements that are available (but not necessarily used every time) at the transmitter and receiver, respectively.

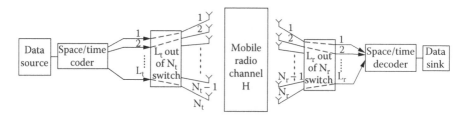

FIGURE 6.1

Block diagram of antenna selection with L_t out of N_t transmit antenna elements selected and L_r out of N_r receive antenna elements selected [4]. (© 2004 IEEE.)

Let L_t and L_r denote the number of RF chains at the transmitter and receiver, respectively. We always have $1 \leq L_t \leq N_t$ and $1 \leq L_r \leq N_r$. A block diagram representation of antenna selection at the transmitter and the receiver is given in Figure 6.1.

The transmitted signal, of size $N_t \times 1$, is denoted by $\mathbf{x}$. The signal, $\mathbf{y}$, of size $N_r \times 1$, received by the N_r antenna elements is given by

$$\mathbf{y} = \mathbf{Hx} + \mathbf{n}, \tag{6.1}$$

where $\mathbf{n}$ represents noise and the matrix $\mathbf{H}$, of size $N_r \times N_t$, denotes the instantaneous channel state. Unless otherwise mentioned, the elements of $\mathbf{H}$ are assumed to be independent of each other. In case of TAS, $N_t L_t$ elements of $\mathbf{x}$, which correspond to the transmit antenna elements not chosen, will be 0. The noise vector, $\mathbf{n}$, of size $N_r \times 1$, is a zero-mean complex white Gaussian random vector. To simplify notation, we shall use $\mathbf{x}$ to also denote the $L_t \times 1$ transmit vector and $\mathbf{y}$ to denote the $L_r \times 1$ receive vector that correspond to only the selected transmit and receive elements, respectively. The notation shall be obvious from the context. The total power transmitted from all the antennas is denoted by ρ. Without loss of generality, we shall assume that each element of $\mathbf{n}$ and $\mathbf{H}$ has unit variance. Therefore, the average SNR of the signal input to any receive antenna element equals ρ / L_t when the power is equally allocated among the L_t transmit elements.

We shall assume that the receiver has perfect channel state information. The receiver can estimate the channel from the training sequences or by using blind estimation techniques. A frequency-flat channel is assumed. The channel is assumed to be block-fading, in that it does not change during the transmission of the space–time codeword and decorrelates thereafter. We shall comment on the impact of the breakdown of these assumptions toward the end of the chapter.

6.2.1   General Notation

For a matrix $\mathbf{A}$, the symbol $\mathbf{A}^\dagger$ denotes the Hermitian transpose of $\mathbf{A}$, $\mathbf{A}^T$ the transpose, $\mathbf{A}^{-1}$ the matrix inverse, $\|\mathbf{A}\|_F$ the Frobenius norm, $|\mathbf{A}|$ the determinant, and Tr$\{\mathbf{A}\}$ the trace. The identity matrix of size $n \times n$ shall be denoted by $\mathbf{I}_n$. The symbol $\mathbf{E}_X[.]$ denotes the expectation with respect to the random variable X.

As we shall see, order statistics, which deals with the statistics of ordered random variables, is often used in the analysis of antenna selection. We shall resort to the following commonly used order statistics notation. When n numbers, $x_1, \ldots, x_n$, are rearranged in descending order, they shall be denoted by $x_{[1]} \geq x_{[2]} \geq \cdots \geq x_{[n]}$, where $[i]$ denotes the index of the ith largest number.

6.3 Spatial Multiplexing

As mentioned, we shall first focus on the case where the transmitter has no knowledge of the channel state $\mathbf{H}$. (In the case of TAS, the transmitter knows, in addition, which antenna elements to use.) This is typically achieved by means of feedback from the receiver in frequency division duplex systems. In time division duplex systems with slowly varying channels, the transmitter can monitor the transmissions from the receiver to determine the channel state without any help from the receiver.

The maximum data transmission rate of any communication system is given by the Shannon capacity [5]. For MIMO systems, the Shannon capacity given the channel state $\mathbf{H}$ takes the form [1]

$$C(\mathbf{H}) = \log_2 \left| \mathbf{I}_{N_r} + \mathbf{H} \mathbf{Q}_X \mathbf{H}^\dagger \right|, \tag{6.2}$$

where $\mathbf{Q}_X$ is the covariance matrix (of size $N_t \times N_t$) of the transmitted signal. The ergodic capacity is defined as the average of $C(\mathbf{H})$ over all the channel realizations. We first discuss the optimal form of $\mathbf{Q}_X$ for antenna selection.

6.3.1 Optimum Signaling with Transmit Antenna Selection

When no CSI is available at the transmitter, it is well known that in a full complexity system it is optimal to allocate the transmit powers uniformly across all the antenna elements and to keep the signals transmitted from different antenna elements uncorrelated, i.e., $\mathbf{Q}_X = \frac{\rho}{N_t} \mathbf{I}_{N_t}$ [6]. In antenna selection, however, this is no longer optimal. This was first highlighted by an example in [7], which showed that at low SNR, a covariance matrix of the form $\mathbf{Q}_X = \frac{\rho}{L_t} \mathbf{1} \mathbf{1}^\dagger$ can outperform $\mathbf{Q}_X = \frac{\rho}{L_t} \mathbf{I}_{L_t}$. Here, $\mathbf{1}$ is a vector of size $L_t \times 1$ and is given by $\mathbf{1} = [11 \cdots 1]^T$. Notice that a covariance of the form $\mathbf{1}\mathbf{1}^\dagger$ also treats all the transmit antenna elements equally and requires no CSI at the transmitter. That $\mathbf{Q}_X = \frac{\rho}{L_t} \mathbf{I}_{L_t}$ is not optimal is because the channel is no longer circularly symmetric Gaussian in the presence of antenna selection, which, depending on the channel realization, changes which transmit antenna elements are chosen.

In fact, the optimum signaling takes the following simple form for all SNR [8]:

$$\mathbf{Q}_x = \frac{\rho}{L_t}\left(\alpha \mathbf{I}_{L_t} + (1-\alpha)\mathbf{11}^\dagger\right), \tag{6.3}$$

where α is a real number that lies between 0 and $L_t/(L_t - 1)$. However, closed-form expressions are not available for α, and a one-dimensional numerical search is required to determine its optimal value.

A large number of papers in the literature do not take the above optimal signaling into account and assume that $\mathbf{Q}_x = \frac{\rho}{L_t}\mathbf{I}_{L_t}$. This is equivalent to setting $\alpha = 1$ in the above equation. We shall also make this assumption henceforth.

6.3.2   Performance of Antenna Selection

When antenna selection is employed at the transmitter or the receiver or both, a subset of the transmit and/or receive antenna elements is chosen. The channel seen by the subset is the sub-matrix, $\tilde{\mathbf{H}}$, that is obtained by selecting only the rows and columns of $\mathbf{H}$ that correspond to the selected receive and transmit antenna elements. The optimal subset is one that leads to the largest mutual information between the antenna elements used for transmission and reception. The capacity with antenna selection is given by

$$C_{\text{sel}} = \max_{S(\mathbf{H})} \log_2 \left| \mathbf{I}_{L_r} + \frac{\rho}{L_t}\tilde{\mathbf{H}}\tilde{\mathbf{H}}^\dagger \right|, \tag{6.4}$$

where $\tilde{\mathbf{H}}$ is an $L_r \times L_t$ matrix obtained by removing $N_r - L_r$ columns and $N_t - L_t$ rows from $\mathbf{H}$ and $S(\tilde{\mathbf{H}})$ denotes the set of all possible $L_r \times L_t$ sub-matrices of $\mathbf{H}$. The optimal channel subset, $\tilde{\mathbf{H}}_{\text{opt}}$, is given by

$$\tilde{\mathbf{H}}_{\text{opt}} = \arg\max_{S(\mathbf{H})} \log_2 \left| \mathbf{I}_{L_r} + \frac{\rho}{N_t}\tilde{\mathbf{H}}\tilde{\mathbf{H}}^\dagger \right|. \tag{6.5}$$

An exact analytical solution for C_{sel} is quite difficult. However, lower and upper bounds are available for C_{sel} for RAS. An upper bound was derived for C_{sel} in [9] by assuming that the signal components are received without any interference from the other components. The receive antenna elements that receive the L_r largest signal components are then selected. Clearly, neglecting the interference from other components is an unrealistic assumption, which is why the formula below is an upper bound:

$$C_{\text{sel}} \leq \sum_{i=1}^{L_r} \log_2 \left(1 + \frac{\rho}{N_t} \gamma_{[i]} \right), \tag{6.6}$$

where $\{\gamma_{[i]}\}_{i=1}^{N_r}$ are ordered chi-square random variables with $2N_t$ degrees of freedom such that $\gamma_{[1]} > \gamma_{[2]} > \ldots > \gamma_{[N_r]}$. The bound is tight for $L_r \leq N_t$, but is rather loose for $L_r > N_t$. For the latter case, the following bound is better:

$$C_{\text{sel}} \leq \sum_{j=1}^{N_t} \xi_j, \tag{6.7}$$

where

$$\xi_j = \log_z \left(1 + \frac{\rho}{L_t} \sum_{i=1}^{L_r} \tilde{\gamma}_{[i]}^{(j)} \right)$$

and, for each j,

$$\left\{ \tilde{\gamma}_{[i]}^{(j)} \right\}_{i=1}^{N_r}$$

are N_r ordered chi-square random variables (ordered with respect to index i) with 2 degrees of freedom.

Using matrix analysis, the following lower bound for the case $L_t = N_t = N_r$ was derived in [10]:

$$C_{\text{sel}} \geq \log_2 \left| \mathbf{I}_{N_t} + \frac{\rho}{N_t} \mathbf{H}^\dagger \mathbf{H} \right| + \log_2 \left| \mathbf{U}_r \mathbf{U}_r^\dagger \right|, \tag{6.8}$$

where $\mathbf{U}$ is the $N_r \times N_t$ orthonormal basis for the column space of $\mathbf{H}$ and $\mathbf{U}_r$ is the $N_t \times N_t$ block of $\mathbf{U}$ corresponding to receive antenna elements chosen. As $\mathbf{U}_r$ is a sub-matrix of the unitary matrix $\mathbf{U}$, we always have $\log_2 \left| \mathbf{U}_r \mathbf{U}_r^\dagger \right| \leq 0$. The capacity loss term $\log_2 \left| \mathbf{U}_r \mathbf{U}_r^\dagger \right|$ has been characterized in detail in [10].

The performance of RAS as a function of L_r for $N_t = 3$ and $N_r = 8$ for $\rho = 20$ dB is shown in Figure 6.2. The figure plots the cumulative distribution function (CDF) of the capacity (in bits/sec/Hz). The CDF is a useful representation because it contains information about not just the first moment (expected value), but also its higher moments. It can be seen that the capacity increases as the number of RF chains, L_r, increases. However, the gains diminish as L_r increases. For $L_r = 5$, the capacity CDF is already very close to that of a full complexity receiver.

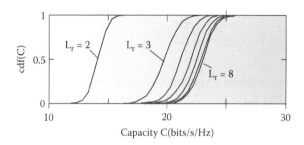

FIGURE 6.2
Capacity of a spatial multiplexing MIMO system with receive antenna selection for different L_r ($N_t = 3$, $N_r = 8$, and $\rho = 20$ dB) [4]. (© 2004 IEEE.)

TAS can even lead to "increased capacity" for low rank channels [11], which arise due to keyhole effects [12] or due to spatial correlation between the transmit and receive antenna elements [13]. This increase is a manifestation of the additional CSI available at the transmitter about which antennas to use; a transmitter without this additional information would have allocated equal powers to all the N_t elements.

6.3.3   Antenna Selection in Presence of Interference

So far, we have assumed that the noise is complex white circular Gaussian. In cellular systems, the transmissions from other cells or from within the cell give rise to co-channel interference, $\mathbf{i}$. The received signal now takes the form

$$\mathbf{y} = \mathbf{Hx} + \mathbf{n} + \mathbf{i}. \tag{6.9}$$

Let $\mathbf{Q}_I$ denote the covariance of the interference plus noise component. The capacity formula for antenna selection changes to

$$C_{\text{sel}} = \max_{S(\mathbf{H})} \log_2 \left| \mathbf{I}_{N_r} + \tilde{\mathbf{H}} \mathbf{Q}_x \tilde{\mathbf{H}}^\dagger \mathbf{Q}_I^{-1} \right|. \tag{6.10}$$

It turns out that this case can be easily mapped to the white Gaussian noise case that we have considered so far. By rearranging the terms in the determinant above, it can be shown that that the scenario above is equivalent to a system with an interference-free MIMO channel [14]. The equivalent channel takes the form $\mathbf{H}_{\text{eq}} = \Lambda_I^{-\frac{1}{2}} \mathbf{U}_I^\dagger \mathbf{H}$, where $\mathbf{U}_I$ is the eigenspace of $\mathbf{Q}_I$ and Λ_I is a diagonal matrix of the eigenvalues of $\mathbf{Q}_I$. When the interference is not Gaussian, the above formula provides a lower bound on the achievable capacity as Gaussian interference is the worst form of interference [5].

We shall not delve into the case where the signaling of the interferers can also be adjusted to maximize global system capacity [15].

6.4 Space–Time Codes

Space–time codes were presented in a series of seminal papers by Tarokh et al. [16,17]. They provide a practical means of exploiting spatial diversity when multiple transmit and receive antenna elements are used. We shall study antenna selection for two different kinds of space–time codes: space–time trellis codes and orthogonal space–time block codes, which offer different trade-offs between performance and complexity. The reader is referred to Chapter 4 for a review of space–time codes.

6.4.1 Space–Time Trellis Codes

Space–time trellis codes (STTC) [16,18] are designed to exploit the spatial diversity present in MIMO systems. Compared to orthogonal space–time block codes, which we discuss in the following section, STTCs also deliver a coding gain. However, the receiver for STTCs is more complicated, with the complexity increasing exponentially with the number of transmit antenna elements.

The performance of STTCs with RAS and a maximum likelihood detector (MLD) was studied in [19]. For STTCs, the expression for the pair-wise error probability of codewords of a given length provides considerable insight into the performance of antenna selection. Performance measures such as the overall error probability of the STTC may then be obtained from the pair-wise error probabilities using classical approaches for trellis codes [20].

Let the ith antenna element transmit a codeword of length l denoted by $\mathbf{C} = [\mathbf{c}_1, \ldots, \mathbf{c}_{N_t}]$. In this notation, the vector $\mathbf{c}_i$ (of size $l \times 1$) is transmitted from the ith antenna element. Given that the receiver knows the channel $\mathbf{H}$ perfectly, the probability that a full complexity receiver erroneously decodes $\mathbf{C}$ to the codeword $\mathbf{E} = [\mathbf{e}_1, \ldots, \mathbf{e}_{N_t}]$ depends on the Euclidean distance between the codewords. It is upper bounded by [16]

$$P(\mathbf{H}) \leq \exp\left(-\frac{\rho}{4N_t} \sum_{j=1}^{N_r} \mathbf{h}_j \mathbf{B} \mathbf{h}_j^\dagger \right), \tag{6.11}$$

where $\mathbf{h}_j$ is the jth row of $\mathbf{H}$ and corresponds to the jth receive antenna element. The matrix $\mathbf{B}$ (of size $N_t \times N_t$) depends on the codewords, with its $(p, q)^{th}$ element given by $\mathbf{B}_{pq} = (\mathbf{c}_p - \mathbf{e}_p)^\dagger (\mathbf{c}_q - \mathbf{e}_q)$. Note that the non-negative term

$$\sum_{j=1}^{N_r} \mathbf{h}_j \mathbf{B} \mathbf{h}_j^\dagger$$

corresponds to the distance between the codewords.

Ideally, the optimal antenna selection algorithm would choose the L_r receive antenna elements with indices $i_1, \ldots, i_{L_r}$ that maximize the sum

$$\sum_{j=1}^{L_r} \mathbf{h}_{i_j} \mathbf{B} \mathbf{h}_{i_j}^\dagger.$$

However, the indices need not be the same for different pairs of codewords. However, at high SNR, it is sufficient to only consider codewords that are separated by the minimum distance of the code because they largely determine the probability of decoding error. For such pairs, it can be argued that the norm-based selection criterion that chooses the antenna elements based on the received signal strength effectively achieves this selection [19]. We shall discuss this further in Section 6.6.2.

Therefore, the pair-wise error probability with RAS can be written as

$$P_{\text{sel}}(\mathbf{H}) \leq \exp\left(-\frac{\rho}{4N_t} \sum_{j=1}^{L_r} \mathbf{h}_{i_j} \mathbf{B} \mathbf{h}_{i_j}^\dagger\right), \tag{6.12}$$

$$\leq \exp\left(-\frac{L_r}{N_r} \frac{\rho}{4N_t} \sum_{j=1}^{N_r} \mathbf{h}_j \mathbf{B} \mathbf{h}_j^\dagger\right). \tag{6.13}$$

The second step follows from the following simple inequality that holds for any sequence of n numbers $x_1, x_2, \ldots, x_n$:

$$\frac{m}{n} \sum_{i=1}^{n} x_i \leq \sum_{i=1}^{m} x_{[i]} \leq \sum_{i=1}^{n} x_i, \quad (1 \leq m \leq n). \tag{6.14}$$

By averaging all the realizations of the channel, $\mathbf{H}$, it can be shown that, for large ρ, the average pair-wise error probability for a full rank STTC takes the form

$$\mathbf{E}_{\mathbf{H}}\left[P_{\text{sel}}(\mathbf{H})\right] \leq |\mathbf{B}|^{-L_r}\left(\frac{L_r}{N_r} \frac{\rho}{4N_t}\right)^{-N_t N_r}. \tag{6.15}$$

Notice that the exponent of ρ is $N_r N_t$ and is independent of L_r. This shows that RAS retains the full diversity order $N_t N_r$. However, compared to a full complexity system (with N_r receive antenna elements and N_r RF chains) this is accompanied by an array gain loss of $10 \log_{10} (N_r/L_r)$. Tighter bounds on the array gain loss are also derived in [19] for the case when only one receive antenna element is selected ($L_r = 1$).

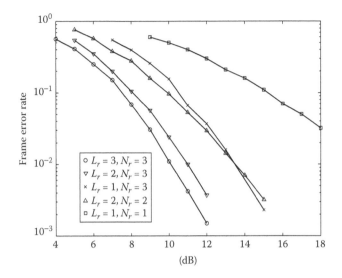

FIGURE 6.3
Pair-wise error probability of STTC with $N_t = 2$ for different L_r and N_r as a function of SNR, ρ [19]. (© 2003 IEEE.)

Figure 6.3 plots the pair-wise error probability as a function of SNR, ρ, for 2 transmit antennas for different numbers of receive antennas and RF chains. It can be seen that when $N_r = 3$, $L_r = 1, 2$, and 3 cases have the same diversity order, which is the slope at high SNR. Moreover, compared to the $L_r = N_r = 3$ case, the array gain losses equal 0.9 dB for $L_r = 2$ and 3.4 dB for $L_r = 1$. Both the losses are considerably less than the respective upper bound values of 1.76 dB and 4.77 dB.

TAS for STTCs was studied in [21]. The analysis turns out to be very similar to the one given above for RAS. TAS also achieves the full diversity order of the system, $N_r N_t$, and incurs an array gain loss that is upper bounded by $10 \log_{10}(N_t/L_t)$. That TAS achieves the full diversity order while transmitting from only L_t antenna elements is noteworthy. This implies that smaller STTCs can be used in conjunction with antenna selection to achieve the full diversity order. This circumvents the exponential increase in the number of trellis states in STTCs that are designed for larger N_t.

6.4.2 Orthogonal Space–Time Block Codes

Orthogonal space–time block codes (OSTBC), introduced in [17], achieve full diversity and yet require only a simple linear receiver. They can be optimally decoded by a low-complexity linear receiver.*

---

* Strictly speaking, the receiver should be called a widely linear receiver as it involves a conjugation operation [22].

Therefore, antenna selection with OSTBCs has received considerable attention in the literature. In OSTBCs, the receive SNR after linear processing can be expressed in terms of the Frobenius norm of the matrix, which is simply the square root of the sum of the squares of the amplitudes of all the elements of the matrix. This simple form for the SNR greatly facilitates analysis. The probability of bit error or codeword error can then be easily written in terms of the SNR depending on the modulation and channel coding scheme under consideration. TAS and RAS have been studied in [23]. The analysis for RAS was generalized to cover several cases in [24]. We illustrate the analysis for RAS below.

Given that the channel subset $\tilde{\mathbf{H}}$ is selected, the SNR, γ_{sel}, of the received data stream can be directly written in terms of the Frobenius norm, $\left\|\tilde{\mathbf{H}}\right\|_F$, of $\tilde{\mathbf{H}}$ [23] as follows:

$$\gamma_{sel} = \frac{\rho}{L_t}\left\|\tilde{\mathbf{H}}\right\|_F^2 = \frac{\rho}{L_t}\sum_{i=1}^{L_r}\sum_{j=1}^{L_t}\left|h_{n_i m_j}\right|^2,\qquad(6.16)$$

where $n_1, \ldots, n_{L_r}$ and $m_1, \ldots, m_{L_t}$ denote the indices of the chosen receive and transmit antenna elements, respectively.

Maximizing the instantaneous SNR is equivalent to maximizing the mutual information or minimizing the BER. The optimal selection criterion for all of these is therefore:

$$\tilde{\mathbf{H}}_{opt} = \arg\max_{S(H)}\left\|\tilde{\mathbf{H}}\right\|_F^2.\qquad(6.17)$$

The above simple criterion is often referred to as the norm-based criterion. In other words, in the case of TAS, the transmitter chooses the antenna elements that correspond to the L_t columns with the largest column-norms. In the case of RAS, the receiver chooses antenna elements that correspond to the L_r rows with the largest row-norms. However, in the case of T-RAS, transmit and receive antenna selection cannot be done independently.

As was the case for STTCs, the analyses for RAS and TAS are very similar. We illustrate the performance of RAS below assuming that uncoded BPSK is used [24]. The bit error probability is given in terms of the Q-function as

$$P_{sel}(\mathbf{H}) = Q\left(\sqrt{2\frac{\rho}{N_t}\sum_{i=1}^{L_r}\sum_{j=1}^{N_t}\left|h_{[i]j}\right|^2}\right),\qquad(6.18)$$

$$\leq Q\left(\sqrt{2\frac{L_r}{N_r}\frac{\rho}{N_t}\sum_{i=1}^{N_r}\sum_{j=1}^{N_t}\left|h_{ij}\right|^2}\right).\qquad(6.19)$$

For large ρ, the average bit error probability simplifies to

$$\mathbf{E_H}\left[P_{sel}(\mathbf{H})\right] \leq \binom{2N_tN_r-1}{N_tN_r}\left(\frac{L_r}{N_r}\frac{4\rho}{N_t}\right)^{-N_tN_r}. \qquad (6.20)$$

Therefore, receive antenna selection maintains the diversity order of N_tN_r with an array gain loss that is upper bounded by $10 \log_{10}(N_r/L_r)$.

As was the case with STTCs, tighter bounds on the array gain loss exist for the case where $L_r = 1$ [24]. When the Alamouti code ($L_t = 2$) is used, exact closed-form expressions for the average SNR and the outage probability have also been derived [23].

6.5 SIMO Systems

In SIMO systems, which have only one transmit antenna, only one data stream can be transmitted. Multiple copies of the transmitted signal arrive at the receiver, but with different amplitudes and phases that depend on the channel. The receiver uses maximum ratio combining (MRC) to combine the multiple received signals and achieve the maximum possible SNR. In MRC, each received signal is weighted with a complex conjugate of the corresponding channel coefficient.

With antenna selection, L_r out of the N_r received signals are selected and maximum ratio combined. Such a receiver is commonly referred to as the hybrid-selection MRC (H-S/MRC) receiver. The SNR at the output of the H-S/MRC receiver is given by

$$\gamma_{H\text{-}S/MRC} = \sum_{i=1}^{L_r} \gamma_{[i]}. \qquad (6.21)$$

Setting $L_r = 1$ corresponds to the special case of a selection combining (SC) receiver that uses only the best receive antenna element.

Closed-form expressions and tight bounds are available for both the SNR and the BER for this case. At first glance, the analysis of HS-MRC receivers seems complicated because the ordered random variables $\gamma_{[1]}, \ldots, \gamma_{[L_r]}$ are correlated even though the underlying variables $\gamma_1, \ldots, \gamma_{N_r}$ are independent of each other. However, the minimum value $\gamma_{[N_r]}$ and the differences $\gamma_{[1]} - \gamma_{[2]}, \ldots,$ $\gamma_{[N_r-1]} - \gamma_{[N_r]}$ are unconstrained, i.e., they can take any value between 0 and ∞. This motivates the following linear mapping, called virtual branch analysis [25], between $\gamma_{[i]}$s and the virtual variables V_js:

$$
\begin{bmatrix} \gamma_{[1]} \\ \vdots \\ \gamma_{[N_r]} \end{bmatrix} =
\begin{bmatrix}
\frac{\rho}{1} & \frac{\rho}{2} & \cdots & \frac{\rho}{N_r} \\
0 & \frac{\rho}{2} & \cdots & \frac{\rho}{N_r} \\
& & \ddots & \\
0 & 0 & \cdots & \frac{\rho}{N_r}
\end{bmatrix}
\begin{bmatrix} V_1 \\ \vdots \\ V_{N_r} \end{bmatrix},
\tag{6.22}
$$

where the V_is are just conveniently scaled versions of the differences. In case of Rayleigh fading, the V_is turn out to be independent and identically distributed exponential random variables with unit mean.

The mean SNR at the output of the H-S/MRC receiver can then be easily shown to be [25,26]

$$
\mathbf{E_H}\left[\gamma_{\text{H-S/MRC}}\right] = \sum_{i=1}^{L_r} \mathbf{E_H}\left[\gamma_{[i]}\right]
\tag{6.23}
$$

$$
= \rho \sum_{i=1}^{L_r} \mathbf{E}_{\delta_i}\left[V_i\right] + L_r \rho \sum_{i=L_r+1}^{N_r} \frac{1}{i} \mathbf{E}_{\delta_i}\left[V_i\right]
\tag{6.24}
$$

$$
= L_r \rho \left(1 + \sum_{i=L_r+1}^{N_r} \frac{1}{i}\right).
\tag{6.25}
$$

And the average symbol error probability (SEP) for MPSK can be shown to be [27]

$$
\mathbf{E_H}\left[P(\mathbf{H})\right] = \frac{1}{\pi} \int_0^{\Theta} \left[\frac{(\sin\theta)^2}{c\rho + (\sin\theta)^2}\right]^{L_r} \prod_{i=L_r+1}^{N_r} \left[\frac{(\sin\theta)^2}{\frac{L_r}{i}c\rho + (\sin\theta)^2}\right] d\theta,
\tag{6.26}
$$

where $\Theta = \pi(M-1)/M$ and $c = (\sin(\pi/M))^2$. For the large SNR case, the formula reduces to:

$$
\mathbf{E_H}\left[P(\mathbf{H})\right] = \rho^{-N_r} \frac{N_r!}{L_r! L_r^{N_r - L_r}} c^{-N_r} \frac{1}{\pi} \int_0^{\Theta} (\sin\theta)^{2N_r} d\theta.
\tag{6.27}
$$

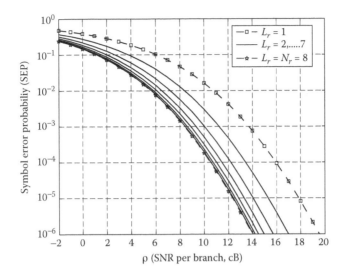

FIGURE 6.4
Symbol error probability as a function of SNR for different L_r ($N_r = 8$) [27]. (© 2001 IEEE.)

Therefore, H-S/MRC achieves a diversity order of N_r, and compared to a full complexity receiver with N_r RF chains, incurs an array gain loss equal to

$$\frac{10}{N_r} \log_{10}\left(\frac{N_r!}{L_r!\,L_r^{N_r-L_r}}\right).$$

Upper and lower bounds for the array gain loss for HS-MRC are derived in [28]. Figure 6.4 plots the SEP for $N_r = 8$ as a function of the SNR, ρ, for different values of L_r. We again see that the SEP decreases as L_r increases. The gains diminish for larger L_r.

The hybrid selection combining that we considered above can also be done at the transmitter in a closed-loop MIMO diversity system, in which signals input to the transmit antenna elements are weighted replicas of the same data stream. The received streams are combined to maximize the instantaneous SNR at the output of the combiner. This is analyzed for the $L_t = 1$ case in [29] and for the general $L_t \geq 1$ case in [30].

6.6 Implementing Antenna Selection: Criteria and Algorithms

In TAS, the transmitter needs to examine $\binom{N_t}{L_t}$ possibilities to choose the best subset of L_t antenna elements out of the N_t available. Similarly, in RAS, the receiver needs to examine $\binom{N_r}{L_r}$ possibilities. In the case of T-RAS, the number of possibilities balloons to $\binom{N_t}{L_t}\binom{N_r}{L_r}$, and coordination between the transmitter and receiver is required to choose the optimal transmit and receive antenna subsets. The combinatorial increase in the number of possibilities makes an exhaustive search impractical even for moderate values of N_t and N_r.

Therefore, a large number of sub-optimal selection algorithms and heuristics with varying levels of complexity have been proposed in the literature. The criteria depend on whether OSTBCs, STTCs, or ideal capacity-achieving space–time codes are assumed, and can also depend on the receiver architecture. We illustrate and compare a few of the proposed approaches for the following cases:

- Capacity-achieving spatial multiplexing with an optimal receiver,
- Spatial multiplexing with linear receivers, and
- Space–time codes with an MLD receiver.

6.6.1 Capacity-Achieving Spatial Multiplexing

In Equation 6.4, the optimal choice of the selection subset is given by a non-linear equation. However, for low SNR, the optimal selection criterion simplifies as follows:

$$C_{sel} \approx \max_{S(\mathbf{H})} \log_2\left(1 + \frac{\rho}{N_t}\left\|\tilde{\mathbf{H}}\right\|_F^2\right). \tag{6.28}$$

Therefore, at low SNR, norm-based selection is optimal [2,31]. Moreover, this is the same as the criterion in Equation 6.17 for antenna selection with OSTBCs.

However, the situation is not as simple for high SNR because of the $\log_2|.|$ form of the capacity formula in Equation 6.4. Therefore, several iterative and computationally simpler criteria have been proposed that exploit basic results from matrix theory [32].

The effect of adding an additional receive antenna element to the capacity formula was considered in [10,31]. A greedy criterion for selecting the next receive antenna element was proposed based on the observation that the capacity with $n + 1$ antenna elements can be written in terms of the capacity with n antenna elements as follows:

$$C(\tilde{\mathbf{H}}_{n+1}) = C(\tilde{\mathbf{H}}_n) + \log_2\left(1 + \frac{\rho}{N_t}\mathbf{h}_j\mathbf{B}_n\mathbf{h}_j^\dagger\right), \tag{6.29}$$

where the matrix $\mathbf{B}_n$ depends on the channel subset chosen so far, and equals

$$\mathbf{B}_n = \left(\mathbf{I}_{N_t} + \frac{\rho}{N_t}\tilde{\mathbf{H}}_n\tilde{\mathbf{H}}_n^\dagger\right)^{-1}.$$

The criterion for choosing the next antenna element then becomes: choose the receive antenna element j corresponding to the row $\mathbf{h}_j$ of the channel matrix, $\mathbf{H}$, that maximizes the following quadratic form:

$$J_n = \arg\max_j \mathbf{h}_j \mathbf{B}_n \mathbf{h}_j^\dagger. \tag{6.30}$$

Calculating the matrix inverse $\mathbf{B}_{n+1} = \left(\mathbf{I}_{N_t} + \frac{\rho}{N_t}\tilde{\mathbf{H}}_{n+1}\tilde{\mathbf{H}}_{n+1}^\dagger\right)^{-1}$ is simplified by using the following iteration:

$$\mathbf{B}_{n+1} = \mathbf{B}_n - \frac{1}{\frac{N_t}{\rho} + \mathbf{h}_{J_n}\mathbf{B}_n\mathbf{h}_{J_n}^\dagger} \mathbf{B}_n \mathbf{h}_{J_n}^\dagger \mathbf{h}_{J_n} \mathbf{B}_n. \tag{6.31}$$

The overall complexity of the algorithm is $O(\max\{N_t, N_r\}N_t L_r)$.

A decremental iterative search algorithm that successively removes antenna elements has also been proposed [10]. The recursion formulae are similar to the ones above, though with the $+$ and $-$ signs reversed. The decremental approach is efficient when L_r is close to N_r, while the incremental approach is efficient for large antenna arrays from which a small subset of antenna elements needs to be chosen ($L_r \ll N_r$). Both approaches achieve performance very close to the optimal selection criterion.

A faster decremental algorithm that avoids matrix inversion and determinants was proposed in [33]. For RAS, it is based on the intuition that a row (receive antenna element) of the channel matrix that is highly correlated with another row adds little additional information about the channel and may be removed. The algorithm calculates the correlation between all the rows of the channel matrix and selects the two rows with the highest correlation. Between the two, it eliminates the row with the lower power. Given that the MIMO transmission rate is better measured by mutual information, an alternate approach is to calculate the mutual information between two rows instead of their correlation. Of the two rows that result in the highest pairwise mutual information, the one with a lower power is eliminated.

Transmit antenna selection criteria have also been motivated by the water-filling principle [6], which optimally allocates more power to the channel modes with better channel quality. For example, [34] proposed finding the subset, represented by the transmit covariance $\mathbf{K}_{sel}$, that is close (in mean square error sense) to the ideal water-filling transmit covariance $\mathbf{K}_w$.

$$\tilde{\mathbf{H}} = \arg\min_{S(\mathbf{H})} |\mathbf{K}_w - \mathbf{K}_{sel}|^2. \tag{6.32}$$

Here, $\mathbf{K}_{sel}$ is a diagonal matrix (of size $N_t \times N_t$) in which the diagonal entries that correspond to the selected antenna elements are set as $\frac{\rho}{L_t}$, while the remaining elements are all set to 0.

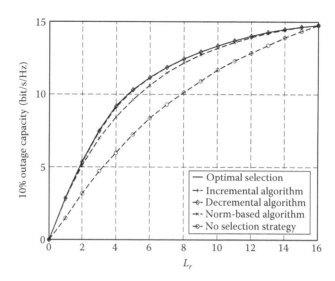

FIGURE 6.5
Outage capacity with a 10% outage probability as a function of L_r for $N_t = 4$, $N_r = 16$, and $\rho = 6$ dB [31]. (© 2004 IEEE.)

Given its simplicity, the norm-based criterion has often been used for selecting antenna subsets even though it is sub-optimal for spatial multiplexing, especially in the presence of spatial correlation [31,33]. Figure 6.5 compares performance of the incremental and decremental algorithms with the norm-based selection criterion. It plots the 10% outage capacity, i.e., the value below which the capacity falls only 10% of the time, as a function of L_r. It can be seen that both the incremental and decremental algorithms perform as well as the optimal selection criterion, while the norm-based selection criterion leads to a lower outage capacity. The differences are even more pronounced in the presence of spatial correlation.

6.6.1.1 When Transmitter Has CSI

So far we have assumed that no CSI is available at the transmitter (except for which antenna elements to use for transmission in the case of TAS). For the case in which CSI is allowed at the transmitter, incremental selection criteria based on heuristic criteria that depend on the signal power and the mutual information between the transmitted (received) signals for TAS (RAS) were proposed in [35], and have been shown to perform well.

6.6.1.2 With Linear Receivers

Linear receivers such as the Zero-forcing (ZF) receiver and the Minimum mean square error (MMSE) receiver, while sub-optimal compared to the

MLD receiver, are popular because they offer a significant reduction in complexity. The ordered successive interference cancellation ZF (OSIC-ZF) receiver, which reorders the streams in terms of their output SNR and decodes the stream with the highest SNR first [1], is also popular. To improve the SNR, the OSIC-ZF receiver cancels the contribution from the streams that have been decoded already.

Antenna selection at the transmitter and/or receiver for linear receivers is discussed in [36–38]. The selection criteria turn out to be different for linear receivers. In a linear receiver, the received signal, $\mathbf{y}$, is passed through a linear spatial filter (equalizer), $\mathbf{W}$. The filter outputs a vector, $\hat{\mathbf{x}}$, that is used to estimate the original data, $\mathbf{x}$:

$$\hat{\mathbf{x}} = \mathbf{Wy} = \mathbf{WHx} + \mathbf{Wn}. \tag{6.33}$$

The SNR of the N_t multiplexed streams is not equal due to interference from the other streams. The stream with the lowest SNR is often the performance bottleneck. It is therefore worthwhile to focus on improving this bottleneck. It can be shown that for ZF, MMSE, and OSIC-ZF receivers, the lowest SNR among all the streams is lower bounded by [37]

$$\text{SNR}_{\min} \geq \frac{\rho}{N_t} \lambda_{\min}(\mathbf{H}^\dagger \mathbf{H}), \tag{6.34}$$

where $\lambda_{\min}(\mathbf{H}^\dagger \mathbf{H})$ is the smallest eigenvalue of the square matrix $\mathbf{H}^\dagger \mathbf{H}$. This motivates the following selection criterion that maximizes the minimum eigenvalue:

$$\tilde{\mathbf{H}} = \arg \max_{S(\mathbf{H})} \lambda_{\min}(\tilde{\mathbf{H}}^\dagger \tilde{\mathbf{H}}). \tag{6.35}$$

However, this criterion does not motivate a low-complexity iterative technique to choose the optimal antenna selection subset. A decremental iterative approach has been suggested in [10] to maximize the throughput of linear receivers such as MMSE and OSIC-ZF. Criteria for joint adaptive modulation and coding and transmit antenna selection are considered in [37].

6.6.2 Space–Time Codes

While the simple norm-based criterion is clearly optimal for OSTBCs, this is not obvious for STTCs. A weak justification for using it for STTCs is presented below. It shows that the norm-based criterion maximizes a lower bound on the pair-wise distance between codewords of the STTC [21]. Other justifications for using the norm-based criterion for STTCs are presented in [19].

In Section 6.4.1, which studied STTCs, the minimum distance between codewords, d, when the transmit antennas with indices $\{i_k\}_{k=1}^{L_r}$ are chosen, was written as

$$d^2 = \sum_{k=1}^{L_r} \mathbf{h}_{i_k} \mathbf{B} \mathbf{h}_{i_k}^\dagger.$$

A lower bound on d^2 can be shown to be

$$d^2 \geq \lambda_{\min}(\mathbf{B}) \sum_{k=1}^{L_r} \sum_{j=1}^{N_t} \left| h_{i_k j} \right|^2 = \lambda_{\min}(\mathbf{B}) \left\| \tilde{\mathbf{H}} \right\|_F^2, \qquad (6.36)$$

where $\lambda_{\min}(\mathbf{B})$ is the smallest eigenvalue of $\mathbf{B}$, which depends entirely on the codewords and, therefore, the STTC structure. The same logic applies to TAS, as well.

6.7 Performance with Non-Idealities

When we considered space–time codes — ideal capacity-achieving codes or practical realizations such as STTCs and OSTBCs — we made the following assumptions:

- *Flat-fading assumption*: We assumed that multipath dispersion is not present.
- *Perfect channel knowledge assumption*: We assumed that the receiver can estimate the channel perfectly.
- *Block-fading channel assumption*: We assumed that the channel does not vary during the transmission of the space–time code.

In a practical system, non-idealities can invalidate these assumptions to varying degrees. We now examine the impact of various non-idealities.

When multipath dispersion is present, antenna selection, while beneficial, is not as effective as in the flat-fading case. In case of Rake receivers for CDMA systems [20,39], which weigh and combine the signals received from the multipaths, different sets of antenna elements may be optimum for receiving different multipaths. In a MIMO-OFDM system, different antenna element subsets may be optimum for different sub-carriers [40]. The performance of various antenna selection criteria in the presence of multipath dispersion is compared in [41]. A genetic-algorithm-based approach for selecting the antenna elements was also proposed.

When the channel state is not estimated perfectly by the receiver, it leads to the selection of sub-optimal subsets of antenna elements at the receiver and at the transmitter, which relies on feedback from the receiver.* Imperfect CSI can also lead to errors in the receive weights in H-S/MRC receivers. Assuming that the receiver always knows which transmit antenna elements are used, the case when the lth best transmit antenna element is selected was investigated in [42]. The analysis showed that the diversity order decreases to lN_r compared to N_rN_t, which is achieved when the best antenna element is chosen. This result shows that selection errors can reduce the diversity order of TAS. The impact of transmit antenna selection errors due to imperfect feedback on the performance of the Alamouti code was considered in [43]. The impact of estimation errors in a diversity system with transmit antenna selection was investigated in [44]. The loss in capacity was found to be tolerable so long as the SNR of the pilot was greater than a threshold value. The capacity degraded significantly for pilot SNRs below the threshold.

In fast fading with STTCs, simulations have shown that the diversity order achieved by receive antenna selection decreases and becomes a function of the number of selected antenna elements [45].

6.8 Antenna Selection with Spatial Correlation

So far, we have assumed that the links between transmit and receive antenna pairs are independent of each other. Moreover, the antenna selection techniques that we considered were based on the instantaneous channel state. In such a set up, a new subset has to be reselected every time the short-term fading variations decorrelate, which happens when the transmitter or the receiver moves over distances as small as half a wavelength. In some cases, such as low pilot power, it might be very difficult to estimate the instantaneous state in a short amount of time. However, it might still be possible to reliably acquire statistical knowledge about the channel, such as spatial correlation, which decorrelates only over distances of tens of meters.

We now consider antenna selection that uses spatial correlation information. The Kronecker model captures the transmit and receive antenna element correlations for many typical channels [46]. The channel matrix takes the form

$$\mathbf{H} = \mathbf{R}_r^{1/2}\mathbf{H}_w\mathbf{R}_t^{1/2}, \tag{6.37}$$

where $\mathbf{R}_t$ and $\mathbf{R}_r$ are the transmit and receive correlation matrices, respectively, and $\mathbf{H}_w$ is a spatially white matrix in which all the elements are mutually independent complex Gaussian random variables. The transmit (receive) covariance depends on the distance between the transmit (receive)

---

* Despite imperfect feedback, it is assumed in [42–44] that the receiver always knows which transmit antenna elements are used.

antenna elements, the mean angle of departure (arrival) of the signal, and the angular dispersion at the transmitter (receiver).

We now analyze the performance of antenna selection when it is based only on the statistical information and not the short-term fading of the channel. Antenna selection based on statistics has been considered in [23,47,48]. Let $\tilde{\mathbf{R}}_t$ and $\tilde{\mathbf{R}}_r$ denote the principal sub-matrices of $\mathbf{R}_t$ and $\mathbf{R}_r$, respectively, that correspond to the antenna elements selected, and have ranks r_t and r_r, respectively. The size of $\tilde{\mathbf{R}}_t$ is $L_t \times L_t$ while that of $\tilde{\mathbf{R}}_r$ is $L_r \times L_r$. Assuming that $\tilde{\mathbf{R}}_t$ and $\tilde{\mathbf{R}}_r$ are full rank, the fading-averaged pair-wise error probability of a space–time code at high SNR can be upper bounded by [23]

$$P \le \frac{1}{\left|\tilde{\mathbf{R}}_r\right|^{N_t} \left|\tilde{\mathbf{R}}_t\right|^{N_r} |\mathbf{B}|} \left(\frac{\rho}{L_t}\right)^{-N_t N_r}, \tag{6.38}$$

where $\mathbf{B}$ depends on the space–time code. For example, it is defined for STTCs in Section 6.4.1. Therefore, the diversity order equals $N_r N_t$. In case $\tilde{\mathbf{R}}_t$ and $\tilde{\mathbf{R}}_r$ are not full-rank, the diversity order scales linearly with the product of the ranks of $\tilde{\mathbf{R}}_t$ and $\tilde{\mathbf{R}}_r$.

For TAS, the optimal criterion is to select the transmit antenna elements that maximize $|\tilde{\mathbf{R}}_t|$, while the optimal criterion for RAS is to select the receive antenna elements that maximize $|\tilde{\mathbf{R}}_r|$. Moreover, the transmit and receive selection processes can occur independently of each other. Note that this is a consequence of assuming a Kronecker structure of the channel.

6.8.1 Alternate Antenna Selection Structures

We have seen that, while antenna selection achieves full diversity, it does suffer from a loss in array gain. Novel antenna selection structures have been recently proposed that involve a joint design of the RF and baseband [47–49]. The joint design introduces an RF pre-processing matrix that processes the signals from the different antennas, and is followed by selection (if necessary), down-conversion, and further processing in the baseband. The scheme is similar to conventional antenna selection because it uses fewer RF chains than the available antenna elements. However, it achieves superior performance by exploiting the spatial correlation of the received signals. For both spatial diversity and spatial multiplexing solutions, the joint-design, unlike conventional antenna selection, does not incur an array gain loss. RF pre-processing is familiar to the microwave community and has been used for applications such as analog beamforming [50]. It is important to understand that while adding a full-rank linear pre-processing block in the receiver cannot increase the capacity in a full complexity system, it does make a substantial difference in the presence of antenna selection, because selection is an inherently lossy operation.

Two designs have been considered. The first design uses an $L \times N_r$ RF pre-processing matrix that outputs only L streams followed by baseband signal

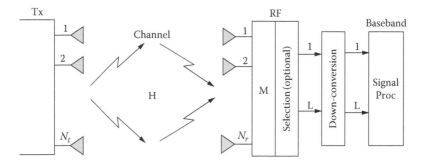

FIGURE 6.6
Block diagram for RF pre-processing and antenna selection at the receiver for spatial multiplexing [48]. (© 2004 IEEE.)

processing — it thus eliminates the need for a selection switch. The second one uses an $N_r \times N_r$ RF pre-processing matrix that outputs N_r streams and is followed by a switch that selects L streams for baseband signal processing. A block diagram of this scheme is shown for a receiver in Figure 6.6.

The optimum RF pre-processing solutions can be tailored to the CSI available for RF pre-processing. The simplest architecture is one in which the RF pre-processing matrix is always fixed. The use of an $N_r \times N_r$ Butler matrix, followed by a selection switch, was proposed in [49], and shown to significantly outperform conventional antenna selection. The other extreme (most complicated, but best) is one in which the RF pre-processing matrix is tuned to the instantaneous channel state [51]. The third and intermediate case is one in which the RF pre-processing solution depends only the slowly varying large-scale statistics of the channel.

We now analyze the case of statistics-based RF pre-processing for spatial multiplexing. The pre-processing matrix, $\mathbf{M}_{L_r}$, is of size $L_r \times N_r$. In baseband notation, the vector, $\tilde{\mathbf{y}}$, of size $L_r \times 1$, at the output of $\mathbf{M}_{L_r}$ now takes the form

$$\tilde{\mathbf{y}} = \mathbf{M}_{L_r}\mathbf{Hx} + \mathbf{M}_{L_r}\mathbf{n}. \qquad (6.39)$$

The ergodic capacity of such a system is given by

$$\mathbf{E_H}\left[C(\mathbf{H}) \right] = \max_{\mathbf{M}_{L_r}} \mathbf{E_H}\left[\log_2 \left| \mathbf{I}_{N_t} + \frac{\rho}{N_t}\mathbf{H}^\dagger\mathbf{M}_{L_r}^\dagger(\mathbf{M}_{L_r}\mathbf{M}_{L_r}^\dagger)^{-1}\mathbf{M}_{L_r}\mathbf{H} \right| \right]. \qquad (6.40)$$

It can be shown that the optimal RF pre-processing matrix that maximizes ergodic capacity is of the form $\mathbf{M}_{L_r} = \mathbf{D}[\mathbf{r}_1, \mathbf{r}_2, ..., \mathbf{r}_{L_r}]^\dagger$, where $\mathbf{r}_i$ is the ith largest eigenvector of the receive covariance matrix $\mathbf{R}_r$, and $\mathbf{D}$ is an arbitrary $L_r \times L_r$ full-rank matrix. When $\mathbf{M}$ is of size $N_r \times N_r$, and is followed by an antenna selection switch, the optimal $\mathbf{M}$ then takes the form $\mathbf{M}_{L_r} = [\mathbf{r}_1, \mathbf{r}_2, ..., \mathbf{r}_{N_r}]$.

Figure 6.7 compares the performance of various RF pre-processing schemes with conventional antenna selection for different spatial correlations. The

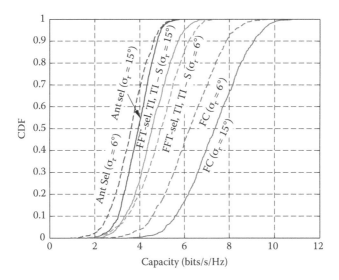

FIGURE 6.7
Performance comparison of RF pre-processing techniques and conventional antenna selection for spatial multiplexing ($N_r = 4$, $L_r = 1$, mean angle of arrival = 60°) [48]. (© 2004 IEEE.)

CDFs of the capacity are plotted for all the solutions. The case in which the RF pre-processing block is of size $L_r \times N_r$ and depends only on the transmit and receive covariances is denoted by TI (time-invariant); the case in which it is of size $N_r \times N_r$ and depends only on the transmit and receive covariances is denoted by TI-S (time-invariant with selection); and the Butler matrix solution is denoted by FFT-sel (fast Fourier transform with selection). Statistics-based and Butler-matrix-based solutions outperform conventional antenna selection even for large angular spreads.*

The optimum RF pre-processing matrices (of sizes $L_r \times N_r$ and $N_r \times N_r$ depending on the design) for a spatial diversity system are derived in [47].

While the elements of the RF pre-processing block can have an arbitrary amplitude, hardware considerations make the implementation of the RF pre-processing matrix using only variable-phase shifters an attractive option. An algorithm for implementing using phase-shifters is proposed in [47]. It has been shown in [52] that the structures are robust to phase quantization and to channel estimation errors.

6.9   Summary

Antenna selection is a promising low-complexity solution that solves the pressing problem of the increased hardware and signal processing complexity

---

* The larger the angular spread, the less the correlation between the antenna elements.

of MIMO systems. This chapter provides an overview of the extensive work that has been done on antenna selection at the transmitter, or at the receiver, or both. We considered antenna selection for SIMO systems and several different MIMO techniques such as capacity-achieving spatial multiplexing, space–time trellis codes, and orthogonal space–time block codes. We saw in several of these cases that antenna selection achieves full diversity order. However, it incurs an array gain loss, which increases as the number of selected antennas decreases. We also developed several criteria for implementing antenna selection that trade off between complexity and performance. The criteria were tailored to the specific system under consideration. We also considered novel structures that use a RF pre-processing block along with the selection switch and, in the presence of spatial correlation, recover the array gain loss of antenna selection to a great extent.

References

1. G.J. Foschini and M.J. Gans. 1998. "On the limits of wireless communications in a fading environment when using multiple antennas," *Wireless Pers. Commun.*, Vol. 6, pp. 311–335.
2. S. Sanayei and A. Nosratinia. 2004. "Antenna selection in MIMO systems," *IEEE Commun. Mag.*, Oct. 2004, pp. 68–73.
3. H. Holma and A. Toskala. 2000. *WCDMA for UMTS*, New York: John Wiley & Sons.
4. A.F. Molisch and M.Z. Win. 2004. "MIMO systems with antenna selection," *IEEE Microwave Mag.*, Vol. 5, March 2004, pp. 46–56.
5. T.M. Cover and J.A. Thomas. 1991. *Elements of Information Theory*, Wiley Series in Telecommunications, New York: John Wiley & Sons.
6. E. Telatar. 1999. "Capacity of multi-antenna Gaussian channels," *European Trans. Telecommun.*, Vol. 10, pp. 585–595.
7. R.S. Blum and J.H. Winters. 2002. "On optimum MIMO with antenna selection," *IEEE Commun. Lett.*, Vol. 6, Aug. 2002, pp. 322–324.
8. P.J. Voltz. 2003. "Characterization of the optimum transmitter correlation matrix for MIMO with antenna subset selection," *IEEE Trans. Commun.*, Vol. 51, Nov. 2003, pp. 1779–1782.
9. A.F. Molisch, M.Z. Win, and J.H. Winters. 2001. "Capacity of MIMO systems with antenna selection," *Proc. ICC*, pp. 570–574; see also A.F. Molisch, M.Z. Win, Y.S. Choi, and J.H. Winters. 2005. "Capacity of MIMO systems with antenna selection," *IEEE Trans. Wireless Commun.*, 4, 142–154.
10. A. Gorokhov, D. Gore, and A. Paulraj. 2003. "Receive antenna selection for MIMO flat-fading channels: theory and algorithms," *IEEE Trans. Inform. Theory*, Vol. 49, pp. 2687–2696.
11. D.A. Gore, R.U. Nabar, and A. Paulraj. 2000. "Selecting an optimal set of transmit antennas for a low rank matrix channel," *Proc. ICASSP*, May 2000, pp. 2785–2788.
12. D. Chizhik, G.J. Foschini, and R.A. Valenzuela. 2000. "Capacities of multi-element transmit and receive antennas: correlations and keyholes," *Electron. Lett.*, Vol. 36, pp. 1099–1100.

13. M.K. Ozdemir, E. Arvas, and H. Arslan. 2004. "Dynamics of spatial correlation and implications on MIMO systems," *IEEE Radio Commun.*, June 2004, pp. S14–S19.

14. T. Shu and Z. Niu. 2003. "A near-optimal antenna selection in MIMO system by using maximum total eigenmode gains," *Proc. Globecom*, pp. 297–301.

15. R.S. Blum. 2003. "MIMO capacity with antenna selection and interference," *Proc. ICASSP*, 2003, pp. IV–824–IV–827.

16. V. Tarokh, N. Seshadri, and A.R. Calderbank. 1998. "Space-time codes for high data rate wireless communication: performance criterion and code construction," *IEEE Trans. Inform. Theory*, Vol. 44, pp. 744–765.

17. V. Tarokh, H. Jafarkhani, and A.R. Calderbank. 1999. "Space-time block codes from orthogonal designs," *IEEE Trans. Inform. Theory*, Vol. 45, July 1999, pp. 1456–1467.

18. A.R. Hammons and H.E. Gamal. 2000. "On the theory of space-time codes for PSK modulation," *IEEE Trans. Inform. Theory*, Vol. 46, pp. 524–542.

19. A. Ghrayeb and T.M. Duman. 2003. "Performance analysis of MIMO systems with antenna selection over quasi-static fading channels," *IEEE Trans. Veh. Technol.*, Vol. 52, March 2003, pp. 281–288.

20. J.G. Proakis. 1989. *Digital Communications.* 2nd ed., New York: McGraw-Hill.

21. Z. Chen, B. Vucetic, J. Yuan, and Z. Zhou. 2004. "Performance analysis of space-time trellis codes with transmit antenna selection in Rayleigh fading channels," *Proc. WCNC*, March 2004, pp. 2456–2462.

22. B. Picinbono and P. Chevalier. 1995. "Widely linear estimation with complex data," *IEEE Trans. Sig. Proc.*, Vol. 43, pp. 2030–2033.

23. D.A. Gore and A. Paulraj. 2002. "MIMO antenna subset selection with space-time coding," *IEEE Trans. Sig. Proc.*, Vol. 50, Oct. 2002, pp. 2580–2588.

24. X.N. Zeng and A. Ghrayeb. 2004. "Performance bounds for space-time block codes with receive antenna selection," *IEEE Trans. Inform. Theory*, Vol. 50, pp. 2130–2137.

25. M.Z. Win and J.H. Winters. 1999. "Analysis of hybrid selection/maximal-ratio combining in Rayleigh fading," *IEEE Trans. Commun.*, Vol. 47, pp. 1773–1776.

26. M.S. Alouini and M.K. Simon. 1999. "Performance of coherent receivers with hybrid SC/MRC over nakagami-m fading channels," *IEEE Trans. Veh. Technol.*, Vol. 48, July 1999, pp. 1155–1165.

27. M.Z. Win and J.H. Winters. 2001. "Virtual branch analysis of symbol error probability for hybrid selection/maximal-ratio combining in Rayleigh fading," *IEEE Trans. Commun.*, Vol. 49, Nov. 2001, pp. 1926–1934.

28. M.Z. Win, N.C. Beaulieu, L.A. Shepp, J. Benjamin F. Logan, and J.H. Winters. 2003. "On the SNR penalty of MPSK with hybrid selection/maximal ratio combining over i.i.d. Rayleigh fading channels," *IEEE Trans. Commun.*, Vol. 51, pp. 1012–1023.

29. S. Thoen, L. Van der Perre, B. Gyselinckx, and M. Engels. 2001. "Performance analysis of combined transmit-SC/receive-MRC," *IEEE Trans. Commun.*, Vol. 49, Jan. 2001, pp. 5–8.

30. A.F. Molisch, M.Z. Win, and J.H. Winters. 2003. "Reduced-complexity transmit/receive-diversity systems," *IEEE Trans. Sig. Proc.*, Vol. 51, Nov. 2003, pp. 2729–2738.

31. M. Gharavi-Alkhansari and A. Gershman. 2004. "Fast antenna subset selection in MIMO systems," *IEEE Trans. Sig. Proc.*, Vol. 52, pp. 339–347.

32. R.A. Horn and C.R. Johnson. 1996. *Matrix Analysis*, Cambridge, U.K.: Cambridge University Press.

33. Y.-S. Choi, A.F. Molisch, M.Z. Win, and J.H. Winters. 2003. "Fast algorithms for antenna selection in MIMO systems," *Proc. VTC,* Fall, pp. 1733–1737.

34. S. Sandhu, R.U. Nabar, D.A. Gore, and A. Paulraj. 2000. "Near-optimal selection of transmit antennas for a MIMO channel based on shannon capacity," *Proc. Asilomar,* pp. 567–571.

35. M. Jensen and M. Morris. 2005. "Efficient capacity-based antenna selection for MIMO systems," *IEEE Trans. Veh. Technol.,* Vol. 54, pp. 110–116.

36. R.W. Heath, S. Sandhu, and A. Paulraj. 2001. "Antenna selection for spatial multiplexing systems with linear receivers," *IEEE Commun. Lett.,* Vol. 5, April 2001, pp. 142–144.

37. R. Narasimhan. 2003. "Spatial multiplexing with transmit antenna and constellation selection for correlated MIMO fading channels," *IEEE Trans. Sig. Proc.,* Vol. 51, Nov. 2003, pp. 2829–2838.

38. H. Zhang and H. Dai. 2004. "Fast transmit antenna selection algorithms for MIMO systems with fading correlation," *Proc. VTC,* pp. 1638–1642.

39. G. Bottomley, T. Ottosson, and Y.-P. Wang. 2000. "A generalized RAKE receiver for interference suppression," *IEEE J. Select. Areas Commun.,* Vol. 8, pp. 1536–1545.

40. G. Stuber, J. Barry, S. McLaughlin, Y. Li, M. Ingram, and T. Pratt. 2004. "Broadband MIMO-OFDM wireless communications," *Proc. IEEE,* Vol. 92, pp. 271–294.

41. P. Karamalis, N. Skentos, and A. Kanatas. 2004. "Selecting array configurations for MIMO systems: an evolutionary computation approach," *IEEE Trans. Wireless Commun.,* Vol. 3, pp. 1994–1998.

42. Z. Chen. 2004. "Asymptotic performance of transmit antenna selection with maximal-ratio combining for generalized selection criterion," *IEEE Commun. Lett.,* Vol. 8, April 2004, pp. 247–249.

43. W.H. Wong and E.G. Larsson. 2003. "Orthogonal space-time block coding with antenna selection and power allocation," *Electron. Lett.,* Vol. 39, pp. 379–381.

44. A.F. Molisch, M.Z. Win, and J.H. Winters. 2002. "Performance of reduced-complexity transmit/receive-diversity systems," *Proc. WPMC,* Oct. 2002, pp. 739–742.

45. A. Ghrayeb, A. Sanei, and Y. Shayan. 2004. "Space-time trellis codes with receive antenna selection in fast fading," *Electron. Lett.,* Vol. 40.

46. J.P. Kermoal et al. 2002. "A stochastic MIMO radio channel model with experimental validation," *IEEE J. Select. Areas Commun.,* Vol. 20, Aug. 2002, pp. 1211–1226.

47. P. Sudarshan, N.B. Mehta, A.F. Molisch, and J. Zhang. 2004. "Spatial diversity and channel statistics-based RF-baseband co-design for antenna selection," *Proc. 60th IEEE Vehicular Techn. Conf.,* invited paper, pp. 1658–1662.

48. P. Sudarshan, N.B. Mehta, A.F. Molisch, and J. Zhang. 2004. "Channel statistics-based joint RF-baseband design for antenna selection for spatial multiplexing," *Proc. Globecom,* Dec. 2004, pp. 3947–3951.

49. A. Molisch and X. Zhang. 2004. "FFT-based hybrid antenna selection schemes for spatially correlated mimo channels," *IEEE Commun. Lett.,* Vol. 8, pp. 36–38.

50. T. Ohira. 2002. "Analog smart antennas: An overview," *Proc. PIMRC,* pp. 1502–1506.

51. A.F. Molisch, X. Zhang, S.Y. Kung, and J. Zhang. 2003. "Antenna selection schemes for spatially correlated MIMO channels," *Proc. PIMRC,* pp. 1119–1123.

52. P. Sudarshan, N.B. Mehta, A.F. Molisch, and J. Zhang. 2004. "Antenna selection with RF pre-processing: robustness to RF and selection non-idealities," *Proc. IEEE Radio & Wireless Conference (RAWCON),* invited paper, pp. 391–394.

7

Performance of Multi-User Spatial Multiplexing with Measured Channel Data

Quentin H. Spencer, Jon W. Wallace, Christian B. Peel, Thomas Svantesson,
A. Lee Swindlehurst, Harry Lee, and Ajay Gumalla

CONTENTS

Abstract

The application of MIMO processing techniques in channels that are shared among multiple users is a relatively new problem that is increasingly important as MIMO transmission is put into practical use. In this chapter we specifically consider the multi-user downlink, where a base station with multiple antennas transmits simultaneously to more than one user. We begin with an overview of some of the multi-user MIMO transmission schemes that have been proposed up to this point, then demonstrate how they might be expected to perform by applying the algorithms to measurement data from indoor and outdoor propagation environments. Specifically, we compare the number of simultaneous users the channel will support for the two different environments, the amount of separation of the users necessary to achieve maximum throughput, and the quality of channel information available to the base station when the users are mobile. In both environments, full multi-user diversity is achieved at relatively short distances on the order of one meter. The total number of simultaneous users in outdoor environments is limited compared to uncorrelated channels due to the relatively sparse multipath structure of the channel. The distances at which channel information becomes too old to be useful to the transmitter appears to be similar for both types of channels.

7.1 Introduction

One of the most important emerging problems for communications system designers is applying multiple-input multiple-output (MIMO) concepts to multi-user environments. The most ubiquitous wireless applications of our time are cellular telephony and wireless LANs — inherently multi-user systems — in which the increasing demand for higher capacity makes them obvious candidates for the capacity improvements promised by MIMO processing.

In order to share a limited amount of frequency spectrum, all multi-user communication systems use one or more of the traditional forms of multiplexing: time-division, frequency-division, and code-division. The best multiplexing scheme for a given application is dependent on the characteristics of the particular channel of interest. The use of antenna arrays in a multi-user channel enables one further type of multiplexing often referred to as spatial multiplexing. Spatial multiplexing is particularly appealing because it can easily be used in conjunction with other forms of multiplexing to dramatically improve the number of users that can share a given channel. In addition to the promise of improved capacity for future communication

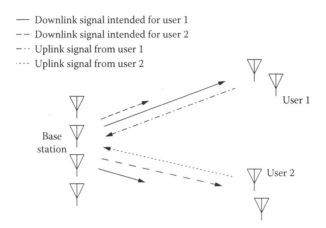

FIGURE 7.1
An illustration of a multiple-user MIMO channel with uplink and downlink. In the uplink, the base station receives multiple interfering signals and uses their spatial properties to separate them. In the downlink, each user often receives data intended for other users, and is not able to coordinate with them.

networks, in some cases these methods can be applied to existing communication protocols, thus extending their usefulness.

The problem of MIMO communications in multi-user environments can be further divided into two distinct problems, each with unique challenges: the "uplink" channel and "downlink" channel, as illustrated in Figure 7.1. The uplink refers to the case where a group of users sharing the same channel (representing a unique time slot, frequency, or code sequence) transmit simultaneously. Among information theorists this is commonly labeled the *multiple access* channel. This scenario requires the multi-antenna base station to successfully separate all of the interfering signals, which can be achieved if the users are transmitting from different locations by exploiting the differing spatial characteristics of the respective channels at the receiver.

The downlink channel (the *broadcast channel* among information theorists) refers to the case where the base station transmits simultaneously to more than one user over a shared channel. This poses challenges that are somewhat different from the uplink channel, because the receivers are unable to cooperate, so the signals at the different receivers cannot be processed jointly. Since the receivers pictured in Figure 7.1 have multiple antennas, they could, in theory, use multiple-user detection (MUD) techniques to avoid the interfering signals. This is typically computationally costly, and in cases where users have only a single antenna, it is not possible without relying on other forms of multiplexing such as CDMA. Ideally, then, we would like to mitigate the multiple-access interference (MAI) at the transmitter by intelligently designing the transmitted signal.

In this chapter, we begin by reviewing the problem of the multi-user MIMO downlink and discuss in general terms several different transmission approaches that have been proposed. We compare the performance of some of the schemes in randomly generated channels under ideal conditions. We then show how the algorithms would perform under real-world conditions using measurement data from an indoor propagation environment and two different outdoor environments. We focus in particular on the achievable capacity, the ability of the channels to support multiple sub-channels per user, the required separation distance between users to maximize available capacity, and the effects of channel estimation error due to user mobility. We begin with a mathematical model used for characterizing multi-user MIMO channels.

7.2   The Multiple-User MIMO Channel

A MIMO channel with n_T transmitters and n_R receivers is commonly represented as a matrix $\mathbf{H}$ of dimension $n_R \times n_T$, where each of the coefficients $[\mathbf{H}]_{i,j}$ represents the complex transfer function from the jth transmitter to the ith receiver. We denote the signal transmitted from the jth transmitter at time t as $x_j(t)$, and represent the total transmitted signal with the vector $\mathbf{x}(t)$ of dimension n_T. Likewise, we represent the received signal with the n_R-dimensional vector $\mathbf{y}(t)$, which can be expressed as a function of the transmitted signal, the channel matrix, and an n_R-dimensional additive noise vector $\mathbf{n}(t)$:

$$\mathbf{y}(t) = \mathbf{H}\mathbf{x}(t) + \mathbf{n}(t). \qquad (7.1)$$

For simplicity, we will omit the dependence on time, referring to the transmitted and received signals instead as $\mathbf{x}$ and $\mathbf{y}$. In a single-user point-to-point MIMO link, all outputs are available to the receiver for processing. In the multiple-user case, the n_R receivers are distributed among different users, so we extend the model to reflect this. Let K represent the number of users sharing a channel, and let n_{R_j} represent the number of antennas for user j, so that the total number of receive antennas

$$n_R = \sum_{j=1}^{K} n_{R_j}.$$

The channel between the base and user j is now the $n_{R_j} \times n_T$ matrix $\mathbf{H}_j$, whose rows we denote by $\mathbf{h}_{ij}^*$ as follows:

$$\mathbf{H}_j = \begin{bmatrix} \mathbf{h}_{1j} & \cdots & \mathbf{h}_{n_{R_j}j} \end{bmatrix}^*,$$

where $(\cdot)^*$ is used to denote the complex conjugate (Hermitian) transpose.

Note that this model is based on the assumption of a flat-fading or narrowband channel, which is not true in many current and next-generation wireless communications applications. This matrix channel model can still be applied to many broadband channels. For example, many modern broadband communication protocols are based on orthogonal frequency division multiplexing (OFDM). Typically, the bandwidth of one OFDM subcarrier is narrow enough that the narrowband model applies, so MIMO processing algorithms could be applied independently for each subcarrier. While we assume a narrowband channel throughout this chapter, the methods discussed here can also be applied to broadband channels using OFDM or similar techniques.

Using this model, consider the signal received by user j. User j not only receives its own signal through the channel $\mathbf{H}_j$, but also contributions from the signals intended for other users:

$$\mathbf{y}_j = \sum_{k=1}^{K} \mathbf{H}_j \mathbf{x}_k + \mathbf{n}_j, \tag{7.2}$$

where $\mathbf{n}_j$ is assumed to be spatially white noise. The transmitted signal $\mathbf{x}$ is then the sum of the transmitted signals for each user:

$$\mathbf{x} = \sum_{j=1}^{K} \mathbf{x}_j.$$

We assume that the transmitted signal $\mathbf{x}_j$ is formed from a vector $\mathbf{d}_j$ of m_j symbols to be transmitted to user j. In a single-user channel, the number of symbols transmitted in parallel is limited by the rank of the channel matrix. Likewise, in a multi-user channel, m_j is limited by the rank of $\mathbf{H}_j$. We use the vector $\mathbf{d}$ to denote the data symbols transmitted to all users:

$$\mathbf{d} = \left[\mathbf{d}_1^T \quad \mathbf{d}_2^T \quad \cdots \quad \mathbf{d}_K^T \right]^T,$$

where the dimension of $\mathbf{d}$ is

$$m = \sum_{j=1}^{K} m_j.$$

It is possible to transmit at different rates to each of the K users by the choice of symbol constellation and channel coding for each user, and by the number of data streams m_j transmitted to each user. Suitable values for $m_1, \ldots, m_K$

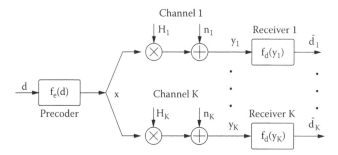

FIGURE 7.2
An illustration of downlink multi-user processing, where each user has multiple antennas and receives parallel data symbols.

will not only depend on the desired data rate for user j, but also on the available transmit power, the achievable SINR, and the number of transmit and receive antennas. Without additional coding or multiplexing, typically $m_j \leq n_{R_j}$, and $\sum m_k \leq n_T$. The choice of m_j to obtain optimal system performance is itself an important problem that has recently been studied in [1].

The transmitted signal $\mathbf{x}$ is formed from $\mathbf{d}$ using some encoding function that we denote by $\mathbf{f}_e$, so that $\mathbf{x} = \mathbf{f}_e(\mathbf{d})$. The receivers use a decoding function $\mathbf{f}_d$ to estimate the data vectors: $\hat{\mathbf{d}}_j = \mathbf{f}_d(\mathbf{y}_j)$. This scheme is illustrated in Figure 7.2. To describe the dimensions of a particular channel, we use the notation $\{n_{R_1}, \ldots, n_{R_K}\} \times n_T$. A system with $K = 4$ users having one antenna each and $n_T = 4$ transmit antennas could be written as a $\{1,1,1,1\} \times 4$ system. Likewise, a $\{1,1,2,2\} \times 4$ describes a similar case where two of the users have two antennas. The challenge of the multi-user MIMO downlink is to choose encoding and decoding functions $\mathbf{f}_e$ and $\mathbf{f}_d$ that optimize the use of channel resources. Specific optimization goals may include maximizing total throughput, ensuring a certain quality of service (QoS) for each user, or minimizing transmit power, among others.

Achieving any of these goals requires that the transmitter have some information about the channel. From an information-theoretic point of view, a MIMO channel where the transmitter has channel state information (CSI) has a different channel capacity than the same channel without CSI. Single-user MIMO systems benefit from having CSI at the transmitter mainly when $n_T \geq n_R$ or at low SNR. On the other hand, CSI in a multi-user MIMO downlink is critical to minimizing inter-user interference under all channel conditions. Obtaining CSI is itself a challenging problem [2]. It can generally be obtained in a two-way communications system by either sending the information over the reverse link, or estimating channel parameters of the reverse link and applying them to the forward link. In this chapter, we assume that CSI is available, and we will later investigate the effects of corrupted channel information on multi-user performance.

Another property of radio channels to consider is how they vary over time. The assumption that CSI is available at the transmitter usually implies that the channel is quasi-static, which has been assumed in much of the research to date on multi-user MIMO channels. This is a reasonable assumption for channel environments such as wireless local area networks (LANs), where users are mobile but do not move rapidly. Cellular telephone applications are more challenging because speeds are much higher. Downlink processing methods for channels that are quickly time-varying with limited CSI is an important problem for future research. However, the results included in Section 7.4 suggest that the prediction horizons for MIMO systems may be much longer than in the SISO case (which has usually proven to be too short to be useful), since multiple antennas reveal more information about the physical structure of the channel [3].

7.2.1   Capacity

Capacity is an important tool for analysis of communication channels. In single-user MIMO channels it is common to assume a constraint on total power broadcast by all transmit antennas. For a multi-user MIMO channel, the problem is more complex. Given a constraint on the total transmitted power, it is possible to allocate varying fractions of that power to different users in the network, so a single power constraint can yield many different information rates. The result is a "capacity region" like that illustrated in Figure 7.3 for a two-user channel. The "corners" of the region represent allocation of 100% of the power to either one of the users. For every possible power distribution in between, there is an achievable information rate, which results in the outer boundary of the region. In Figure 7.3, two regions are

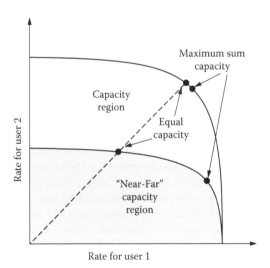

FIGURE 7.3
An illustration of a multi-user capacity region. The sum capacity may penalize certain users, depending on the shape of the capacity region.

shown: one for the case where both users have similar maximum capacity, and one for the case where they are different (due, for example, to user 2's channel being attenuated relative to user 1, sometimes referred to as the "near-far" problem). For K users, the capacity region is characterized by a K-dimensional volume.

In Figure 7.3, two points are indicated on the boundary of each of the two regions. One point represents the maximum achievable throughput of the entire system, or the point on the curve that maximizes the sum of all users' information rates. It is clear that this *sum capacity* point does not always represent a fair distribution of resources among the users. The second point on the two curves is located where the curves intersect with the line $C_2 = C_1$, and represents the maximum achievable rate such that both users have equal rates. The problem in this case is that the total throughput is substantially reduced in the "near-far" case. While the sum capacity point clearly does not convey all the relevant information about a multi-user MIMO channel, it is nevertheless a useful tool for understanding the relative capabilities of a particular transmission algorithm or channel, and will be used extensively in this chapter for that purpose.

The capacity of the multi-user MIMO channel is achieved by applying a concept that originates from a paper by M. Costa [4] known as "writing on dirty paper." Costa studied communication channels with interference and proved the somewhat surprising result that if a received signal y is defined as

$$y = s + i + w, \tag{7.3}$$

where s is the transmitted signal, i is interference known deterministically to the transmitter, and w is additive white Gaussian noise, the capacity of the system is the same as if there were no interference present, regardless of how strong the interference is and whether or not it is known to the receiver. Using the dirty paper analogy, the capacity of a dirty sheet of paper is the same as that of a clean sheet if the location of the dirt is known.

The implications of this result for multi-user MIMO channels with CSI available at the transmitter are clear: since the channel and the transmitted signals are all known, the transmitter knows how a signal designed for one user interferes with other users and can design the signals for the other users to compensate. This is the basis for many of the results on capacity of the MIMO broadcast channel [5–9]. While the initial capacity results characterize the achievable rate, they only prove achievability, but do not describe how this rate is achieved. More recently, some practical transmission schemes [10–12] have been proposed that use dirty paper codes to approach the capacity of the scalar interference channel. To date, no practical application of dirty-paper codes to the downlink problem has appeared, though the techniques of [13] are related to dirty-paper coding (DPC).

A simplified approach to transmission in multi-user channels is to treat all interference as noise. Clearly, this is suboptimal, but it also results in much simpler implementations. The result is that the encoder $\mathbf{f}_e$ and decoder $\mathbf{f}_d$ are linear functions of the data to be transmitted and the received signals, respectively. The linear class of transmission schemes can generally be viewed as a type of beamforming. Linear transmission schemes exist for a wide variety of channel configurations, while so far, dirty-paper schemes have only been proposed for the special case where all receivers have one antenna $n_{R_j} = 1$. However, the existing dirty-paper schemes can be applied to the more general case where $n_{R_j} > 1$ by combining them with linear processing methods. In the next section we review some of these schemes.

7.3   Multi-User MIMO Transmission Schemes

As noted above, the existing schemes for transmitting from an antenna array to a group of users can be put in two broad categories: linear and non-linear. They can also be categorized as algorithms for single-antenna receivers and multi-antenna receivers. In this section we review some of the linear methods that have been proposed, and briefly discuss non-linear methods.

7.3.1   Linear Processing, Single-Antenna Receivers

We begin with linear transmission schemes for the case where each user has only one receive antenna: $n_{R_j} = 1$. With only one antenna, the receiver is unable to perform any spatial interference suppression of its own, so the transmitter is responsible for precoding the data in such a way that the interference seen by each user is tolerable. We consider three techniques for solving this problem: channel inversion, regularized channel inversion, and optimal beamforming.

7.3.1.1   *Channel Inversion*

Perhaps the simplest way of managing the inter-user interference in a multi-user downlink is using the (pseudo-) inverse of the channel matrix $\mathbf{H}$ [14–16]. For non-square channels where $n_T \geq K = n_R$, the transmitted signal $\mathbf{s}$ is

$$\mathbf{s} = \mathbf{H}^* \left(\mathbf{H}\mathbf{H}^* \right)^{-1} \Gamma \mathbf{d} . \qquad (7.4)$$

The matrix Γ is a diagonal matrix used to scale the power transmitted to each user. The channel inversion nulls out all inter-user interference, reducing the problem to K independent scalar channels, so the amount of power allocated to one user does not affect the others. Given a constraint on the total transmitted power ρ, Γ can be chosen in different ways to achieve different goals: allocate equal power to all users, allocate equal capacity to all users, or maximize sum capacity. Sum capacity can be maximized by computing the gain of each of the independent channels and using the water-filling algorithm to distribute the available power.

A very simple way of allocating the power is to set $\Gamma = \gamma\mathbf{I}$, where $\gamma = 1/\rho$. One problem with channel inversion arises when $\mathbf{H}$ is ill-conditioned. In such cases, at least one of the singular values of $(\mathbf{HH}^*)^{-1}$ is very large, γ will be large, and the SNR at the receivers will be low. It is interesting to note the similarity between channel inversion and least-squares or "zero-forcing" (ZF) receive beamforming, which applies a dual of the transformation in Equation 7.4 to the receive data. Such beamformers are known to cause noise amplification when the channel is nearly rank deficient. On the transmit side, ZF produces signal attenuation instead. In fact, it has been shown that in the ideal case where the elements of $\mathbf{H}$ are independent complex Gaussian random variables, the probability density of γ has an infinite mean [17]. It is also shown in [17] and the simulation results section of this chapter that the capacity of channel inversion does not grow linearly with K.

7.3.1.2   *Regularized Channel Inversion*

When rank-deficient channels are encountered in ZF receive beamforming, one technique to reduce the effects of noise amplification is to regularize the inverse in the ZF filter. If the noise is spatially white and an appropriate regularization value is chosen, this approach is equivalent to using a minimum mean-squared error (MMSE) criterion to design the beamformer weights. Applying this principle to the transmit side suggests the following solution:

$$\mathbf{s} = \frac{1}{\sqrt{\gamma}}\mathbf{H}^*\left(\mathbf{HH}^* + \zeta\mathbf{I}\right)^{-1}\mathbf{d}, \qquad (7.5)$$

where ζ is the regularization parameter. When $\zeta \neq 0$, the transmitter does not perfectly cancel out all interference. The key is to define a value for ζ that optimally trades off the numerical condition of the matrix inverse against the amount of interference that is produced. It has been shown that choosing $\zeta = K/\rho$ approximately maximizes the SINR at each receiver, and leads to linear capacity growth with K [17]. Because each user sees some interference from other users, this scheme does not allow the same flexibility as exact channel inversion in adjusting the power transmitted to each user, because a change to the power weighting for one user changes the interference seen by all other users.

7.3.1.3   *Optimal Linear Precoders*

Regularized channel inversion demonstrates that perfectly canceling out all inter-user interference is not optimal, and provides a good solution in closed form at low computational cost. However, it is still not necessarily the optimal linear beamformer. Attempting to design a set of transmit beamformers without any constraints on inter-user interference is a very challenging problem because they are all interdependent. If an optimal beamformer is designed for one user, it will produce some interference for the other users. If the interference is taken into account in designing an optimal beamformer for a second user, it will emit interference that makes the first user's beamformer suboptimal. This suggests that the optimal solution can not be computed in closed form but requires an iterative approach.

With a zero-forcing solution, a set of independent channels is created, so the power transmitted to each user can readily be adjusted to achieve a variety of different goals, such as maximizing sum capacity or insuring equal capacity for all users given a power constraint, or minimizing power given a capacity constraint for each user. In the case where inter-user interference is allowed, the solution to each of these problems will be different. Optimal transmit beamformers have been found for a variety of different optimization criteria [18–23]. We give as an example the linear precoder that optimizes sum capacity [18], which takes the form

$$s = \left(H^* D H + \zeta I_{n_T} \right)^{-1} H^* \Delta d. \tag{7.6}$$

This is similar to regularized channel inversion, but we have introduced two diagonal matrices, D and Δ, which are used respectively to weight the rows of H inside the inverse and weight the columns of the resulting beamformer. The optimal values for these matrices, and the scale constant ζ, can be computed using the iterative algorithm given in Table 7.1.

The sum capacity as a function of the channel matrix size for the linear precoders we have discussed so far is compared with the sum capacity of the channel and capacity of an equivalent single-user channel in Figure 7.4.

TABLE 7.1

Linear Precoding for Maximum Sum Capacity

---

1. Initialize $\rho_j(1) = 1$ for $j = 1...K$, $D = I$, $\Delta = I$
2. Repeat until convergence

   a. $M = \left(H^* D H + \mathrm{tr}(D)/\rho I \right) H^* \Delta$

   b. $W = H M \sqrt{\rho / \mathrm{tr}(M^* M)}$

   c. $n_j = \left| [W]_{j,j} \right|^2$

   d. $d_j = 1 + \sum_{i=1, i \neq j}^{K} \left| [W]_{i,j} \right|^2$

   e. $[D]_{j,j} = \dfrac{n_j}{d_j(d_j + n_j)}$

   f. $[\Delta]_{j,j} = [W]_{j,j}/d_j$

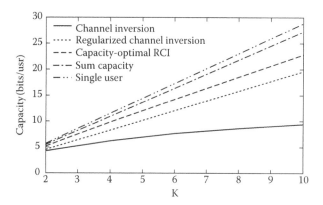

FIGURE 7.4
Mean sum-capacity of various precoders for uncorrelated Gaussian channels with K users and $n_T = K$ transmitters at a SNR of 10 dB.

Of all the precoders, only channel inversion fails to achieve a sum capacity that increases linearly with K and n_T. Regularized channel inversion offers a substantial improvement in performance, and the optimal linear precoder (labeled capacity-optimal RCI) is even better, achieving most of the sum-capacity of the channel that is achievable using DPC.

As we noted earlier, there are many situations in which optimizing sum capacity is problematic because it does not guarantee a minimum level of signal to any user. There are many other optimizations that have appeared recently in the literature that may be of greater practical interest. The "power control" problem was the first of these [19,20], and can be stated as follows: given a set of SINR requirements for each user, compute the set of beam-formers $\mathbf{b}_1, \ldots, \mathbf{b}_k$ such that the SINR requirements are met and total trans-mitted power is minimized. We define γ_j as the SINR for user j, which can be expressed as:

$$\gamma_j = \frac{\mathbf{b}_j^* \mathbf{H}_j^* \mathbf{H}_j \mathbf{b}_j}{\sum_{k \neq j} \mathbf{b}_k^* \mathbf{H}_j^* \mathbf{H}_j \mathbf{b}_k + 1}, \tag{7.7}$$

where we have assumed that the noise has unit variance. The power mini-mization problem can be stated mathematically as

$$\min_{\mathbf{b}_1, \cdots, \mathbf{b}_K} \sum_{k=1}^{K} \mathbf{b}_k^* \mathbf{b}_k$$

$$\text{s.t.} \quad \frac{\mathbf{b}_j^* \mathbf{H}_j^* \mathbf{H}_j \mathbf{b}_j}{\sum_{k \neq j} \mathbf{b}_k^* \mathbf{H}_j^* \mathbf{H}_j \mathbf{b}_k + 1} \geq \gamma_j, \quad j = 1, \cdots, K. \tag{7.8}$$

Solutions to this problem have been proposed in [19–21]. Other optimizations for which solutions have recently been proposed include the maximization of the SINR margin for all users given a minimum SINR requirement for each user and a total power constraint [22,23]. In addition to the sum-capacity solution, [18] also proposes a solution for maximizing the minimum capacity for each of the individual users. This whole class of solutions requires more computation than plain or regularized inversion, but since beamformers are typically computed once for an entire block of transmitted symbols, this is still a practical solution.

7.3.2   Linear Processing, Multi-Antenna Receivers

With only a single antenna, the receivers are not able to perform spatial interference suppression of their own, so they can only receive data over a single spatial channel. With multiple antennas, these restrictions are removed, provided that the transmitter and receiver can coordinate their spatial processing, and appropriately allocate the available spatial resources. A simple approach to this would be to apply the single-antenna techniques just described, provided that $n_R \leq n_T$, where n_R is the total number of receive antennas summed over all users. This effectively treats each receive antenna as if it were a separate user, so no joint processing among the receive antennas is required, but performance is limited because the problem is overly constrained, and the number of users is limited more than necessary.

Both channel inversion and regularized channel inversion limit the number of users to $K \leq n_T$. For optimal beamforming, it is technically possible to support cases where $K > n_T$, but realistically this can occur only when the SINR requirements are very low. So, it is reasonable to consider n_T to be the practical upper bound on the number of users. If the receivers have multiple antennas, it is still possible to support up to $K = n_T$ users by using the concept of coordinated beamforming. To illustrate this, consider the block diagram of Figure 7.2. Assume that the decoding function $f_d(\mathbf{y}_j)$ for user j is a linear operator $\mathbf{w}_j$, so that $\hat{d}_j = \mathbf{w}_j^* \mathbf{y}_j$. If the beamformers were known to the transmitter in advance, then the *virtual channel*, which represents the transfer function from the transmitter to the output of the beamformer of user j, is

$$\overline{\mathbf{h}}_j^* = \mathbf{w}_j^* \mathbf{H}_j .$$

If we collect the virtual channels for each user, we can define a virtual channel for the entire system:

$$\overline{\mathbf{H}} = \begin{bmatrix} \overline{\mathbf{h}}_1 & \overline{\mathbf{h}}_2 & \cdots & \overline{\mathbf{h}}_K \end{bmatrix}^* $$

As long as $K \leq n_T$, it possible to apply any of the single-antenna algorithms described earlier to the virtual channel $\overline{\mathbf{H}}$. The remaining problem is determining the receive beamformers $\mathbf{w}_j$. This information could be obtained if the transmitter were to assume a specific approach to designing the beam-formers. For example, both MMSE and MRC designs for $\mathbf{w}_j$ are functions of only the channel and the transmit beamformers, so $\mathbf{w}_j$ could be computed from information available to the transmitter. However, this results in a situation where the solutions to the transmit and receive beamformers are dependent on each other. This suggests the following iterative approach:

1. Assume an initial set of $\mathbf{w}_j$ values.
2. Compute the virtual channel $\bar{\mathbf{H}}$ and the transmit beamformers.
3. Update the receive beamformers $\mathbf{w}_j$.
4. Repeat steps 2 and 3 until convergence.

The convergence properties of this approach will depend, in general, on what algorithms are used on both the transmitter and receiver side to determine the beamforming weights. The concept of beamforming that is coordinated between the transmitter and receiver is the basis for several recent multi-user transmission schemes [1,24–29].

In single-user MIMO channels with CSI available to the transmitter, capacity is achieved by spatial multiplexing, where a number of independent sub-channels are created that carry independent streams of data. In a multi-user MIMO downlink where the receivers have multiple antennas, it is also possible to transmit multiple data streams to each user. We define m_j to be the number of sub-channels allocated to user j, and

$$m = \sum_{j=1}^{K} m_j$$

to be the total number of sub-channels. The restrictions on these values are that $m_j \leq \tilde{L}_j$, where $\tilde{L}_j$ is the rank of $\hat{\mathbf{H}}_j = \begin{bmatrix} \mathbf{H}_1^T \dots \mathbf{H}_{j-1}^T \ \mathbf{H}_{j+1}^T \dots \mathbf{H}_k^T \end{bmatrix}^T$, and $m \leq n_T$. This means that allocating multiple sub-channels to individual users limits the total number of users that can be served. The problem of choosing a good value of m_j has not yet been studied extensively, but the simulation results presented later in this chapter illustrate the trade-offs involved. In the case where $m_j > 1$, define the receiver for user j as the $m_j \times n_{R_j}$ matrix $\mathbf{W}_j$, and the linear precoder as the $n_T \times m$ matrix,

$$\mathbf{B} = \begin{bmatrix} \mathbf{B}_1 & \mathbf{B}_2 & \dots & \mathbf{B}_k \end{bmatrix},$$

where $\mathbf{B}_j$ is the $n_T \times m_j$ precoder for user j. As in the single sub-channel case, the transmit precoders $\mathbf{B}_j$ can be derived by selecting an initial set of receivers $\mathbf{W}_j$, and alternately updating $\mathbf{B}_j$ and $\mathbf{W}_j$ until convergence is reached.

In this section, we discuss two general approaches to this problem. The first is a coordinated zero-forcing approach that is a generalization of channel inversion. The second is a framework for applying other methods like regularized channel inversion or optimal beamforming in a context where users have multiple antennas.

7.3.2.1 Coordinated Zero-Forcing

As noted previously, in cases where the receivers have multiple antennas, it is possible to use channel inversion at the transmitter if $n_r \leq n_T$, but this over-constrains the problem, by forcing **HB** to be completely diagonal. In fact, all inter-user interference can be eliminated by constraining **HB** to be block-diagonal (i.e., $\mathbf{H}_i\mathbf{B}_j = 0$, for $i \neq j$). A procedure for computing the optimal **B** given this constraint has been proposed [24,30–34], but it imposes restrictions on the channel configurations that can be accommodated. In order to accommodate all possible receiver sizes, those restrictions can be eliminated using the coordinated beamforming approach: estimate the receivers $\mathbf{W}_j$ and force $\mathbf{W}_i^*\mathbf{H}_i\mathbf{B}_j$ to be zero. A method for computing this iteratively, referred to as the "coordinated zero-forcing" algorithm, is listed in Table 7.2.

TABLE 7.2

Coordinated Zero-Forcing Algorithm

---

1. For each user, initialize $\mathbf{W}_j$ as the m_j dominant left singular vectors of $\mathbf{H}_{jj}$, and define $\bar{\mathbf{H}}_j = \mathbf{W}_j^*\mathbf{H}_j$.

2. For each user, define

$$\tilde{\bar{\mathbf{H}}}_j = \left[\bar{\mathbf{H}}_1^T \ \cdots \ \bar{\mathbf{H}}_{j-1}^T \ \ \bar{\mathbf{H}}_{j+1}^T \ \cdots \ \bar{\mathbf{H}}_K^T \right]^T,$$

let $\tilde{\mathbf{V}}_j^{(0)}$ represent an orthogonal basis for the right null space of $\tilde{\bar{\mathbf{H}}}_j$, and compute the SVD

$$\mathbf{H}_j\tilde{\mathbf{V}}_j^{(0)} = \left[\mathbf{U}_j^{(1)} \ \ \mathbf{U}_j^{(0)} \right] \Sigma_j \left[\mathbf{V}_j^{(1)} \ \ \mathbf{V}_j^{(0)} \right]^H,$$

where $\mathbf{U}_j^{(1)}$ and $\mathbf{V}_j^{(1)}$ represent the first m_j left and right singular vectors. Update the transmitter and receiver beamformers: $\mathbf{W}_j = \mathbf{U}_j^{(1)}$ and $\mathbf{B}_j = \mathbf{V}_j^{(1)}$, and define

$$\mathbf{S} = \begin{bmatrix} \mathbf{W}_1^*\mathbf{H}_1 \\ \vdots \\ \mathbf{W}_K^*\mathbf{H}_K \end{bmatrix} \begin{bmatrix} \mathbf{B}_1 & \mathbf{B}_K \end{bmatrix}.$$

3. Repeat step 2 until

$$\min_{i=1,..K} \frac{[\mathbf{S}]_{i,i}}{\sum_{j \neq i}[\mathbf{S}]_{i,j}} < \varepsilon$$

for some value of ε.

4. Use water-filling to determine power allocation given the diagonal values of the Σ_j matrices as the channel gains.

The result of the coordinated zero-forcing algorithm is a set of non-interfering virtual channels. One advantage of this approach is that, since the channels do not interfere, the solution is independent of the power allocation to each channel, and therefore the power allocation can be performed independently from the computation of the beamformers.

There are a few special cases of the coordinated zero-forcing algorithm worth noting. First, if $n_{R_j} = 1$ for all users, the solution is equivalent to channel inversion with optimal power allocation. Second, if $n_T > \max\{\mathrm{rank}(\tilde{\mathbf{H}}_1, \ldots, \mathrm{rank}(\tilde{\mathbf{H}}_K)\}$, the convergence criterion is reached at the first step, and the solution is equivalent to the block-diagonalization solution of [24]. Third, if $m_j = 1$ for all users, the receiver beamformers $\mathbf{W}_j$ are equivalent to maximal ratio combiners, and the solution for $\mathbf{B}$ is equivalent to channel inversion of $\bar{\mathbf{H}}$ (this allows for some computational savings over the generalized implementation).

7.3.2.2 General Coordinated Beamforming

As noted in the discussion of channel inversion, the use of zero-forcing at the transmitter has some disadvantages, so there are good reasons to use other beamforming methods at the transmitter. This can be done in channels where the receivers have multiple antennas by applying the same general approach as in coordinated zero-forcing. A general algorithm for doing this is listed in Table 7.3.

There are two reasons that computing the zero-forcing solution makes a good initialization point for the algorithm in step 1. The first is that as SNR increases, the difference between the zero-forcing solution and other beamforming algorithms will become increasingly small, so starting with the zero-forcing solution can significantly reduce the number of iterations to convergence [29]. The second reason is that zero-forcing is the only way the beamforming weights and power allocation can be decided independently, so initializing with zero-forcing is a means of intelligently estimating how many bits should be allocated to each sub-channel before proceeding with beamformer optimization. In [29] this approach was used with MMSE receivers and optimal beamforming for minimum power at the transmitter.

TABLE 7.3

Coordinated Transmitter/Receiver Beamforming Algorithm

---

1. Assume an initial set of receiver weights $\mathbf{W}_1, \ldots, \mathbf{W}_K$. Two good candidates for this are to use the dominant left singular vectors of the respective channel matrices $\mathbf{H}_j$, or to compute the full coordinated zero-forcing solution and use the resulting values of $\mathbf{W}_j$.
2. Given $\mathbf{W}_1, \ldots, \mathbf{W}_K$, calculate $\bar{\mathbf{H}}$ and find $\mathbf{B}$ using any of the algorithms discussed earlier (regularized channel inversion, optimal beamforming).
3. Given $\mathbf{B}$, recalculate the receiver beamformers $\mathbf{W}_1, \ldots, \mathbf{W}_K$ according to some assumed receiver design (MMSE, MRC, etc).
4. If the SNR or sum rate achieved by $\mathbf{B}$ and $\mathbf{w}_j$ has changed from the last iteration, go to step 2; otherwise, stop.

7.3.3   Non-Linear Processing Methods

All of the transmission schemes discussed so far use linear processing at the transmitter and receiver. However, as noted previously, channel capacity in a multi-user environment depends on the use of DPC techniques, which are inherently non-linear in nature. DPC techniques have been demonstrated to outperform linear methods [13], but implementation is more expensive. Some efficient DPC precoders have computational complexity similar to that of linear precoders, but the computation must be performed separately for each transmitted symbol, while linear precoders can be computed once for an entire block of transmitted symbols.

Another limitation of DPC it that none yet exists that is designed for multiple-antenna receivers. However, there are straightforward ways of combining DPC with linear processing methods to make them usable for multi-antenna receivers. One simple example is using coordinated zero-forcing to compute a set of transmit and receive vectors that allow one sub-channel per user ($m_j = 1$). After this is computed, the linear beamformers could be replaced by a DPC encoder that uses the channel matrix $\bar{\mathbf{H}}$.

It is reasonable to assume that in a real multi-user environment it will be common to have a mixture of users with single and multiple antennas. In this type of environment, one proposal for obtaining the benefits of non-linear precoding is to use block-diagonalization for the users with multiple antennas and non-linear DPC methods for the users with only one antenna [35]. In this approach, the beamformers for the multiple-antenna users are chosen to lie in the null space of the channel matrices of the other users including those with single antennas. The equivalent channel for the single-antenna users looks as if there are no multiple-antenna users present, which improves diversity for those users. The data transmitted to the multiple-antenna users are also precoded using a linear precoder in order to eliminate the multi-user interference, which in this case only originates from the single-antenna users. This approach significantly improves the performance of the single-antenna users and, hence, also that of the overall system.

7.4   Channel Measurements

In the results that follow, we examine the performance of linear precoding schemes in realistic environments using channel measurements from both indoor and outdoor propagation environments. The measurements were taken from a narrowband channel sounding system designed and built at Brigham Young University (BYU). The transmitter of the system modulates the chosen carrier frequency using BPSK modulation with a unique pseudo-random binary sequence for each of the antennas in the transmit array. In the receiver, the signals from each of the elements of the receive array are down-converted to an intermediate frequency and sampled using a high frequency, multi-channel, analog-to-digital converter. The sampled signals are stored and processed off-line to extract the complex gain from each of the transmit antennas to each of the receive antennas. The frequency with which the channel can be sampled is a function of the length of binary sequence used for modulation. For a more detailed description of the channel sounder and the post-processing, see [36].

All of the results included here were collected at a carrier frequency of 2.43 GHz, with a bandwidth of 25 KHz. This frequency is used by some of the popular wireless LAN standards and is close to the 1.9 GHz frequency used in some mobile telephone networks. In the measurement results presented here, the transmitter was kept at a fixed location and the receiver moved while sampling the channel every 2.5 ms. Multi-user channels are created by selecting samples from multiple points along the measurement path. Because the average number of samples per wavelength is as high as 30 for most of the cases considered here, relatively small separations between users can be simulated.

The measurements used here come from three different sets. The first is a set of indoor measurements taken inside a typical university building.* The measurements were taken with the transmitter in a fixed location and the receiver moving in a straight path with an approximate length of about 40 meters along a long corridor at constant speed. The measurement path is illustrated in Figure 7.5. All channels were non-line-of-sight (NLOS), which would typically lead to reduced power but enhanced multipath diversity relative to line-of-sight (LOS) channels. Both the transmitter and receiver used 10 monopole antennas arranged in a circular pattern with a radius of 0.86 wavelengths, equivalent to a spacing of approximately 0.5 wavelengths between adjacent elements.

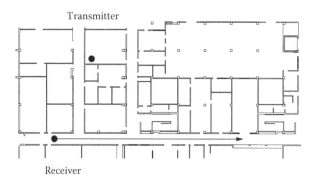

FIGURE 7.5
Illustration of the measurement path and part of the building used for the indoor channel measurements.

---

* The building was the Clyde Engineering Building on the BYU campus, which has steel-reinforced concrete structural walls and cinder-block partition walls.

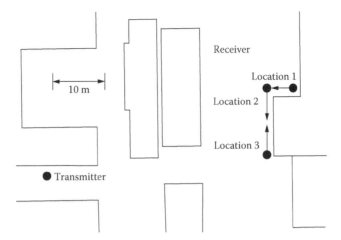

FIGURE 7.6
Illustration of the measurement paths and neighboring buildings for outdoor channel A. These channels are almost all non-line-of-sight (NLOS).

Two outdoor data sets are also considered here. The first, referred to as outdoor channel A, placed the transmitter between two buildings on the BYU campus, and the receiver behind a neighboring building, creating a NLOS channel similar to that often seen in urban environments. These measurements were collected with 8-element uniform linear arrays of monopole antennas at both transmitter and receiver, with a spacing of 0.3 wavelengths. The receiver was placed at three different locations and moved along a straight path with a length of about 10 meters. The measurement paths and neighboring buildings for outdoor channel A are illustrated in Figure 7.6. The results in the next section derived from these measurements are averaged over the three different locations.

The second outdoor environment, referred to in the next section as outdoor channel B, contained mostly LOS channels. The transmitter was placed in two locations a few meters from the wall of a building. The receiver was placed at four different locations near the same building, and moved distances of 10–12 meters. These measurements were collected using uniform linear arrays of seven antennas at both transmitter and receiver with a spacing of 0.39 wavelengths. The building and the measurement paths for this channel are illustrated in Figure 7.7. The composite results for outdoor channel B also are averaged over the four measurement locations.

Most of the test cases considered scenarios with fewer antennas than the original data set. Appropriate antenna subsets were selected as follows. On the transmit side, antennas with maximal separation were chosen to mimic a base station that uses the entire array aperture. For example, the 4-element transmitter that is used in many of the results is taken from the 7-element linear array by choosing 4 elements with uniform separation of 0.78 wavelengths. A mobile receiver, on the other hand, would be expected to have limited

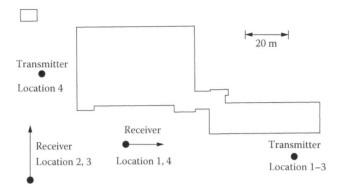

FIGURE 7.7
Illustration of the measurement paths and neighboring building for outdoor channel B. These channels are almost all line-of-sight (LOS).

size, and thus only adjacent antennas were used when simulating end users with multiple antennas.

An important issue that arises in MIMO channel data sets is how the various channels are normalized prior to processing. There are two common approaches. One approach is to scale each individual channel sample to have the same received power (measured using the Frobenius norm). This approach ignores power fluctuations due to multipath fading or shadowing and does not allow channels with "near-far" effects, but it allows for more consistent comparisons with simulated propagation environments and makes the results less dependent on the specific physical environment. This type of normalization was used in all of the results measuring capacity of various precoding algorithms. A second approach is to preserve the power relationships, normalizing so that the average Frobenius norm has a fixed value. This approach can be subject to large power fluctuations. We used this type of global channel normalization when measuring the effects of channel estimation error.

7.5 Performance Results

In this section, we compare the performance of some of the various downlink multiplexing schemes we have described. We begin with results derived from simulation of random channels, and then show how the algorithms perform with measurement data. We focus in particular on three processing schemes whose capacity can be computed easily. Note that capacity here refers to the maximum achievable rate for the given transmission scheme, which is different from the true channel capacity. The first is coordinated zero-forcing, referred to in many of the plots as zero-forcing or ZF. The second is coordinated beamforming using regularized channel inversion on the transmit side and MMSE beamformers at the receivers when multiple antennas are present. This is labeled in many of the plots as regularized channel inversion or RCI. The third scheme is coordinated beamforming using MMSE receivers and the iterative RCI algorithm for achieving maximum capacity given in Table 7.1, referred to in the plots as capacity-optimal RCI.

7.5.1   Multi-User Performance in Randomly Generated Channels

We begin by illustrating the performance of different multiplexing methods in uncorrelated Gaussian channels, which is usually the best-case scenario. Figure 7.8 shows complementary cumulative density functions (CCDF) of sum capacity for coordinated zero-forcing, coordinated beamforming using regularized channel inversion at the transmitter and MMSE receivers, and coordinated beamforming using capacity-optimal RCI at the transmitter with MMSE receivers. Note that for $m_j = 1$, and in about 40% of cases where $m_j = 2$, zero-forcing has slightly better performance than RCI. The reason for this is that with the ZF solution the power for each user was adjusted to maximize sum capacity, while this is not possible with RCI without using the iterative algorithm. The capacity is quite similar for all algorithms for $m_j = 1$, but for $m_j = 2$, it is clear that the capacity-optimal RCI method makes far better use of the second spatial sub-channel than either of the others. In some cases, as will be seen with the data-derived results, adding the second sub-channel does not provide any additional benefit because of the channel characteristics. Another reason it may be preferable to use less than all available sub-channels is that this allows additional degrees of freedom in optimizing the transmission to other users.

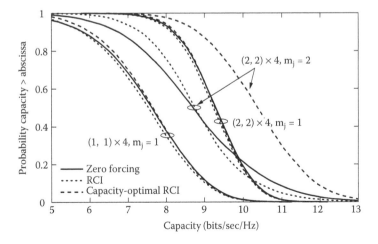

FIGURE 7.8
CCDFs of capacity for zero-forcing and regularized channel inversion in uncorrelated Gaussian channels at a SNR of 10 dB.

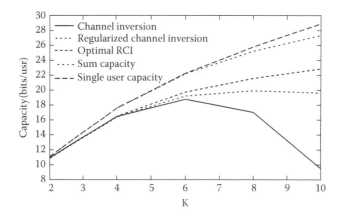

FIGURE 7.9

Comparison of capacity for channel inversion and regularized channel inversion with the sum capacity of the channel and capacity of the equivalent single-user channel for K users and $n_T =$ 10 at a SNR of 10 dB.

In Figure 7.4 (see also [17,37]) is a comparison of capacity for channel inversion and regularized channel inversion for channels where $K = n_T$ revealed that RCI and capacity-optimal RCI achieve linear growth in capacity as a function of K, while channel inversion does not. In Figure 7.9, we assume that n_T is fixed at ten antennas, and we compare the sum capacity of channel inversion and RCI with the theoretical limits as a function of K. In this case, the theoretical limits increase with the number of users, but the linear processing schemes achieve maximum capacity at six users, and channel inversion actually loses a substantial amount of capacity as $K \rightarrow n_T$. This is due to the fact that the power scaling in Equation 7.4 is limited by singular values of $(\mathbf{H}\mathbf{H}^*)^{-1}$, which, for random complex Gaussian matrices, are not well conditioned when $\mathbf{H}$ is square [17].

Figure 7.10 compares the same three linear precoders in the context of multi-antenna receivers, with the number of data streams per user m_j fixed at 1. While there is a sizable gap between zero-forcing and the RCI regularized channel inversion when the receivers have only one antenna, the difference becomes much smaller as a function of the number of receive antennas, to the point that it becomes almost negligible for $n_{R_j} = 3$.

In multi-user MIMO channels, optimal transmission schemes depend heavily on the availability of CSI at the transmitter. In practice, CSI will likely be corrupted by noise. Figure 7.11 shows the effects of channel estimation error on performance. In this case, the error is modeled as an additive error matrix $\mathbf{N}$ such that the estimated channel $\hat{\mathbf{H}} = \mathbf{H} + \mathbf{N}$. The error is quantified by the ratio of the total power in $\mathbf{H}$ to the total power in $\mathbf{N}$: $\|\mathbf{H}\|_F^2 / \|\mathbf{N}\|_F^2$. It is apparent from the curves in Figure 7.11 that receivers with additional antennas reach their maximum capacity with more error in their channel estimates than those with only one antenna. With one antenna per user, not only do

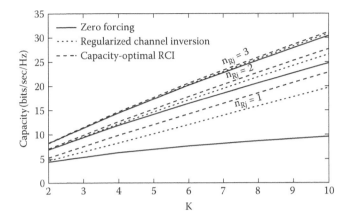

FIGURE 7.10
Capacity comparison of coordinated zero-forcing and regularized channel inversion for $m_j = 1$ and varying values of n_{R_j} for K users at a SNR of 10 dB.

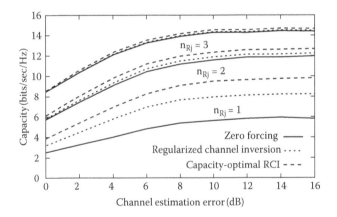

FIGURE 7.11
Capacity comparison as a function of channel estimation error.

the RCI precoders significantly outperform zero-forcing, they are much more robust in the presence of channel estimation error.

7.5.2 Multi-User Performance in Channels Derived from Measurements

While uncorrelated Gaussian channels are useful as an analysis tool, this assumption is a best-case scenario, and it is important to also consider the specific channel conditions in which these algorithms are likely to be used. For example, if two users are located close together, their channels will likely be highly correlated, which will affect the transmitter's ability to achieve signal separation with precoding. An important question then is what physical spacing is required to achieve the capacity levels of uncorrelated users.

For indoor environments, this problem was recently studied using both channel measurements and statistical models [38,39]. In this section, we compare performance derived from the indoor measurements with outdoor measurements and random Gaussian channels. The transmission scheme in all cases is coordinated beamforming with MMSE receivers and capacity-optimal RCI at the transmitter, so the capacity referred to here is the maximum achievable throughput given linear precoding and decoding. We consider three important questions. First, we test how closely two users can be located in space before a significant reduction in spatial multiplexing performance is observed. Second, we test channels with many users to see how much multipath is present in the channel and how many users can be supported, and third, we test how far a receiver terminal can move before updated CSI is required.

7.5.2.1 Effects of Inter-User Separation

We begin by examining the performance of two-user channels. Figure 7.12 shows the capacity as a function of separation distance for coordinated beam-forming using regularized channel inversion for indoor channels. The cases shown are for $n_{R_j} = 1$, and $n_{R_j} = 2$ with $m_j = 1$ and 2. As a reference, the mean capacity was computed for uncorrelated Gaussian channels and for measured channels where the users' locations were chosen randomly from anywhere in the data set, and those values are shown along the left and right sides of the plot, respectively. While random spacing allows slightly higher capacity than fixed spacing, capacity for fixed spacing in all three cases appears to reach its maximum at a separation of 5 wavelengths. For the

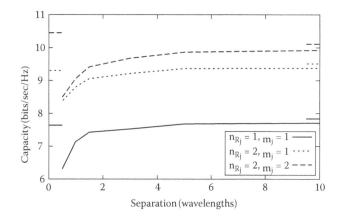

FIGURE 7.12
Mean system capacity as a function of separation distance for a two-user MIMO channel derived from indoor measurement data. Markers on the right side are capacity for random separation, and along the left side are for uncorrelated Gaussian channels.

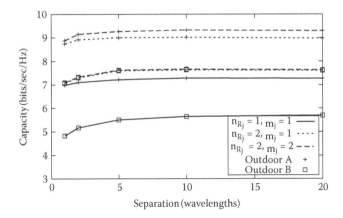

FIGURE 7.13
Mean system capacity as a function of separation distance for a two-user MIMO channel derived from outdoor measurement data.

2.43 GHz channel we are considering, this is equivalent to about 60 cm. For both cases where $m_j = 1$, the capacity from measured data is very close to that of uncorrelated Gaussian channels, but there is a gap for $m_j = 2$. This is an indication that there is likely a greater spread in the singular values of the channel matrices for the measured data than for simulated random channels. Causes of this include correlation between antenna elements and a dominant path in the multipath environment.

Figure 7.13 shows the mean capacity as a function of user separation for the two outdoor channels. Outdoor channel A, which is almost entirely NLOS with multiple buildings in the vicinity, achieves higher overall capacity and the maximum capacity appears to be reached at 5 wavelengths, as in the indoor case. Channel B, on the other hand, which consists of LOS channels, requires about 10 wavelengths to reach maximum capacity and has lower overall capacity. Even a distance of 10 wavelengths is relatively small, considering it is equivalent to about 1.2 meters at our measurement frequency. Since the channels have all been normalized, relative attenuation between the different propagation environments is not considered here. It is also interesting to note that when $n_{R_j} = 2$, outdoor channel A achieves a small increase in capacity when m_j is increased from 1 to 2 (but a smaller increase than the indoor environment), but channel B does not. This is an indicator that channel B has virtually no multipath diversity — the channel is almost always dominated by a single multipath component — and Channel A has less multipath diversity than the indoor channel.

In Figure 7.14, we consider the performance of a system with a larger number of transmit antennas as a function of the number of users sharing the channel. We compare the performance of uncorrelated Gaussian channels with locations randomly selected from the data sets as a function of the number of users for $n_{R_j} = 1$ and 3, with $m_j = 1$. The random Gaussian channel

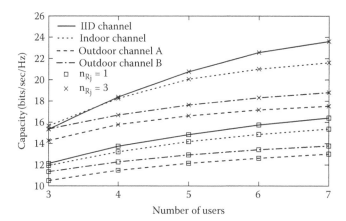

FIGURE 7.14
Sum capacity of coordinated beamforming with optimal RCI for multi-user channels with 7 transmit antennas and users placed at random locations.

outperforms the measured channels by a larger margin for multi-antenna receivers than for single-antenna receivers. For multi-antenna receivers, the indoor channel achieves performance close to that of random channels except for six and seven users. For both outdoor channels, there is significantly less capacity than random channels for as few as four users. This illustrates that regardless of the number of transmitters and users, the system capacity in real propagation environments may be limited by the multipath structure. Beyond a certain limit (defined by the multipath diversity of the channel) the addition of transmit and receive antennas results mainly in beamforming gain rather than diversity or multiplexing gain.

7.5.2.2 Effects of User Motion

In Figure 7.11, we demonstrated that noise in the CSI does not measurably degrade performance if the energy in **H** is greater than the noise in the channel estimate by about 12 dB. In this section, we assume that the noise effects are negligible and measure the effects of errors in the CSI due to user motion. We assume that CSI is made available to the transmitter via a feedback channel, but the receiver may have moved by the time the transmitter has processed the channel estimate. A similar case was considered in [40]. In [39], this scenario was studied for a $\{2,2\} \times 4$ channel with $m_j = 2$ in an indoor environment. Because we observe better overall performance with $m_j = 1$, we consider here the case of $\{1,1\} \times 4$ channels and $\{2,2\} \times 4$ channels with errors in CSI.

The sum capacity at 10 dB SNR as a function of separation distance from CSI measurement to CSI usage is shown for the indoor and both outdoor channels in Figure 7.15. In the first 0.25 wavelengths, the degradation is

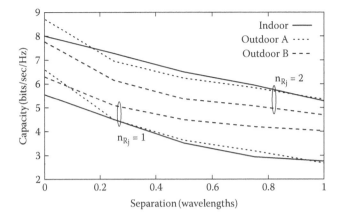

FIGURE 7.15
Mean capacity as a function of distance between a user's actual location and the location of the available channel estimate.

steeper for outdoor channels than for indoor channels, while the rate is similar for all of the channels from 0.25 to 1 wavelength. The slope of the curves for larger separations appears to be less steep for $n_{R_j} = 2$. This can be attributed to the fact that the dominant eigenvector of each user's channel is closely related to the angle of strongest propagation path, which will tend to change more slowly than the individual channel coefficients. This is another advantage of adding additional antennas at the receivers.

In general the effects of CSI error appear to be similar for both the indoor and outdoor environments studied here. It is possible to compensate for these errors up to a certain point by adding additional SINR margin to the design requirements for transmit beamformers. This type of CSI error is a limited problem in indoor environments because mobility speeds are quite low. The slightly higher sensitivity to CSI error in outdoor environments combined with higher speeds imply that CSI will not be useful for long periods of time, making the problem of obtaining relevant CSI at the transmitter a more significant challenge for outdoor channels than indoor.

7.6 Summary

Spatial multiplexing algorithms in multi-user MIMO systems can substantially increase the capacity of a wireless network, assuming that accurate CSI is available and that the channels for different users are uncorrelated. In this chapter, we have reviewed some of the available algorithms for the downlink, and demonstrated the expected performance in both ideal uncorrelated

Gaussian channels and channels derived from measurements of both indoor and outdoor propagation environments. While many of the multiplexing algorithms proposed so far are designed for the case where each user has only one antenna, they can be readily adapted to cases where the users have more than one antenna by using linear processing at the receiver to reduce the dimension of the channel. The results demonstrate a clear benefit from adding a second or third antenna to each user and using simple processing schemes at the receiver, such as MMSE beamforming. When users have more than one antenna and coordinated beamforming is used, the performance gaps between transmission approaches such as channel inversion and regularized channel inversion become much smaller.

In practice, achieving spatial multiplexing requires that users' channels be sufficiently uncorrelated, which implies a certain minimum separation between them. For indoor propagation measurements tested here, a spacing of about 5 wavelengths (approximately 60 cm at the measurement frequency) appears to be sufficient, regardless of the number of users, while in outdoor environments with significant multipath, a spacing of 5–10 wavelengths (0.6–1.2 m) is sufficient, depending on the environment.

While in uncorrelated Gaussian channels, system capacity can grow linearly with the number of transmit antennas and number of users, the multipath structure encountered in measured channels imposes limits on how many users can be multiplexed spatially. In the particular indoor environment measured here, with a base station array of 7 elements, most of the available capacity is reached at around five to six users, while the number is three to four for the outdoor environments that were studied here. This limit is a function of the number of significant multipath components for each user and the difference in multipath structures for users located near each other. All of the measurement results were taken over limited paths, so it is possible that more users can be supported if they are scattered over a wider area.

The use of CSI at the transmitter in multi-user MIMO downlinks appears to be particularly challenging in outdoor propagation environments because performance quickly becomes degraded when the channel has changed relative to the available CSI. Protocols that use spatial multiplexing in outdoor MIMO downlinks will, therefore, use CSI for very short time horizons.

7.7 Acknowledgments

This work was supported by the U.S. Army Research Office under the Multi-University Research Initiative (MURI) grant W911NF-04-1-0224, by the DARPA Advanced Technology Office, and by San Diego Research Center, Inc.

References

1. R.L.-U. Choi, M.T. Ivrlač, R.D. Murch, and J.A. Nossek. 2003. "Joint transmit and receive multi-user MIMO decomposition approach for the downlink of multi-user MIMO systems," in *Proceedings of the IEEE 58th Vehicular Technology Conference*, Orlando, FL: IEEE, Oct. 6–9, 2003.

2. D.J. Love, R.W. Heath, Jr., W. Santipach, and M.L. Honig. 2004. "What is the value of limited feedback for MIMO channels?" *IEEE Communications Magazine*, Vol. 42, No. 10, Oct. 2004, pp. 54–59.

3. T. Svantesson and A.L. Swindlehurst. 2003. "A performance bound for prediction of a multipath MIMO channel," in *Proc. 37th Asilomar Conference on Signals, Systems, and Computers, Session: Array Processing for Wireless Communications*, Pacific Grove, CA, November 2003.

4. M. Costa. 1983. "Writing on dirty paper," *IEEE Transactions on Information Theory*, Vol. 29, May 1983, pp. 439–441.

5. G. Caire and S. Shamai. 2003. "On the achievable throughput of a multi-antenna Gaussian broadcast channel," *IEEE Transactions on Information Theory*, Vol. 43, July 2003, pp. 1691–1706.

6. W. Yu and J. Cioffi. 2002. "Sum capacity of a Gaussian vector broadcast channel," in *Proceedings IEEE International Symposium on Information Theory*, July 2002, p. 498.

7. P. Viswanath and D. Tse. 2003. "Sum capacity of the vector Gaussian broadcast channel and uplink-downlink duality," *IEEE Transactions on Information Theory*, Vol. 49, No. 8, Aug. 2003, pp. 1912–1921.

8. S. Vishwanath, N. Jindal, and A. Goldsmith. 2003. "Duality, achievable rates and sum capacity of Gaussian MIMO broadcast channels," *IEEE Transactions on Information Theory*, Vol. 49, No. 10, Aug. 2003, pp. 2658–2668.

9. H. Weingarten, Y. Steinberg, and S. Shamai. 2004. "The capacity region of the Gaussian MIMO broadcast channel," in *Proceedings Conf. on Information Sciences and Systems (CISS)*, Princeton, NJ, March 2004.

10. G. Caire and S. Shamai. 2002. "LDPC coding for interference mitigation at the transmitter," in *Proc. 40th Annual Allerton Conference on Communication, Control, and Computing*, October 2002.

11. G. Caire and S. Shamai. 2002. "Writing on dirty tape with LDPC codes," in *Multiantenna Channels: Capacity, Coding and Signal Processing*, DIMACS Series in Discrete Mathematics and Theoretical Computer Science, Vol. 62, pp. 123–140.

12. T. Philosof, U. Erez, and R. Zamir. 2002. "Precoding for interference cancellation at low SNR," in *22nd Convention of IEEE Israel Section*, Tel-Aviv University, December 2002.

13. B.M. Hochwald, C.B. Peel, and A.L. Swindlehurst. 2005. "A vector-perturbation technique for near-capacity multi-antenna multi-user communication — part II: perturbation," *IEEE Transactions on Communications*, Vol. 53, No. 3, March 2005, pp. 537–544.

14. J.H. Winters, J. Salz, and R.D. Gitlin. 1994. "The impact of antenna diversity on the capacity of wireless communication systems," *IEEE Transactions on Communications*, Vol. 42, No. 2, Feb./Mar./Apr. 1994, pp. 1740–1751.

15. D. Gerlach and A. Paulraj, "Adaptive transmitting antenna arrays with feedback," *IEEE Signal Processing Letters*, Vol. 1, No. 10, October 1994, pp. 150–152.
16. T. Haustein, C. von Helmolt, E. Jorwieck, V. Jungnickel, and V. Pohl. 2002. "Performance of MIMO systems with channel inversion," in *Proceedings of the IEEE 55th Vehicular Technology Conference*, Vol. 1, Birmingham, AL, May 2002, pp. 35–39.
17. C.B. Peel, B.M. Hochwald, and A.L. Swindlehurst. 2005. "A vector-perturbation technique for near-capacity multiantenna multiuser communication-part I: channel inversion and regularization," *IEEE Transactions on Communications*, Vol. 53, No. 1, January 2005, pp. 195–202.
18. M. Stojnic, H. Vikalo, and B. Hassibi. 2004. "Rate maximization in multi-antenna broadcast channels with linear preprocessing," in *Proceedings of IEEE Globecom*, November 2004, pp. 3957–3961.
19. F. Rashid-Farrokhi, K.R. Liu, and L. Tassiulas. 1998. "Transmit beamforming and power control for cellular wireless systems," *IEEE Journal on Selected Areas in Communications*, Vol. 16, No. 8, October 1998, pp. 1437–1450.
20. E. Visotsky and U. Madhow. 1999. "Optimum beamforming using transmit antenna arrays," in *Proceedings of the IEEE Vehicular Technology Conference*, Vol. 1. Houston, TX: IEEE, May 16–20, 1999, pp. 851–856.
21. M. Bengtsson and B. Ottersten. 2001. "Optimal and suboptimal beamforming," in *Handbook of Antennas in Wireless Communications*, L. Godara, ed., Boca Raton, FL: CRC Press.
22. H. Boche and M. Schubert. 2002. "Multi-antenna downlink transmission with individual SINR receiver constraints for cellular wireless systems," in *Proceedings of the 4th International ITG Conference on Source and Channel Coding*, Informationstechnische Gesellschaft im VDE (ITG). Berlin: VDE Verlag GmbH, January 2002, pp. 159–166.
23. M. Schubert and H. Boche. 2004. "Solution of the multiuser downlink beamforming problem with individual SINR constraints," *IEEE Transactions on Vehicular Technology*, Vol. 53, No. 1, January 2004, pp. 18–28.
24. Q.H. Spencer, A.L. Swindlehurst, and M. Haardt. 2004. "Zero-forcing methods for downlink spatial multiplexing in multi-user MIMO channels," *IEEE Transactions on Signal Processing*, Vol. 52, No. 2, February 2004.
25. J.-H. Chang, L. Tassiulas, and F. Rashid-Farrokhi. 2002. "Joint transmitter receiver diversity for efficient space division multiaccess," *IEEE Transactions on Wireless Communications*, Vol. 1, No. 1, January 2002, pp. 16–27.
26. Z. Pan, K.-K. Wong, and T. Ng. 2003. "MIMO antenna system for multi-user multi-stream orthogonal space division multiplexing," in *Proceedings of the IEEE International Conference on Communications*, Vol. 5, Anchorage, AK: IEEE, May 2003, pp. 3220–3224.
27. K.-K. Wong, R. Murch, and K.B. Letaief. 2003. "A joint-channel diagonalization for multiuser MIMO antenna systems," *IEEE Transactions on Wireless Communications*, Vol. 2, No. 4, July 2003, pp. 773–786.
28. Q.H. Spencer, A.L. Swindlehurst, and M. Haardt. 2003. "Fast power minimization with QoS constraints in multi-user MIMO downlinks," in *Proceedings of the IEEE International Conference on Acoustics, Speech, and Signal Processing*, IEEE, April 2003.

29. Q.H. Spencer and A.L. Swindlehurst. 2004. "A hybrid approach to spatial multiplexing in multi-user MIMO downlinks," *EURASIP Journal on Wireless Communications and Networking*, Vol. 2004, No. 2, December 15, 2004, pp. 236–247. Available: http://wcn.hindawi.com/volume-2004/issue-2.html.

30. R.L.-U. Choi and R.D. Murch. 2002. "A downlink decomposition transmit pre-processing technique for multi-user MIMO systems," in *Proceedings of IST Mobile & Wireless Telecommunications Summit*, June 2002.

31. A. Bourdoux and N. Khaled. 2002. "Joint Tx-Rx optimization for MIMO-SDMA based on a null-space constraint," in *Proceedings of the IEEE Vehicular Technology Conference*, September 2002, pp. 171–174.

32. Q.H. Spencer and M. Haardt. 2002. "Capacity and downlink transmission algorithms for a multi-user MIMO channel," in *Conference Record of the 36th Asilomar Conference on Signals, Systems and Computers*. IEEE, November 2002.

33. M. Rim. 2002. "Multi-user downlink beamforming with multiple transmit and receive antennas," *Electronics Letters*, Vol. 38, No. 25, December 5, 2002, pp. 1725–1726.

34. R. Choi and R. Murch, "A transmit preprocessing technique for multiuser MIMO systems using a decomposition approach," *IEEE Transactions on Wireless Communications*, Vol. 3, No. 1, January 2004, pp. 20–24.

35. V. Stankovic, M. Haardt, and M. Fuchs. 2004. "Combination of block diagonal-ization and THP transmit filtering for downlink beamforming in multi-user MIMO systems," in *Proc. European Conference on Wireless Technology (ECWT 2004)*, Amsterdam, the Netherlands, Oct. 2004, pp. 145–148.

36. J.W. Wallace, M.A. Jensen, A.L. Swindlehurst, and B.D. Jeffs. 2003. "Experimental characterization of the MIMO wireless channel: data acquisition and analysis," *IEEE Transactions on Wireless Communications*, Vol. 2, No. 2, March 2003, pp. 335–343.

37. Q.H. Spencer, C.B. Peel, A.L. Swindlehurst, and M. Haardt. 2004. "An introduction to the multi-user MIMO downlink," *IEEE Communications Magazine*, Vol. 42, No. 10, October 2004, pp. 60–67.

38. Q. Spencer and T. Svantesson. 2003. "MIMO downlink spatial multiplexing algorithms applied to channel measurements," in *Proceedings of the IEEE 58th Vehicular Technology Conference*, Orlando, FL: IEEE, October 6–9 2003.

39. Q.H. Spencer, T. Svantesson, and A.L. Swindlehurst. 2004. "Performance of MIMO spatial multiplexing algorithms using indoor channel measurements and models," *Wireless Communications and Mobile Computing*, Vol. 4, No. 7, November 2004, pp. 739–754,

40. P. Zetterberg, M. Bengtsson, D. McNamara, P. Karlsson, and M.A. Beach. 2002. "Performance of multiple-receive multiple-transmit beamforming in WLAN-type systems under power or EIRP constraints with delayed channel estimates," in *Proceedings of the 55th Vehicular Technology Conference*, Vol. 4, IEEE, pp. 1906–1910.

8

Multiuser MIMO for UTRA FDD

Jyri Hämäläinen, Risto Wichman, Markku Kuusela, Esa Tiirola,
and Kari Pajukoski

CONTENTS

Abstract

Link capacity of cellular networks has become an issue with forecasted growing public demand for high data rate services. Recently, multiple-input multiple-output (MIMO) antenna techniques have been received enthusiastically within wireless communications due to their potential to increase link capacities in a spectrally efficient manner. Despite their great theoretical promise, the introduction of MIMO transceivers to commercial wireless communication systems faces several challenges.

In this chapter, we introduce recent developments of MIMO techniques within a third generation cellular system known as universal terrestrial radio access (UTRA), which is standardized by the third generation partnership project (3GPP). Rather than discussing general MIMO algorithms, we will concentrate on MIMO techniques that can be implemented without major revisions to the current UTRA air interface, wideband code division multiple access (WCDMA). The first full specification of WCDMA was already completed at the end of 1999, and backward compatibility to the legacy system is one of the most important design issues when considering future enhancements to WCDMA. Thus, small modifications to air interface can be accepted more easily in standardization than schemes that would require large redesign efforts. We will concentrate on UTRA frequency division duplex (FDD) mode, because it is a more widely deployed system than UTRA time division duplex mode.

8.1   Introduction

Multiple-input multiple-output (MIMO) transceivers have great potential to improve the performance of wireless systems. Compared to wireline systems, the inferior capacity of wireless cellular systems is caused by several different physical constraints like co-channel and adjacent channel interference, channel propagation loss, and flat or multipath fading channels. Multiantenna transmission and reception techniques are currently regarded as one of the most promising approaches for significantly increasing the coverage, capacity, and spectral efficiency of wireless systems.

The multiple degrees of freedom offered by multiple transmit and receive antennas can be used for diversity or for spatial multiplexing. In the former case, a single data stream is transmitted and multiple antennas are used to decrease the variance of the received signal and thereby improve the quality of the radio link. In the latter case, multiple transceiver antennas are used for parallel multiplexing, i.e., to transmit several data streams simultaneously to a user to increase peak data rates. In this study, these two approaches will be referred to as diversity MIMO and information MIMO, respectively.

In terms of channel capacity, information MIMO is a much more impressive concept than diversity MIMO. With parallel multiplexing, the capacity increases linearly with the number of transmit and receive antennas [1,2] when the number of transmit antennas equals the number of receive antennas and when the channels between the antennas are independent and identically distributed. This observation has stimulated great interest in MIMO transceivers, and in addition to academic research, MIMO is actively being studied in standardization for different wireless systems, such as UTRA, IEEE 802.11n and IEEE 802.16e.

Multiple antenna transceivers promise high spectral efficiencies to wireless systems, and spectral efficiencies as large as 20 to 40 bps/Hz have been reported in laboratory environments [3]. This would present a significant improvement to current cellular systems that typically operate with 0.05 to 2 bps/Hz range. However, the introduction of MIMO to commercial systems still faces several challenges, especially when a system has not been designed for MIMO from the beginning. For example, channel estimation in GSM relies on seven different midamble sequences that have been carefully selected to possess optimal autocorrelation properties. If one base station (BS) uses multiple midamble sequences, the seven sequences quickly run out, because frequency reuse limits the usage of the same sequences in nearby base stations. Furthermore, cross-correlation properties of the GSM training sequences are poor, making their use in the same base station suboptimal. Thus, the introduction of coherent information MIMO techniques would require the redesign of the entire GSM channel estimation scheme. At the same time hundreds of millions of legacy GSM handsets should be compatible with the redesigned system. In principle, differential modulation or blind detection and estimation could be applied to relax the requirement of multiple training sequences in the case of MIMO. However, blind algorithms are not considered viable in commercial cellular systems due to their complexity of implementation. Differential modulation suffers from 3 dB SNR penalty when compared to coherent signaling, and the use of differential modulation and increased power levels would increase the interference in wireless networks.

The first release of universal terrestrial radio access (UTRA) frequency division duplex (FDD) air interface, referred to as wideband code division multiple access (WCDMA), was completed in 1999 by the third generation partnership project (3GPP). At that time, research on information MIMO techniques was in the initial stage, but the release already specified two-antenna open-loop and closed-loop transmit diversity modes that are part of the most recent Release 6 [4] as well. Still, the introduction of more advanced MIMO techniques to WCDMA has not been straightforward. A great deal of standardization effort has been dedicated to extend the specification of diversity MIMO to more than two transmit antennas [5], and information MIMO techniques to WCDMA high-speed downlink packet access (HSDPA) [6] have been extensively studied as well [7]. However, these techniques were not accepted before Release 6 and the standardization for Release 7 is still ongoing.

In this chapter, we concentrate on MIMO transceiver techniques that do not require major modifications to the current UTRA FDD specifications. Instead of assessing the performance of single user MIMO systems we discuss MIMO transceivers in system level within UTRA FDD. This is because the gains of a single link do not necessarily translate into gains in a cellular system where users are subject to intracell and intercell interference. To this end, it is first necessary to understand the limitations and characteristics of UTRA FDD air interface. Section 8.2 highlights the important features in WCDMA uplink and downlink, and Section 8.3 presents multiantenna transceiver techniques that can be applied within the current WCDMA framework.

The salient difference between the uplink and downlink in UTRA FDD is that the uplink is interference limited while the downlink is code limited. Thus, code reuse and spectrally efficient MIMO techniques are in great demand in downlink, while in uplink each user can employ the entire channelization code set and MIMO schemes relying on well-known multiuser detection algorithms [8,9]. MIMO in UTRA FDD uplink is discussed in Section 8.4.

Section 8.5 presents various information MIMO algorithms [10] that have been proposed for WCDMA HSDPA mode. Requirements for MIMO performance according to [10] are discussed and their meaning to the development of MIMO algorithms is outlined. It is interesting to note that while the discussion on MIMO transceivers usually focuses on high data rates, increasing peak data rates are only considered of secondary importance in 3GPP, the most important factors being coverage and cell throughput. At the end of this section we present a multiuser MIMO scheme for WCDMA downlink that increases cell throughput without requiring any modifications to the present UTRA FDD specification. Finally, discussion and visions on future developments of MIMO transceiver techniques within UTRA FDD framework are presented in Section 8.6.

8.2   UTRA Framework

In WCDMA, narrowband user information is spread over a wide bandwidth by modulating a low-rate user data sequence with a high-rate spreading code (channelization code). The length of the spreading codes varies from 4 to 512 chips (in the uplink the maximum code length is 256 chips) and the chip rate is 3.84 Mcps (Mega chips per second), which together with transmit pulse shape filter leads to the carrier bandwidth of approximately 5 MHz.

Variable spreading factors and multicode transmission are used in order to support a wide scale of different data rates.

Scrambling codes of length 38400 chips are employed on top of channelization codes. While channelization codes are used in the downlink to separate different intracell users and separation of intercell users is based on different scrambling codes, in the uplink the scrambling codes separate different users and channelization codes separate different physical data and control channels of a user. This difference between the code usage in the downlink and uplink is essential: The limited number of orthogonal channelization codes imposes a strict upper bound to the achievable downlink capacity unless information MIMO techniques reusing the channelization codes are used. The limitation is not strict in the uplink, because one user may use all channelization codes and the set of long scrambling codes contains several million codes.

The basic UTRA FDD mode supports user data rates up to 2.3 Mbps both in uplink and downlink. In downlink, transmission on the peak data rate requires the use of three parallel channelization codes with spreading factor 4, which allocates 75% of code resources to a single user. In uplink, terminal transmitting with the peak data rate is seen as a large interference source in the network but available code resources of other intracell users are not affected. High peak-to-average ratio of multicode signals sets stringent requirements to transmitters, which are particularly critical to mobile stations.

8.2.1 UTRA FDD Downlink

Figure 8.1 presents a simplified downlink channel and frame structure. Each frame is divided into 15 time slots, with the frame length being 10 ms. The radio frame carries a dedicated physical channel (DPCH) that further divides into dedicated physical data channel (DPDCH) and dedicated physical control channel (DPCCH). The control channel contains transmit power control (TPC) bits for fast power control, transport format combination indicator

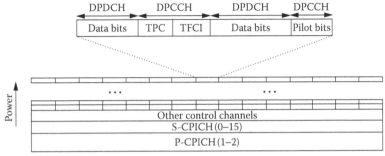

FIGURE 8.1
Downlink frame structure for dedicated physical channel.

(TFCI) bits that inform the receiver which transport channels are active and the pilot bits enabling the channel and signal to interference and noise ratio (SINR) estimation from DPCH.

Figure 8.1 further shows the primary common pilot channel (P-CPICH) and the secondary common pilot channel (S-CPICH). The numbers in parentheses indicate that one sector in downlink may contain 1–2 parallel P-CPICHs and 0–15 parallel S-CPICHs. The downlink pilot structure is introduced here, because it is needed to understand system specific problems of different multiantenna processing methods. While the receiver may estimate the channel from the dedicated channel, a better result is usually obtained when employing P-CPICH or S-CPICH, because these common pilot channels, especially P-CPICH, have higher transmit power. In practice, P-CPICH usually takes 10% of transmit power resources of the base station, while power on S-CPICHs is lower. This high transmit power on P-CPICH is due to the fact that P-CPICH is always transmitted to the whole cell enabling handover and cell selection/reselection measurements. By adjusting power differences between P-CPICHs of adjacent cells, the network load can be balanced. The power of S-CPICH can be lower because it has to cover only part of the cell. Usually S-CPICH is applied in connection with beamforming.

According to present UTRA FDD specification only two P-CPICHs are available at maximum while multiple S-CPICHs can be introduced, for example, for beamforming purposes. This introduces a serious backward compatibility problem for the design of MIMO algorithms. If the power in P-CPICH is kept fixed but the power is divided to more than two transmit antennas, the channel estimation in legacy mobile terminals is deteriorated. If the power of P-CPICH is increased, interference to network increases as well.

8.2.1.1 High-Speed Downlink Packet Access

Starting from Release 5, WCDMA contains a highly optimized downlink data transmission concept referred to as high speed downlink packet access (HSDPA). The HSDPA concept introduces a new type of transport channel, referred to as the high speed downlink shared channel (HS-DSCH), with a fixed spreading factor of length 16. HSDPA applies several advanced physical layer solutions to enable high throughput and reduced service delays. The new physical layer techniques in HSDPA are outlined below.

- Adaptive modulation and coding (link adaptation)
- High order modulation
- Hybrid automatic repeat request (ARQ) solution
- Reduced length (2 ms) transport time interval (TTI)
- Fast cell selection (FCS)
- Physical layer scheduler located in base station

The control of HS-DSCH is terminated in the base station to adapt rapidly to changing channel conditions. Adaptation rate is increased by shortening transport time interval TTI from 10 ms in Release '99 and Release 4 to 2 ms. The same TTI can be used to transmit with multiple channelization codes to the same user equipment (UE), depending on terminal capability. In addition, it is possible to multiplex multiple UEs in the code domain within one HS-DSCH TTI.

Link adaptation is implemented with a large set of possible transport format combinations, each associated with a unique combination of modulation, coding, and block size parameters. Release '99 and Release 4 support only QPSK modulation, but an HSDPA terminal also supports 16-QAM. This increases the peak date rates in good channel conditions. Fast power control, which aims to keep the received SINR constant, is not applied. This is because with non-real time data transmission it is more efficient to allow the received SINR to vary and change modulation and coding according to the channel state. Furthermore, fast power control used together with a high date rate and high power transport channel would generate large interference peaks to neighboring cells.

Data rates are assigned to a UE based on channel state information. Channel state information is measured at the UE and signaled to the base station in the form of a channel quality indicator (CQI). Thus, CQI feedback provides the necessary information for efficient real-time link adaptation, and this enables a form of multiuser diversity when applied together with physical layer scheduling. In a single-input single-output (SISO) scenario, throughput is maximized by maximum SNR scheduler transmitting to the user that reports the best SNR to the base station [11]. When the distances of the users from the serving base station are different, the scheduler is not fair, because the users on cell edge rarely get any transmission. Therefore, different fair schedulers have been developed, which aim to equalize transmission periods among the users. This kind of physical layer scheduling mechanism is part of the 3GPP WCDMA HSDPA specification, but its implementation is a vendor-specific option. Instead of SNR, it is also possible to signal data rates or capacities by mapping the SNR values appropriately.

Fast cell selection is used to select the transmitting base station from the set of active base stations. Hybrid ARQ (HARQ) combines retransmission with the previous transmissions to improve reliability, while ARQ ignores erroneously received packets. HARQ provides implicit rate matching, while adaptive modulation and coding (AMC) is used to maximize instantaneous throughput given the instantaneous CQI.

A number of contributions have emerged within 3GPP that propose different MIMO and MISO transceivers to further increase instantaneous data rates or spectral efficiency of HSDPA. These proposals are presented in more detail in Section 8.2. In addition, the enhancements to the physical layer techniques already presented are considered as well [13].

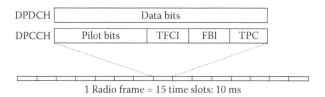

FIGURE 8.2
Uplink frame structure for DPDCH/DPCCH.

8.2.2 UTRA FDD Uplink

Figure 8.2 presents uplink frame and time slot structures for DPDCH and DPCCH. In uplink, user and control data are I/Q code multiplexed and several DPDCHs can be associated to a single DPCCH. In addition to pilot bits, the TFCI and TPC uplink DPCCH slot contains feedback information bits (FBI) used to convey partial channel state information when UTRA two-antenna closed-loop transmit diversity modes are applied. Only user-specific dedicated pilot channels are available in uplink.

Uplink users are not synchronized, and due to the non-orthogonality of users' channelization codes, multiuser interference cannot be avoided. Accurate and fast TPC is indispensable to uplink performance, because otherwise users in the vicinity of the base station would completely mask intracell users on cell edges. Fast power control is an inherent characteristic of CDMA systems, and it is applied in most of the downlink and uplink data channels in UTRA FDD.

8.3 Current Multiantenna Methods in UTRA FDD

Conventionally, multiantenna techniques have mainly been considered within base stations, because deploying multiple antennas in the user equipment is not straightforward due to cost, complexity of signal processing, and power consumption. This is still true in the present terminals that may support several radios like WCDMA, GSM, GPS, Bluetooth, WLAN. Thus, handsets that support only SISO processing may still require several antennas for different frequency bands, although the current trend is to handle the different frequencies with a single multifrequency antenna.

Multiantenna signal processing is useful in base station receivers, but the use of multiple antennas for transmission requires careful study. All precoding methods that modify the transmitted signal constellation must be standardized to ensure that all mobile receivers are able to estimate and detect the signals from different antennas. An extensive standardization process has been required even in the case of transmit beamforming.

Present UTRA FDD specification supports two-antenna transmit diversity and transmit beamforming. Such MISO systems for downlink have been intensively studied, because at the time when the 3GPP standardization was initiated, most scenarios predicted that the capacity of wireless networks would be limited by the downlink connection. Although diversity MISO does not promise such high peak data rates as information MIMO, transmit diversity improves fading resistance and beamforming increases the system capacity while its impact to individual user data rates is small. Information MIMO techniques are not included in the UTRA FDD specification, mostly because the initial standardization work was carried out during the 1990s when the development of practical information MIMO algorithms was still only beginning.

8.3.1 Beamforming

Conventional transmit beamforming provides the simplest approach to increase system capacity and coverage in downlink. The current 3GPP specification supports both fixed beamforming (FBF) and user specific beamforming (USBF) modes in downlink. Fixed beamforming introduces additional sectors to the cell, because the short term variation of the beams is small and a handover between the beams is required for mobile users. Each beam is associated with a unique S-CPICH while P-CPICH is transmitted to the whole cell.

Fixed beamforming is able to relax the code limitation in downlink, because orthogonal spreading codes can be reused within the same cell due to spatial filtering. It is also possible to use different scrambling codes in different beams. In the presence of flat fading, the channelization codes of different users within the same beam remain orthogonal because they are scrambled with the same code. Different beams interfere with one another and increase intracell interference, but when angular spread (AS) is reasonably small as in macrocells, this interference remains acceptable. In addition to larger interference levels, control signaling overhead increases due to inter-beam handovers, because logically the fixed beams behave in the same way as different sectors in base stations. The overall effect on system capacity is positive, though, and according to [14], a 2.4-fold system capacity gain can be achieved with a four-element antenna array when compared to a single antenna BS.

Instead of fixed transmit beams, USBF generates individual beams to each user. More sophisticated signal processing algorithms can be used, and consequently, USBF is able to provide higher individual data rates than FBF. In practice, USBF cannot be utilized in most cases within UTRA FDD downlink, because it prevents the user from employing the P-CPICH as a phase reference, and according to the current specification, S-CPICH cannot be employed in individual beams. Thus, channel estimation must be based on dedicated pilot channels whose transmit power is low, which may seriously corrupt channel estimation in UE. Then only high data rate users near the base station may take full advantage from USBF. User-specific beamforming is optional for UE in HS-DSCH and, therefore, the network cannot assume that all UE users are able to support it.

Downlink beamforming changes the statistics of the fading signal, and in environments with small angular spread, array gain dominates diversity gain. Therefore, the estimation of downlink beamforming gain over single antenna transmission is rather straightforward: The gain is approximately the same as the array gain. This is not as simple, though, in uplink.

In uplink, both UTRA FDD beamforming techniques can be implemented in a straightforward manner in the receiver. The simplest approach is to combine the selected signal paths using maximal ratio combining (MRC) in the beam space. This leads to a standard Rake-receiver concept. The most challenging practical problems are related to beam selection (direction of arrival (DoA) estimation in USBF) and cost-efficient receiver structures. The former problem is related to the large variety of physical environments with different channel profiles. Furthermore, mobile's transmit power is typically low for low data rate connections, making the channel estimation difficult. The latter problem arises from the fact that baseband complexity increases rapidly when spatio-temporal estimation processes are introduced. For example, preambles of random access channel (RACH) need to be monitored simultaneously for each beam in order to avoid additional delays in connection setup time.

Transmit beamforming improves the downlink capacity in UTRA FDD, but it is important to note that receive beamforming is not necessarily the best solution in uplink due to the lack of diversity gain [15]. This gives rise to a tradeoff between uplink and downlink design targets. If downlink capacity is the bottleneck, FBF provides a good solution in macrocells. On the other hand, if good uplink coverage is the primary target, uncorrelated antennas with suitable receiver algorithms provide a better solution.

8.3.2 Transmit Diversity

According to [15], basic receive diversity solution outperforms receive beamforming performance even when angular spread is small. Moreover, transmit beamforming in downlink loses its good performance in terms of capacity and coverage when AS becomes large, so that beams cannot be accurately pointed to users anymore. Hence, there is a need for a multiantenna transmission method in downlink that performs well when correlation between the transmit antennas is low. In general, low correlation can be achieved when the distance between antenna elements in the base station is several wavelengths. Conventional beamforming algorithms do not apply anymore, and to this end, several open-loop transmit diversity techniques have been developed in recent years. The simplest space–time block code [16] has been adopted into 3GPP specification as a two-antenna open-loop transmit diversity method. In UTRA FDD parlance, the scheme is referred to as space–time transmit diversity (STTD).

Space–time codes provide diversity reducing the variance of the received signal, which further translates into reduced transmit power in downlink. However, space–time codes do not increase the received SNR when compared to single antenna transmission. With uncorrelated transmit antennas, received SNR can be improved when short-term channel state information (CSI) is made available in the transmitter. Such transmit diversity algorithms are referred to as closed-loop transmit-diversity methods. The current UTRA FDD specification contains two closed-loop modes [4], where the relative phase and power between the two transmit antennas is adjusted according to feedback information bits signaled from mobile to a base station through a dedicated control channel. (See Figure 8.2.)

UTRA FDD supports two P-CPICHs in order to enable efficient channel estimation of the two channels when transmit diversity is in use. In case of closed-loop modes, feedback commands are determined from the P-CPICHs and transmit weights are applied on dedicated traffic channels. Common pilots are transmitted with higher power than dedicated pilots, and therefore, common pilots should be used for channel estimation whenever possible. With closed-loop modes, dedicated pilots are used for the verification of the transmit weights (feedback commands may be erroneously decoded in the base station, and the mobile cannot automatically assume that the transmitter uses the weights that the mobile has instructed), and channel estimates can still be calculated from P-CPICHs.

Closed-loop techniques outperform open-loop ones, particularly within low-mobility environments, when the delay of the feedback signaling does not exceed channel coherence time. Moreover, contrary to conventional beamforming techniques, closed-loop algorithms do not require accurate calibration of the transmit antennas due to feedback information. Only coarse calibration of antenna elements is necessary in order to avoid spurious antenna gain patterns and unexpected interference to the network.

The discussion on transmit diversity extensions to more than two transmit antenna systems is ongoing within 3GPP [5,7]. The main concern is that the benefits from additional transmit antennas may not justify the additional complexity of transceiver and system design. Firstly, transmit diversity increases the robustness of the link, but it does not increase data rate. Secondly, a difficult backward compatibility with former releases is encountered if transmit power of common pilot channels is divided between, say, four transmit antennas to support channel estimation in UEs. Then legacy UEs that are able to receive only two pilot channels lose performance, because they lose half of the available power of the common pilot signals.

8.4 MIMO in UTRA FDD Uplink

MIMO discussion in 3GPP has focused downlink, but the introduction of new services such as videophones will make it extremely important to reach high spectral efficiency also in the uplink direction. Besides this service-based demand, there are also two apparent reasons why MIMO in UTRA FDD uplink is attractive and, thus, worthy of a closer study. First, base stations will have two or more antennas in the near future, and at the same time, two-antenna mobiles, which are at least capable of interference cancellation, will become more common. Second, in contrast to UTRA FDD downlink, only minor changes to the UTRA FDD specifications are needed to enable simple uplink MIMO approaches.

8.4.1 MIMO Algorithms

We divide MIMO techniques into two main classes — diversity and information MIMO. UTRA FDD uplink provides a good means to implement diversity and information MIMO techniques, because uplink users are separated by different scrambling codes and, therefore, the whole channelization code space is available for any one user.

8.4.1.1 Diversity MIMO

In diversity MIMO, M_t replicas of the same data stream are transmitted by using different orthogonal channelization codes but the same scrambling code. This is depicted in Figure 8.3, which shows the system model of diversity MIMO for two transmit antennas. The received signal in kth base station antenna becomes

$$y_k(t) = \sum_{i=-\infty}^{\infty} \sum_{m=1}^{M_t} h_{m,k}(t) * \frac{\sqrt{P_{tx}}}{M_t} \left\{ S_{d,m}(t) + j \cdot S_{c,m}(t) \right\},$$

(8.1)

$$S_{x,m}(t) = b_x[i] \cdot \beta_x \cdot c_{x,m}(t - iT_x - \delta) \otimes s_{dpch}(t - iT_s - \delta),$$

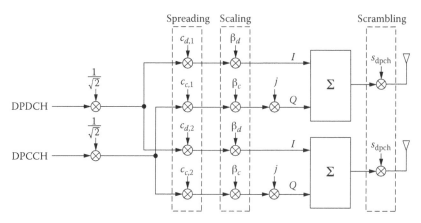

FIGURE 8.3
System model for diversity MIMO.

where the subscript $x \in \{d, c\}$, * refers to convolution with multipath channel $h_{m,k}$, and $\otimes$ refers to chip-by-chip multiplication of user's scrambling code s_{dpch} and channelization codes $c_{d,m}$ and $c_{c,m}$ of dedicated physical data channel (DPDCH) and dedicated physical control channel (DPCCH), respectively. Total transmit power is given by P_{tx}, and power ratio between DPDCH and DPCCH is adjusted by scaling factors β_d and β_c, and user's transmitted bits in DPDCH and DPCCH are denoted by b_d and b_c. Users are not synchronized in uplink, and the delay is denoted by δ. Finally T_d, T_c, and T_s refer to the lengths of the two channelization codes and the scrambling code.

Each transmit antenna uses different channelization codes $c_{d,m}$ and $c_{c,m}$ so that the base stations can separate the signals from different antennas and later combine them using maximal ratio combining. Thus, the scheme doubles the usage of uplink channelization codes when compared to single antenna transmission. The achievable bit rate with diversity MIMO is at maximum 2 Mbps/M_t. This reduction of maximum bit rate could be partly avoided by puncturing the applied error correcting codes, but information MIMO provides a more attractive choice since the reduced code rate quite rapidly destroys system performance. However, for low and medium bit rate services, this does not present a problem.

We note that in the case of flat fading and perfect channel state information in the receiver, the link performance of the proposed two-antenna diversity MIMO algorithm is the same as with STTD, because the diversity order of both schemes is two. However, the system utilizing orthogonal channelization codes is more robust, because orthogonality of the received signals does not depend on the channel estimation, as in the case of space–time coding.

8.4.1.2 Information MIMO

In the case of information MIMO of Figure 8.4, a composite data stream is multiplexed into two or more independent substreams that are transmitted from separate antennas by employing different scrambling codes. All streams contain DPDCH and DPCCH so that in the base station they can be interpreted as signals from different independent users. The received wideband signal in kth antenna is given by

$$y_k(t) = \sum_{i=-\infty}^{\infty} \sum_{m=1}^{M_t} h_{m,k}(t) * \frac{\sqrt{P_{tx}}}{M_t} \{S_{d,m}(t) + j \cdot S_{c,m}(t)\},$$

$$S_{x,m}(t) = b_{x,m}[i] \cdot \beta_x \cdot c_x(t - iT_x - \delta) \otimes s_{dpch,m}(t - iT_s - \delta),$$

(8.2)

where again $x \in \{d,c\}$. We note that now information bits as well as scrambling code chips vary between antennas. In contrast, the channelization code can be the same in all antenna branches.

The present UTRA FDD specification allows the mobile to use only a single scrambling code as well as a single DPCCH and DPDCH. For the proposed

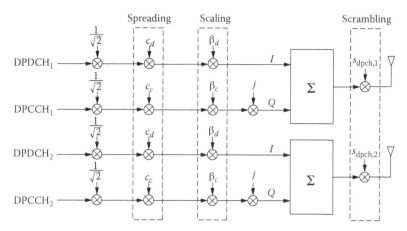

FIGURE 8.4
System model for information MIMO.

information MIMO, the specification should be changed such that the use of multiple scrambling codes is allowed in the mobile end. Then, it would be possible to independently apply DPCCH and DPDCH to each scrambling code. This does not represent a big change to radio interface, because it only requires that a single user can set up M_t different links. For the effective use of several simultaneous links, some new code puncturing sets would also be needed.

8.4.1.3   Advanced Receiver Techniques

As mentioned before, UTRA FDD uplink is interference limited. The interference-limited nature of CDMA systems results from the receiver design: the reception is typically based on a matched filter that does not take into account multiple access interference (MAI). However, MAI arises in multipath channels or with asynchronous communications when spreading codes do not usually stay completely orthogonal anymore. This limitation can be relaxed by employing advanced multiuser receiver algorithms, which aim to reduce MAI. Information MIMO increases MAI in the base station, and therefore advanced receivers become important to the system performance when implementing information MIMO techniques.

In order to avoid the overwhelming complexity of multiuser detection, various suboptimal multiuser receivers have been developed to cope with MAI. The multiuser receivers can be categorized in several ways. One possibility is to classify the receivers to linear (decorrelating detector, minimum mean square error (MMSE) detector) and non-linear receivers. Subtractive interference cancellation (IC) is one example of the latter category, which first estimates the MAI component and then subtracts it from the received signal to improve the reliability of the symbol decisions.

Cancellation of MAI can be carried out in parallel to all users, i.e., with parallel interference cancellation (PIC) receivers, or in a serial fashion, i.e., using serial interference cancellation (SIC) receivers. In un-decoded (conventional) PIC receivers, IC is based on tentative symbol decisions. The quality of the tentative decisions is essential, because incorrect decisions increase interference instead of decreasing it. In practice, the error rate of un-decoded bits varies between 5 and 20% in UTRA FDD uplink. The error rate depends mainly on the effective coding rate (i.e., rate of the channel encoder taking into account possible repetition and puncturing), the radio environment, and the quality of service requirements.

Unreliable tentative bit decisions compromise the performance of un-decoded PIC. One possibility to improve the reliability of the decisions is to first decode the received symbols, re-encode the decoded symbols again, and finally subtract the contribution of the re-encoded symbols from the received signal. For more details we refer to [17], where this kind of decoded PIC is applied to the convolutionally coded CDMA system. In [18], this idea is extended to the Turbo-coded CDMA system.

8.4.2 Uplink Load Equations

Load equations provide a useful tool in both initial planning and performance analysis of the network. Here we briefly recall the load equations [19,20] that will serve as a basis for further studies.

8.4.2.1 Basic Load Equations

In the following N_{own} and N_{other} refer to the number of own cell and other cell users, respectively. The received wideband signal in baseband at time instant t can be expressed in the form

$$r(t) = \sum_{n=1}^{N_{\text{own}}} y_n(t) + \sum_{n=1}^{N_{\text{other}}} z_n(t) + n(t),$$

(8.3)

where $n(t)$ is complex zero-mean Gaussian, $y_n(t)$ is the interference from nth own cell user, and $z_n(t)$ is the interference from nth other cell user. The system load can be estimated from the sum of received wideband powers of different users. Signals from different users are mutually independent and from Equation 8.3 the received wideband power is given by

$$E\left\{ |r(t)|^2 \right\} = I_{\text{total}} = I_{\text{own}} + I_{\text{other}} + P_N,$$

where E refers to expectation, P_N is the noise power, and

$$I_{\text{own}} = \sum_{n=1}^{N_{\text{own}}} E\left\{|y_n(t)|^2\right\}, \quad I_{\text{other}} = \sum_{n=1}^{N_{\text{other}}} E\left\{|z_n(t)|^2\right\}.$$

Received energy per data bit divided by the noise spectral density is an important variable in the analysis of UTRA FDD uplink. For user n it is given by

$$E_n = \frac{1}{\kappa_n} \cdot \frac{P_{\text{rx},n}}{I_{\text{total}} - P_{\text{rx},n}}, \tag{8.4}$$

where $P_{\text{rx},n}$ refers to the received power and κ_n is the ratio between bit rate and chip rate of user n. In UTRA FDD uplink, fast power control tries to keep E_n at a target level. Therefore, changes in the interference do not reflect to E_n, as long as power control is able to compensate the increased interference through $P_{\text{rx},n}$. From Equation 8.4 we easily find that

$$P_{\text{rx},n} = \frac{\kappa_n E_n}{1 + \kappa_n E_n} \cdot I_{\text{total}} =: \eta_n \cdot I_{\text{total}},$$

where η_n is the load from the nth user. For the received wideband power in the own cell there holds

$$I_{\text{own}} = \sum_{n=1}^{N_{\text{own}}} P_{\text{rx},n} = \sum_{n=1}^{N_{\text{own}}} \eta_n \cdot I_{\text{total}}. \tag{8.5}$$

Furthermore, the total wideband power can be written in the form

$$I_{\text{total}} = I_{\text{own}} + I_{\text{other}} + P_N = (1 + v)I_{\text{own}} + P_N, \tag{8.6}$$

where $v = I_{\text{other}}/I_{\text{own}}$ is the intercell to intracell interference ratio seen by the base station receiver. After combining Equation 8.5 and Equation 8.6 and solving I_{total} from the resulting formula, we obtain the following expression for the noise rise

$$\mu = \frac{I_{\text{total}}}{P_N} = \left(1 - (1 + v)\sum_{n=1}^{N_{\text{own}}} \eta_n\right)^{-1}, \tag{8.7}$$

where the sum term defines the own cell load η. In UTRA FDD uplink, load control drives noise rise to a target level. The more loading is allowed in the system, the larger is the required interference margin max{μ} and the smaller is the coverage area. Interference margin defines the maximum allowed noise rise, and typically values 1.0 to 3.0 dB are used for coverage-limited cases with 20 to 50% load, and in capacity-limited cases, higher interference margins up to 6 dB can be used [19].

8.4.2.2 Load Equations for MIMO

In diversity MIMO with Rake receiver, the load from the nth user is of the form

$$\eta_n^D = \frac{\kappa_n E_n^D}{1 + \kappa_n E_n^D},$$

where the superscript D refers to the diversity MIMO. Diversity MIMO reduces the required energy per data bit and there holds

$$E_n^D = \delta_{M_t,M_r} E_n / M_r, \tag{8.8}$$

where the received energy per data bit in single antenna system E_n is divided by the number of receiver antennas, because Rake collects energy from M_r independent antenna branches. Factor δ_{M_t,M_r} depends on the channel diversity and system-related issues such as the accuracy of channel estimation. Only in simple channels, such as single path and ITU Pedestrian A channels, may the effect of δ_{M_t,M_r} be noticeable, while in more realistic channels, such as ITU Vehicular A and ITU Pedestrian B, δ_{M_t,M_r} is round 0 dB. The ITU channel profiles [21,22] are listed in Table 8.1.

In information MIMO with Rake, the uplink load of a single user is of the form

$$\eta_n^I = \sum_{m=1}^{M_t} \frac{\kappa_{n,m} E_{n,m}^I}{1 + \kappa_{n,m} E_{n,m}^I}, \tag{8.9}$$

where $\kappa_{n,m}$ is the ratio between bit rate and chip rate of the mth data stream and $E_{n,m}^I$ is the received energy per data bit divided by the noise spectral

TABLE 8.1

ITU Multipath Channel Profiles

Model	Delay Profile [ns]	Power Profile [dB]
Pedestrian A	0 110 190 410	0,–9.7, –19.2, –22.8
Pedestrian B	0 200 800 1200 2300 3700	0, –0.9, –4.9, –8.0, –7.8, –23.9
Vehicular A	0 310 710 1090 1730 2510	0, –1, –9, –10, –15, –20

density for the mth data stream. We note that with M_r receive antennas and the Rake receiver there it holds that

$$E_n^l = \delta_{M_r} E_n / M_r, \tag{8.10}$$

where δ_{M_r} depends on the diversity gain and is around 0 dB in case of multipath diversity. In contrast to diversity MIMO, δ_{M_r} depends only on the number of receive antennas because each transmit antenna applies a different scrambling code. Without CSI in the transmitter data, rates in different transmit antenna branches are equal, and Equation 8.9 becomes

$$\eta_n^l = M_t \cdot \frac{\kappa_n E_n^l}{1 + \kappa_n E_n^l}. \tag{8.11}$$

Equation 8.11 indicates that the base station receiver sees the usage of information MIMO as an increased number of SISO users. This additional load can be effectively reduced by applying PIC, which removes part of the own cell interference in the detection process. Let us denote the PIC efficiency by β, $0 \le \beta \le 1$. Then the load of a single user can be written as

$$\eta_n = (1 - \beta) \frac{\kappa_n E_n}{1 + \kappa_n E_n}.$$

Since PIC is not able to remove the intercell interference, the total wideband power seen by the base station receiver is of the form

$$I_{\text{total}} = (1 - \beta) I_{\text{own}} + I_{\text{other}} + P_N = (1 - \beta + v) I_{\text{own}} + P_N \tag{8.12}$$

and the corresponding cell load is given by

$$\eta = (1 - \beta + v) \sum_{n=1}^{N_{\text{own}}} \eta_n. \tag{8.13}$$

In case of PIC diversity, MIMO operates in a similar manner as in case of SISO. For a single user load we have

$$\eta_n^D = (1 - \beta) \frac{\kappa_n E_n^D}{1 + \kappa_n E_n^D}. \tag{8.14}$$

Furthermore, for information MIMO the single user load is given by

$$\eta_n^I = (1-\beta) \sum_{m=1}^{M_t} \frac{\kappa_{n,m} E_{n,m}^I}{1+\kappa_{n,m} E_{n,m}^I}. \tag{8.15}$$

8.4.3 Performance Comparisons

In the following, we show performance results for UTRA FDD uplink and discuss the connection between the results and the load equations. Simulations use full 3GPP link level modeling with inner and outer loop power control and realistic channel and interference estimation algorithms. The radio channel is modeled according to ITU models in Table 8.1. Our system model follows accurately the present UTRA FDD specifications and simulations are done following strictly the recommendations given in [21].

Let us consider two systems where the number of users is the same and all users employ the same service. Assume that the number of receive antennas is the same in both systems, but the first system applies the diversity MIMO while the second one employs SIMO. In both systems, the base station applies the Rake receiver. Then, according to Equation 8.8, the ratio between cell loads is given by

$$\frac{\eta_{\text{sys}_2}}{\eta_{\text{sys}_1}} = \frac{\delta_{1,M_r} E/M_r (1+\kappa \cdot \delta_{M_t,M_r} E/M_r)}{\delta_{M_t,M_r} E/M_r (1+\kappa \cdot \delta_{1,M_r} E/M_r)} \cdot \frac{1+v_2}{1+v_1}, \tag{8.16}$$

where E refers to the received energy per bit in SISO system, and v_1 and v_2 denote intercell interference ratios in systems 1 and 2. Since δ_{M_t,M_r} and δ_{1,M_r} are around 0 dB, we can approximate

$$\eta_{\text{sys}_2} \approx \frac{1+v_2}{1+v_1} \cdot \eta_{\text{sys}_1}. \tag{8.17}$$

Thus, there is no noticeable capacity gain from additional transmit antennas in an isolated cell where $v_1 = v_2 = 0$. Assume that we know the intercell to intracell interference ratio v_1 and the original load η_{sys_1}. Then we can compute the capacity gain, provided that the ratio v_2/v_1 can be deduced. We have

$$\frac{v_2}{v_1} = \frac{I_{\text{own},1}}{I_{\text{own},2}} \cdot \frac{I_{\text{other},2}}{I_{\text{other},1}} = \frac{I_{\text{other},2}}{I_{\text{other},1}}, \tag{8.18}$$

where the latter equality holds because the number of own cell users is the same in both systems and power control drives SIR (Equation 8.4) to the same target value. Hence, v_2/v_1 depends only on the ratio between other cell interferences that are proportional to the average mobile transmission

TABLE 8.2

Gain of Diversity MIMO Relative to SIMO in Terms of Reduced Cell Load

$\eta = 0.75$	$M_r = 2$	$M_r = 4$
$v_1 = 0.5$, $M_t = 2$	11.1%	4.8%
$v_1 = 0.5$, $M_t = 4$	14.3%	6.7%
$v_1 = 2.0$, $M_t = 2$	22.2%	9.5%
$v_1 = 2.0$, $M_t = 4$	28.6%	13.3%

power. After computing the mean transmit powers in diversity MIMO and SIMO systems it is found that [23–25]

$$v_2 = \frac{M_t(M_r - 1)}{M_t M_r - 1} \cdot v_1, \quad M_r > 1. \tag{8.19}$$

This equality holds in flat fading. Table 8.2 shows the relative reduction $(\eta_{sys_1} - \eta_{sys_2})/\eta_{sys_1}$ of the cell load for diversity MIMO when the reference system applies SIMO. The results show that the gain from diversity MIMO can be noticeable especially when the initial intercell interference ratio v_1 is large.

While the gain of diversity MIMO strongly depends on the intercell interference level, the gain from additional receive antennas is remarkable also in isolated cells. Applying Equation 8.8, the required energy per data bit and noise spectral density is approximately inversely proportional to the number of receive antennas M_r. Another measure for the cell coverage is the power per data bit that is required to compensate the path loss and multiuser interference. Assume that all users apply the same service. Then by Equation 8.4, Equation 8.5, and Equation 8.10 we have in isolated cells

$$\frac{P_{rx}}{P_N} = \frac{\delta_{M_r} \cdot \kappa \cdot E}{M_r - (N_{own} - 1) \cdot \delta_{M_r} \cdot \kappa \cdot E}, \tag{8.20}$$

where E is the energy per data bit and noise spectral density for SISO. We note that although Equation 8.10 is given for MIMO system, it is applicable also in the case of SIMO. Equation 8.20 shows that required transmit power, which is directly proportional to the received power, does not depend only on the number of receive antennas, but it also depends on the number of users. It is found that the range gain from additional receive antennas increases with the load while the absolute cell radius is decreasing when the load is growing.

In the following, we study the system performance also by simulations. The main simulation parameters and assumptions are shown in Table 8.3.

TABLE 8.3

Simulation Parameters

Carrier frequency	1940 MHz
Chip rate	3.840 Mchips
Sampling rate	1 sample/chip
Power control	ON, both inner and outer loops
BLER target (QoS)	10%
Rake finger allocation	Known delays
Maximum number of allocated Rake fingers	Five per receive antenna
Channel estimation	Estimated (DPCCH)
Signal-to-interference estimation	Estimated (DPCCH)

Full 3GPP link level modeling was used with inner and outer loop power control and realistic channel and interference estimation algorithms. For more details, see [26].

Figure 8.5 depicts the received power per bit and antenna for SIMO in terms of the number of 0.96 Mbps users assuming isolated cell and Pedestrian B channel with 3 km/h mobile speed. The results in Figure 8.5 were obtained through multiuser simulations, but they are well in line with Equation 8.20. It is found that the range gain from two additional receive antennas is of the order of 3 dB in the single user case, but the gain grows rapidly with additional users. If, however, noise rise is fixed, the range gain from additional antennas is inversely proportional to M_r. This is seen by studying the intersection points between power and noise rise curves.

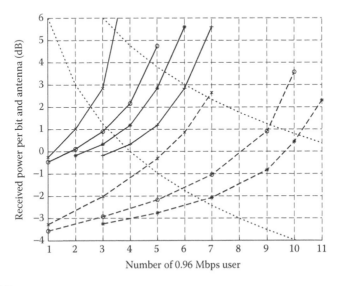

FIGURE 8.5
Received power per bit and antenna as a function of 0.96 Mbps users assuming two (solid curves) and four (dashed curves) receive antennas in Pedestrian B channel with 3 km/h mobile speed. Receivers are Rake (x), conventional PIC (o), coded turbo PIC with one iteration (*), and coded turbo PIC with three iterations (+). Dotted curves show the 3 dB and 6 dB noise rise levels.

Cell capacity also can be estimated from the results in Figure 8.5. We recall that the load control of the network drives the noise rise to a predefined target value that is usually between 3 dB and 6 dB, depending on the network configuration. Intersection with 6 dB level curve shows that two-antenna and four-antenna Rake receivers reap around 3.3 Mbps and 6.5 Mbps cell capacities, respectively. In addition to Rake curves, Figure 8.5 shows performance results for conventional PIC and coded turbo PIC. Although the largest gains are obtained by increasing the number of receive antennas, PIC provides an efficient way to improve cell coverage and capacity when the number of receive antennas is fixed. In two-antenna case coded turbo PIC with three iterations results in cell throughput up to 5.8 Mbps and the range gain against Rake is around 2.5 dB for 6 dB noise rise. Assuming four receive antennas and coded PIC with single iteration allows almost 10 Mbps throughputs for 6 dB noise rise.

We recall from Equation 8.14 and Equation 8.15 that the effect of PIC can be modeled through the efficiency β. This parameter is not constant but depends on cell conditions such as the number of users and channel model. According to simulations the efficiency for uncoded PIC is around 0.4 at maximum while coded PIC with turbo receiver can provide maximum efficiency of around 0.7.

Figure 8.5 shows the performance for SIMO configuration. However, the performance for information MIMO can be deduced from the same figure by setting the data rate to $M_t \times 0.96$ Mbps and scaling the number of users accordingly. In the analysis, this equivalence is seen in Equation 8.11. Thus, there is no throughput gain at the cell level from information MIMO for low user data rates. Instead, the performance may suffer, because the user's pilot power is divided between multiple transmit antennas, which deteriorates channel estimation in the receiver.

The benefits of information MIMO become visible with high user data rates. The main reason for the superiority of information MIMO to SIMO and diversity MIMO is that in information MIMO high data rates can be obtained without heavy code puncturing because independent data streams are transmitted from separate antennas. Both SIMO and diversity MIMO face serious performance degradation when effective code rate increases due to puncturing. We note that adaptive modulation and coding in uplink is not supported in UTRA FDD specification, and therefore code puncturing cannot be avoided when SIMO and diversity MIMO are used on high data rates.

Figure 8.6 shows the received power per bit and antenna in Vehicular A channel with 30 km/h mobile speed assuming information MIMO and Rake receiver. Six multicodes of spreading factor 4 were applied for data rates higher than 2 Mbps. Data rates higher than $M_t \times 2$ Mbps were generated by using the code puncturing. It is found that additional transmit antennas provide a remarkable performance gain when the base station has four or eight receive antennas. Especially in the case of eight receive antennas, the gain from the second and the third data stream is large, and data rates up

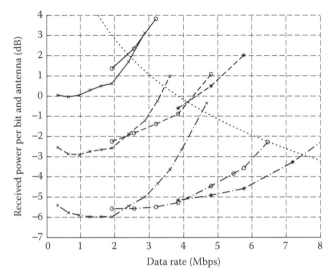

FIGURE 8.6
Received power per bit and antenna as a function of data rate assuming Rake with two (solid curves), four (dashed curves), and eight (dashed and dotted curves) receive antennas in Vehicular A channel with 30 km/h mobile speed. Numbers of transmit antennas are one (x), two (o), and three (*). Dotted curve shows the 3 dB noise rise level.

to 7.5 Mbps are obtained with 3 dB noise rise. On the other hand, with two receive antennas, channel estimation errors slightly degrade the system performance when the second data stream is added. For low data rates, this phenomenon is also visible when $M_r > 2$. Besides the channel estimation errors, the poor orthogonality properties of different scrambling codes also have a negative impact to the system performance. Figure 8.6 also shows the negative effect of code puncturing. Especially in SIMO curves, rapid degradation in performance is clearly noticeable when the data rate exceeds 2 Mbps.

8.5 MIMO in UTRA FDD Downlink

In the previous section, we showed that in UTRA FDD uplink, spectral efficiency is not the main issue because the uplink is not code limited but interference limited. However, UTRA FDD downlink is code limited, and therefore it is necessary to pursue spectrally efficient MIMO transceiver solutions.

In this section, we first review basic MIMO algorithms whose derivatives have been presented within 3GPP standardization [10]. Many good research papers and tutorials on MIMO exist (see, e.g., [27]), and we only recall the fundamental results in order to illustrate potential benefits of MIMO. Then we will discuss the 3GPP requirements for MIMO performance [10] and explain their impact to algorithm design. At the end of the section we present an alternative scheme that increases cell throughput by utilizing the existing CSI signaling structure without additional complexity in UE.

8.5.1 Theoretical Background

8.5.1.1 Single-User MIMO

For simplicity, we present the capacity expressions in single path channels to avoid the notational complexity of capacities in multipath channels. Capacities in multipath (wideband) channels can be expressed by dividing the channel into several narrowband channels and integrating over the frequency band to obtain the sum capacity.

Let $\mathbf{H}$ be a $M_r \times M_t$ matrix that consists of normalized complex channel coefficients $h_{m,n}$. The bound for the information rate of a memoryless diversity MIMO system in additive white Gaussian noise (AWGN) channel is given by

$$C = \log_2\left(1 + \frac{\rho}{M_t}\sum_{m=1}^{M_r}\sum_{n=1}^{M_t}|h_{m,n}|^2\right) \quad \text{bits/s/Hz,} \tag{8.21}$$

where ρ is the average SNR, which is divided by M_t because the total transmit power is equally divided between the M_t transmit antennas. Ergodic capacity in fading channels is then obtained by averaging the capacity (Equation 8.21) over all possible channel states.

For the information MIMO the bound for the information rate is of the form

$$C = \sum_{l=1}^{L} \log_2\left(1 + \frac{\rho}{M_t}\lambda_l\right); \quad L = \min\{M_r, M_t\}, \tag{8.22}$$

where λ_l is an eigenvalue of $\mathbf{HH}^\dagger$. We note that Equation 8.21 and Equation 8.22 apply when the transmitter has no CSI and the total transmit power is fixed. Ergodic capacities in fading channels are obtained from Equation 8.21 and Equation 8.22 by averaging over channel states.

Wireless channels are often slow fading and coding over all channel states is not possible within the transmission of a data packet. Thus, it is important to evaluate the performance of suboptimal transceivers that can be implemented in practice. In vertical encoding (also referred to as V-BLAST [1]), the input data stream is demultiplexed into as many streams as there are transmit antennas. Each substream undergoes independent encoding, interleaving, and symbol mapping before it is transmitted from its designated antenna. This simplifies the receiver design, because the substreams can be decoded independently. However, when the substreams make up a composite data packet, the capacity is limited by $C = M_t \log_2\left(1 + \min(\eta_1, \cdots, \eta_{M_t})\right)$, where

η_i refers to the received SINR in independently coded stream i, because the composite packet is decoded correctly only if the stream with the worst SINR is decoded correctly.

The performance of V-BLAST can be improved if CQI feedback for each substream is made available in the transmitter. Then the ergodic open-loop capacity can be obtained by transmitting independent data streams from different antennas with equal power but with different rates [28,29] and using the MMSE receiver and SIC. This is useful, because the complexity of the MMSE-SIC receiver is smaller than that of maximum-likelihood decoding. The capacity becomes

$$
\begin{aligned}
C &= \log_2 \det\left(\mathbf{I}_{M_t} + \frac{\rho}{M_t} \mathbf{H}^\dagger \mathbf{H} \right) \\
&= \sum_{i=1}^{M_t} \log_2\left(1 + \frac{\rho}{M_t} \mathbf{h}_i^\dagger \left[\mathbf{I}_{M_t} + \frac{\rho}{M_t} \mathbf{H}_{(i)}^\dagger \mathbf{H}_{(i)} \right]^{-1} \mathbf{h}_i \right) \\
&= \sum_{i=1}^{M_t} C_i
\end{aligned}
$$

where the channel matrix $\mathbf{H}_{(i)}$ is obtained by removing data streams 1, ..., $i - 1$. Thus, the receiver first demodulates and deinterleaves the substream from transmit antenna 1, decodes the data, subtracts its contribution from the received signal, and proceeds to substream 2, etc. The rates that each substream can support are different, and therefore, these rates must be signaled to the transmitter. In practice, the MMSE-SIC suffers from error propagation, because if one substream is incorrectly decoded, all subsequent substreams are affected. Therefore, decoding should start from the stream with the largest received SINR.

Instead of using CQI feedback, the open-loop capacity with MMSE-SIC receiver can be achieved by coding across transmit antennas referred to as diagonal encoding or D-BLAST [1]. In diagonal encoding, the composite data packet undergoes horizontal encoding after which the codewords are split into blocks. The blocks of any one codeword pass through a stream rotator, which rotates the blocks in a round-robin fashion so the mapping from the substreams to the antennas is periodically altered. Thus, each codeword spans multiple antennas, but the symbols sent simultaneously from different transmit antennas belong to different codewords and substreams. Demodulation and decoding are similar to those of V-BLAST and MMSE-SIC. With V-BLAST the decoding is performed for codewords that are transmitted from one antenna only, but now decoding is performed over a codeword consisting of M_t blocks, where each block is transmitted from a different antenna. The capacity of any one substream in D-BLAST is the mean of the capacities of the blocks transmitted from different antennas,

$$C_{stream} = \frac{1}{M_t} \sum_{i=1}^{M_t} C_i,$$

and when there are M_t such streams in parallel the total channel capacity in Equation 8.22 follows.

D-BLAST requires a more complicated transceiver algorithm than V-BLAST because of the diagonal encoding and decoding. It also suffers from a rate loss, because in the initial phase, transmission from some of the antennas has to be suspended in order to facilitate the initialization of the MMSE-SIC receiver. On the other hand, D-BLAST does not use any CSI in the transmitter, and in theory, system design is simple, because no feedback channel is required.

The capacity can be further improved if the transmitter knows the eigenmodes of the channel. This does not increase the degrees of freedom of the channel, but the power efficiency is improved because the transmitter may pour power to different eigenmodes instead of allocating power to different transmit antennas. This is the celebrated waterfilling principle. Moreover, receiver design is highly simplified, because SIC is not needed. Instead, a matched filter, matched to the singular vectors of the MIMO channel, is able to orthogonalize received substreams. When the number of transmit and receive antennas is equal, the gap between the capacities with and without CSI shrinks as a function of SNR [30], but when the number of transmit antennas exceeds the number of receive antennas, the differences may be large. Unfortunately, waterfilling requires full CSI in the transmitter, which is not realistic in FDD systems. Therefore, suboptimal solutions, which communicate only partial CSI to the transmitter, have been extensively studied.

Antenna selection requires relatively small amount of CSI and different selection algorithms for MIMO have been developed (see, e.g., [31]). When the number of available transmit antennas is larger than the number of substreams to be transmitted, the transmitter may choose a subset of antennas for transmission according to CSI feedback. Thus, the signaling of eigenvectors is avoided by using a fixed set of basis vectors, e.g., the transmit antennas, and CSI feedback consists of the indices of the basis vectors.

Antenna selection improves the quality of the received signal but requires CSI in the transmitter, although signaling overhead is much less than the overhead required to signal the singular vectors of MIMO channel. Another way to improve the robustness of the link is to add spatial redundancy to the transmitted substreams. Diversity MIMO (Equation 8.21) uses all degrees of freedom in the MIMO channel for this redundancy. Information MIMO achieves higher data rates (Equation 8.22) and has no spatial redundancy between the substreams, but with practical receivers pure information MIMO approach leads to poor link-level performance at low SNR. This motivates the search for schemes that achieve a tradeoff between data rate and diversity [32]. One such 2×2 MIMO scheme is presented in [33], which transmits two STTD-coded substreams and rotates the other substream to avoid pathological error events for some combinations of transmitted symbols. In [34] two STTD-coded substreams are transmitted within a 4×2 setup, which is referred to as double STTD within 3GPP standardization [35].

8.5.1.2 Multiuser MIMO

The capacity expressions above assume a point-to-point MIMO link where the interference consists of AWGN. However, cellular systems supporting high data rates are subject to non-Gaussian MAI, which should be taken into account in the design of MIMO transceivers. The downlink channel is a point-to-multipoint channel, which is referred to as broadcast channel in information theory. Characteristic of the problem is that the transmitter is able to coordinate its transmissions to different users, but receivers are not able to cooperate. Moreover, transmit power resources are shared among the users.

The capacity region for MIMO broadcast channels has been studied in [36,37,38], and it has been shown that the sum-rate capacity is achieved by dirty-paper coding (DPC) [39]. DPC allows the base station to efficiently transmit data to multiple users simultaneously and avoid the impact of MAI by jointly encoding the transmitted signals. This is possible when the base station knows the MIMO channels of the users so that it is able to control the multiuser interference caused by the simultaneous transmissions. Even though the DPC achievable region is the largest known achievable region for the multiple-antenna broadcast channel, it is difficult to implement in practical systems. Different techniques have been investigated to achieve the gains promised by DPC [40,41], but these techniques are still in the development stage. Another nonlinear technique is to use decision feedback equalization in the transmitter, i.e., Tomlinson-Harashima precoding for MIMO channels [42].

Instead of non-linear precoding, multiuser beamforming (MUB) techniques can serve multiple users simultaneously with reduced complexity. With multiuser beamforming, each user stream is coded independently and multiplied by a beamforming weight vector for transmission through multiple antennas. In block diagonalization [43], the multiuser MIMO channel is effectively converted into multiple point-to-point MIMO channels.

Allowing the scheduler to select the served user among a pool of users brings another dimension to the problem. When the number of users is large, the base station can schedule its transmission to those users with good channel conditions to achieve higher date rates. This form of selection diversity is referred to as multiuser diversity, and it can be exploited to increase the system throughput in delay tolerant data applications [44,45]. The authors of [46] indirectly show that the combination of beamforming and multiuser diversity asymptotically achieves the optimal sum-rate of DPC when the number of users is large. However, finding the optimal beamforming weight vector is a difficult non-convex optimization problem. All these multiuser MIMO schemes assume perfect CSI at the transmitter. Under partial CSI as in UTRA FDD, the performance of multiuser MIMO is still an open problem.

8.5.2 Proposed MIMO Algorithms within 3GPP Standardization

The standardization of MIMO for HSDPA is still ongoing, and many algorithms have been proposed by different standardization bodies within 3GPP. In the following, we briefly summarize the algorithms proposed so far; their acronyms are presented in Table 8.4. All concepts adjust the modulation and coding of independent substreams using CQI feedback, and D-BLAST-type processing is not applied. In addition, some schemes employ space–time coding and four transmit and two receive antennas to gain diversity and improve link-level performance (DSTTD-SGRC and RC-MPD). Alternatively, link performance is improved by linear precoding in the transmitter (SS CL MIMO, PU²RC, S-PARC, D-TxAA).

The first two proposals, PARC and DSTTD-SGRC, are well documented, but documentation on the other six algorithms is sparse. Therefore, we will introduce them only briefly. A more detailed description and performance analysis can be found in [10] and numerous standardization contributions showing performance results for the algorithms of Table 8.4. However, there is no widespread agreement concerning the ranking of the candidate algorithms, and therefore, we refrain from comparing them here.

All proposals invariably increase the overhead of CSI feedback when compared to HSDPA in Release 5. Multiple data streams require multiple CQIs, and additional transmit antennas require additional CSI when compared to 2×1 MISO configuration supported in UTRA FDD Release 5. However, reducing the overhead due to feedback signaling has been identified as one of the goals in enhanced HSDPA [13]. Uplink interference due to feedback signaling may cause problems, especially when the mobile is located on the cell edge.

TABLE 8.4

Proposed MIMO Algorithms within 3GPP Standardization

MIMO Scheme	Space–Time Coding	Linear Precoding
PARC		
DSTTD-SGRC	✓	
SS CL MIMO		✓
RC-MPD	✓	
PU2 RC		✓
TPRC for CD-SIG		
S-PARC		✓
D-TxAA		✓

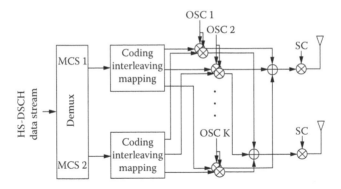

FIGURE 8.7
Transmitter structure for PARC.

Per-Antenna Rate Control (PARC). Figure 8.7 shows a block diagram of PARC architecture for two transmit antennas. The HS-DSCH data packet is demultiplexed into two low-rate substreams. Both substreams are turbo encoded, interleaved, and mapped to either QPSK or 16 QAM symbols. Code rates and symbol mappings between the substreams may vary, and therefore the number of information bits assigned to the streams can be different. Furthermore, symbols in each spatial substream are divided into K subchannels, where K is the maximum number of HS-PDSCHs defined by the UE capability. After spreading these subchannels by different orthogonal spreading codes (denoted by OSC 1-OSC K in Figure 8.7), the subchannels are summed up and modulated by a scrambling code and are finally transmitted from their designated antennas. Thus, PARC includes multiplexing in spatial and code domains.

The base station selects data rates and corresponding modulation, coding, and channelization codes for different antennas based on antenna-specific CQI feedback. For this purpose, UE estimates SINR for all antennas, maps SINR values to CQI values, and sends this information to the base station through a feedback channel. The number of possible transport combinations is large, and CQI must be quantized to avoid excessive feedback overhead. In order to guarantee the performance of PARC at poor channel conditions, the lower end of the CQI values should correspond to selective transmit diversity (STD) transmission.

To emphasize the importance of STD feature for the PARC performance, probabilities for dual stream and STD transmission for 2×2 PARC in a macrocell environment are presented in Figure 8.8. Probabilities are presented both in ITU Pedestrian and Vehicular A channels. It should be noted that the probability for the single stream transmission is as high as 75 to 80%. Moreover, Vehicular A has higher dual stream transmission probability as it provides more multipath diversity than Pedestrian A.

Double STTD with Subgroup Rate Control (DSTTD-SGRC). This scheme assumes that $M_t = 2M_r$. Transmit antennas are divided into M_r two-antenna subgroups that apply STTD and AMC. Both antennas within the subgroup

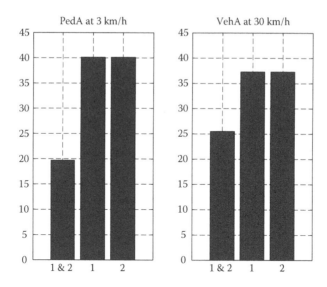

FIGURE 8.8
Stream distribution for 2–2 PARC in ITU Pedestrian A and Vehicular A channels. Round robin scheduler is used and 70% of the total transmit power in the base station is allocated to HS-DSCH.

use the same modulation and coding scheme (MCS), but the data rates of different groups can be adjusted independently or jointly by selecting a suitable MCS. In WCDMA, the maximum number of transmit antennas is expected to be four, and thus, at maximum, two independent data streams can be transmitted.

DSTTD-SGRC is a natural extension to the conventional STTD supported by UTRA FDD Release'99. While conventional STTD employs two transmit antennas and a single data stream, DSTTD-SGRC doubles the number of transmit antenna and data streams, provided that UE is equipped with at least two antennas. Figure 8.9 shows the structure of the DSTTD-SGRC with four transmit antennas. The incoming HS-DSCH data are divided into two substreams. Selected MCS and the number of orthogonal spreading codes (OSC 1-OSC K) define the number of information bits allocated to each substream. For both substreams, information bits are coded, interleaved, and modulated according to the MCS scheme. After STTD encoding, the two substreams are split into K parallel streams corresponding to K spreading codes. In the last stage, streams are combined, scrambled, and transmitted.

Rate-Control Multipaths Diversity (RC-MPD). Here each data stream is transmitted from at least two antennas, and the number of data streams is equal to the number of transmit antennas. A pair of data streams that share the same two antennas apply the same MCS. The basic idea is to transmit a second copy of the signal after one chip delay by using the STTD encoding. Hence, in the case of two transmit antennas, two data streams and corresponding symbols s_1 and s_2, the transmitted signal consists of symbols s_1 and

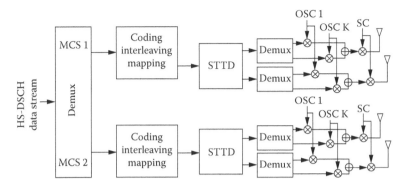

FIGURE 8.9
Transmitter structure for DSTTD-SGRC.

s_2 at time T and symbols $-s_2^*$ and s_1^* at time $T + T_c$, where T_c is the chip interval. The aim of the method is to achieve multipath diversity that is orthogonalized through STTD encoding.

Single Stream Closed-Loop MIMO (SS CL MIMO). This is a four-antenna extension of the two-antenna closed-loop Mode 1 that is supported by UTRA FDD Release'99. The scheme supports only a single data stream but requires more CSI than Mode 1 due to additional transmit antennas.

Per-User Unitary Rate Control (PU²RC). This algorithm is based on the singular value decomposition of the MIMO channels. Precoding in the transmitter is based on the unitary matrix that is a combination of the selected unitary basis vector from all UEs. The aim is to utilize multiuser diversity on top of MIMO transmission.

Tx Power Ratio Control for Code Domain Successive Interference Cancellation (TPRC for CD-SIC). Here the receiver applies code-domain SIC to suppress the impact of code domain interference in addition to space–time interference. System performance is further boosted by employing a code-domain transmit power ratio control that requires additional feedback signaling. Power control allocates different powers to different spreading codes according to the order of successive interference cancellation in code domain.

Selective Per-Antenna Rate Control (S-PARC). The aim is to improve the performance of conventional PARC by combining PARC with antenna selection.

Double Transmit Antenna Array (D-TxAA). Composite data stream is split into two substreams and each substream is transmitted from two antennas by applying either of the closed-loop methods in UTRA FDD Release'99. Hence, the total number of transmit antennas is four.

8.5.3 Practical Considerations

The high speed downlink packet access in Release 5 contains several enhancements to UTRA FDD downlink compared to earlier releases as discussed in Section 8.1. MIMO concepts presented above show one possibility to further improve the performance of HSDPA. Naturally, the improvement due to MIMO has to be substantial when compared to the complexity of the implementation. We recall the most important requirements that should be taken into account when evaluating different candidate MIMO techniques. In the following, each paragraph starts with a direct quotation from [10].

MIMO proposals shall be comprehensive to include techniques for one, two, and four antennas at both the base station and UE. Deploying multiple antennas in the user equipment or base stations to support MIMO techniques is not straightforward due to concerns of cost, implementation complexity, and visual impact. This is especially true in the present mobile terminals, where basic products with large production volumes may have, at most, two antennas. Present macro base stations typically employ two or four antennas, and it is expected that two-antenna base stations will dominate in number in the near future.

For each proposal, the transmission techniques for the range of data rates from low to high SIR shall be evaluated. This requirement is due to the fact that the gain from information MIMO greatly depends on the SINR. For example, in macrocell environments the operating SINR in HSDPA is less than 10 dB most of the time. Therefore, practical performance differences between various diversity and information MIMO techniques are not great because the differences become only significant at high SNR region.

Operation of the MIMO technique should be specified under a range of realistic conditions. The requirement sparked the development of a new MIMO channel model for standardization purposes [47]. The model is a geometry-based stochastic model, which replaced the earlier stochastic channel model [48,49]. Link-level simulations with the spatial channel model are only used for calibration purposes, and system-level simulations are required to verify the performance of MIMO proposals. Moreover, implementation un-idealities should be taken into account when modeling realistic operation environments.

The MIMO technique shall have no significant negative impact on features available in earlier releases. This requirement makes the design of competitive four-antenna MIMO algorithms difficult, because the present specification contains only two primary common pilot channels. To support four-antenna MIMO, a straightforward solution would be to define two additional primary common pilot channels. However, since the total transmit power in the base station cannot increase — otherwise the base station would generate more interference to the network — the transmit power per antenna should be halved. Legacy UEs can receive only two common pilot channels so that, in case of a four-antenna cell, they would be able to gather only half of the pilot power when compared to a two-antenna cell. This would seriously limit the coverage and service availability within the four-antenna cell.

MIMO techniques should demonstrate significant incremental gain over the best performing systems supported in the current release with reasonable complexity. Theoretical capacity curves suggest that information MIMO may give remarkable gains when compared to diversity MIMO. It has been observed, however, that with two transmit and receive antennas, practical gains from information MIMO can be small [50]. Increasing the number of antennas increases the gain of information MIMO, but at the same time, the implementation complexity grows rapidly, and backward compatibility, such as the pilot design problem, becomes an issue.

The focus shall be on strengthening the UTRA system as a reliable and cost-effective access technique in urban and suburban areas. This means that the goal is to increase the number of users, and/or to increase their coverage compared to earlier releases. In other words, the improvement of the service availability as compared to earlier releases shall be used as a primary evaluation criterion. The increase in maximum data rate per cell is also of interest. In WCDMA downlink, capacity and coverage are highly interdependent. Multiple intracell users, which share the same transmit power resource, interfere with each other. In general, the system can support more users at low power than at high power, and the availability of services in a cell depends on required transmit powers. This requirement suggests that instead of increasing data rates with parallel multiplexing, multiple transceiver antennas should be used for diversity or beamforming to improve the quality of the received signal.

8.5.4 Multiuser Beamforming

In this section we present a concept that fulfills the requirements cited in the previous section without additional feedback overhead or receiver complexity when compared to WCDMA Release 5. This can be accomplished by utilizing the existing closed-loop modes and CQI feedback in UTRA FDD together with physical layer scheduling.

UTRA FDD two-antenna closed-loop modes [51] aim to maximize the received SNR by signaling transmit weights from UE to BS. Due to the limited overhead of the feedback channel, the transmit weights must be quantized. Mathematical formulation of the problem for finding the optimal transmit weight $\mathbf{w}_0$ is given by

$$\text{Find } \mathbf{w}_0 \in \mathbf{W}: \ \left|\mathbf{h}\mathbf{w}_0\right| = \max_{\mathbf{w} \in \mathbf{W}} \left|\mathbf{h}\mathbf{w}\right|,$$

where $\mathbf{W} = \{\mathbf{w} = (w_1, w_2)^T : \|\mathbf{w}\| = 1 \text{ and } w_1, w_2 \in C\}$, and in the presence of flat fading $\mathbf{h} = (h_1, h_2)$ consists of complex impulse responses corresponding to the first and the second channel. It can be assumed without loss of generality that w_1 is real. Therefore, the solution can be characterized by a single complex coefficient $w_2 = ze^{j\phi}$ and

$$(z, \phi) = \text{argmax}\left\{\left|\sqrt{1 - z^2}\,h_1 + ze^{j\phi}h_2\right|^2 : z \in [0,1], \phi \in [0, 2\pi)\right\}.$$

With UTRA FDD closed-loop Mode 1 and Mode 2, the feedback weights are quantized to follow a time-varying QPSK and 16-QAM signal constellation, respectively.

Assume now that the base station transmits to two users simultaneously employing either of the present closed-loop modes, and let the channelization and scrambling codes be the same for both users. Such assumptions are reasonable in HSDPA, where the shortage of channelization codes puts strict limits on the cell throughput. After despreading, the received signals of the two users become

$$r_1 = \mathbf{h}_1 \mathbf{w}_1 s_1 + \mathbf{h}_1 \mathbf{w}_2 s_2 + n_1,$$

$$r_2 = \mathbf{h}_2 \mathbf{w}_1 s_1 + \mathbf{h}_2 \mathbf{w}_2 s_2 + n_2,$$

where r_i is the received signal of user i and $\mathbf{h}_i = (h_{i,1}, h_{i,2})$ is the impulse response vector between the transmitter and the user i, $\mathbf{w}_i$ is the transmit weight vector requested by user i, s_i is the transmitted symbol, and n_i is the noise term.

Mutual interference between simultaneously scheduled users that use the same channelization code can be suppressed by selecting users i and j in such a way that their transmit weights are orthogonal, i.e., $\mathbf{w}_i^\dagger \mathbf{w}_j = 0$. Now the expectation $\gamma_{i,i}$ for the desired power of user i, $(i \in \{1,2\})$ is given by

$$\gamma_{i,i} = \mathrm{E}\left\{ |\mathbf{h}_i \mathbf{w}_i|^2 \right\} = \mathrm{E}\left\{ \max\left\{ |\mathbf{h}_i \mathbf{w}|^2 : \mathbf{w} \in \mathbf{W} \right\} \right\} \tag{8.23}$$

and the expected undesired interference powers between the users are

$$\gamma_{1,2} = \mathrm{E}\left\{ |\mathbf{h}_1 \mathbf{w}_2|^2 \right\} = \mathrm{E}\left\{ \min\left\{ |\mathbf{h}_1 \mathbf{w}|^2 : \mathbf{w} \in \mathbf{W} \right\} \right\},$$

$$\gamma_{2,1} = \mathrm{E}\left\{ |\mathbf{h}_2 \mathbf{w}_1|^2 \right\} = \mathrm{E}\left\{ \min\left\{ |\mathbf{h}_2 \mathbf{w}|^2 : \mathbf{w} \in \mathbf{W} \right\} \right\}.$$

Moreover, assuming that the channels corresponding to separate users are uncorrelated, we find that

$$\mathrm{E}\left\{ (\mathbf{h}_1 \mathbf{w}_1)^\dagger \mathbf{h}_1 \mathbf{w}_2 \right\} = w_{1,1}^* w_{2,1} \mathrm{E}\left\{ |h_{1,1}|^2 \right\} + w_{1,2}^* w_{2,2} \mathrm{E}\left\{ |h_{2,1}|^2 \right\}. \tag{8.24}$$

When the base station transmits with a fixed power, as in HSDPA, the expectation of Equation 8.25 vanishes due to orthogonal transmit weights.

Computation of SIR can be carried out provided that the expectations of Equation 8.23 and Equation 8.24 can be evaluated. This can be done by

TABLE 8.5

Simulation Assumptions

Site-to-site distance	2.8 km
Pathloss model	3GPP UMTS
Number of sectors	Three/cell
Power delay profile	ITU Pedestrian A, ITU Vehicular A
Number of HS-DSCH codes	10
MCS set	{QPSK (1/2), QPSK (3/4), 16QAM (1/2),16QAM (3/4)}
Receiver	Time-domain LMMSE equalizer
Power of HS-DSCH	70% of the total BS power
Power of overhead channels	30% of the total BS power
Feedback delay	1 slot
Feedback error rate	4%

following the method of [52], and in the case of Rayleigh fading, the resulting SIR values corresponding to Mode 1 and Mode 2 are given by

$$\text{SIR} = \frac{\gamma_{1,1}}{\gamma_{1,2}} = \frac{\gamma_{2,2}}{\gamma_{2,1}} = \begin{cases} 7.65\text{dB} & \text{for Mode 1} \\ 13.5\text{dB} & \text{for Mode 2} \end{cases}$$

Performance of the proposed multiuser beamforming (MUB) technique is simulated with a quasi-static system level simulator assuming UTRA FDD closed-loop Mode 1. Average throughput per sector is used as a performance measure and 2×2 PARC provides a reference case. The simulation parameters are listed in Table 8.5. The primary user is chosen by using a round robin (RR) scheduler, which does not take into account CSI. The secondary user with the same spreading code and orthogonal transmit weight vector is chosen by either RR or max SINR scheduler. If there is no user having orthogonal transmit weights with the primary user, full HS-DSCH power is allocated to the primary user. Otherwise, HS-DSCH power is evenly divided between the primary and the secondary user. Thus, the proposed scheduling is a simple extension of the conventional RR providing a fair share of transmission resources to primary users in a RR fashion while additional resources are given to secondary users depending on the orthogonality of transmit weights.

The average sector throughput values in Pedestrian A channel are presented in Figure 8.10. When the secondary user is scheduled by using RR scheduler (legend "MUB, RR"), the gain in throughput over single-user transmission is approximately 40%, whereas for max SINR scheduling (legend "MUB, max SINR") the gain is 70%. Multiuser beamforming with RR scheduler provides almost the same average sector throughput as 2×2 PARC, whereas with max SINR scheduler MUB provides 15% better average sector throughput than PARC. We emphasize that the MUB schemes employ only one receive antenna in UE. It has been shown that the performance of the MUB can be further improved by slightly modifying the calculation of CQI in user equipment [53]. However, here the simulations assume that

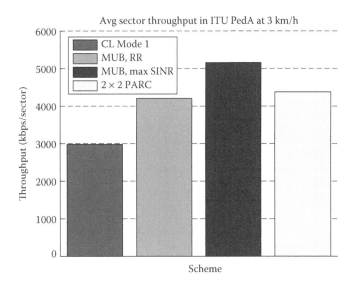

FIGURE 8.10
Average sector throughputs in ITU Pedestrian A channel with 3 km/h mobile speed.

UEs are perfectly backward compatible to earlier UTRA FDD releases and CQI calculation is not altered.

Cumulative distribution functions (CDF) for user throughputs are presented in Figure 8.11. For MUB with RR scheduler, the shape of the distribution is maintained but the distribution is shifted to the right when compared to the single-user distribution. The shape of the distribution of MUB with max SINR scheduling is less steep than the CDF of MUB with

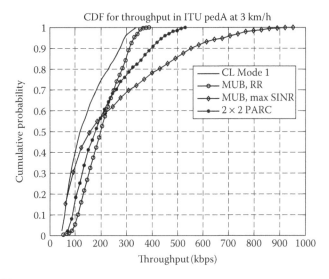

FIGURE 8.11
User throughput CDFs, ITU Pedestrian A channel with 3 km/h mobile speed.

RR scheduling. Multiuser beamforming with max SINR scheduling increases the probability of serving good users when compared to MUB with RR, which explains the higher rates. On the other hand, users with a low SINR are only scheduled as primary users in MUB with max SNR scheduling, whereas MUB with RR schedules weak users as primary and secondary users alike. This explains the shape of the CDF at low data rates. When comparing the average user throughput CDFs of MUB against the PARC, we note that MUB with RR provides best average user bit rates for users in poor channel conditions, whereas PARC provides better average bit rates for users in good channel conditions than MUB with RR scheduler. The max SINR scheduler provides clearly the best average bit rates, but on the other hand, it also provides the smallest average bit rates for the users in poor channel conditions.

Average sector throughputs in ITU Vehicular A channel at 30 km/h are presented in Figure 8.12, and the corresponding CDFs for average user throughputs are shown in Figure 8.13. Comparing the results in Pedestrian A and in Vehicular A channels reveals the following differences: First, the average sector throughputs in Vehicular A channel are smaller than in Pedestrian A channel for all simulated schemes. For the closed-loop Mode 1 and the MUB schemes the reduced throughputs are explained by reduced beamforming gains. For PARC the performance reduction is due to the interference between the streams, which increases in multipath channels. Second, in Vehicular A channel MUB with RR scheduler outperforms PARC, while the opposite is true in Pedestrian A channel. This is due to the increase in interference between users and substreams, the increment being larger for PARC than for MUB. This is based on the fact that MUB minimizes the interference between

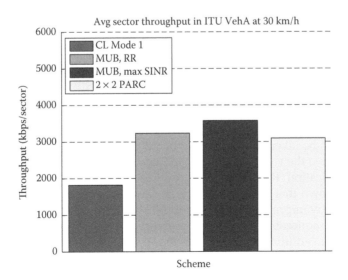

FIGURE 8.12
Average sector throughputs in ITU Vehicular A channel with 30 km/h mobile speed.

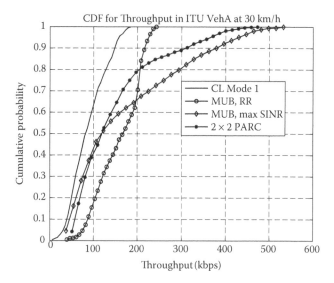

FIGURE 8.13
User throughput CDFs, ITU Vehicular A channel with 30 km/h mobile speed.

the simultaneously scheduled users given the CSI feedback, whereas for PARC there is no protection between the streams. Third, due to the reduced variance of received SINR, gains from max SINR scheduling are smaller in Vehicular A than in Pedestrian A channel. The smaller variance in SINR can also be seen from the CDF curves of average user bit rates, the slopes being steeper in Vehicular A channel.

8.6 Conclusions and Discussion

We studied different options for introducing multiple-input multiple-output (MIMO) transceivers to 3G WCDMA uplink and downlink. Basic universal terrestrial radio access (UTRA) frequency-division duplex (FDD) parameters were discussed in order to illustrate the boundary conditions in UTRA FDD that need to be taken into account when designing MIMO transceivers. We also presented the various MISO techniques supported by the present UTRA FDD specification, Release 6.

It was shown that the uplink coverage and capacity of UTRA FDD mode can be greatly improved by SIMO and MIMO. While the gain from transmit diversity is noticeable only in the presence of heavy intercell interference, the gain from additional receive antennas is remarkable also in isolated cells. Information MIMO introducing additional scrambling codes should be used with high data rates instead of diversity MIMO. This is mostly due to the fact that heavy code puncturing can be avoided by using the information MIMO. UTRA FDD uplink is not code limited and therefore diversity and information

MIMO can be implemented by allocating additional channelization or scrambling codes to different transmit antennas. Thus, the receiver in the base station can separate the signals from different antennas in code domain, and it is not necessary to rely on spatial signatures and space–time code design. This makes the system more robust and receivers can utilize well-known multiuser and interference cancellation receivers developed for CDMA systems. Furthermore, these MIMO schemes can be implemented with only minor changes to the present specification.

In contrast to uplink, UTRA FDD downlink is code limited and it is necessary to search for spectrally efficient MIMO techniques that are able to reuse channelization codes. Therefore, similar MIMO techniques as in UTRA FDD uplink cannot be employed. The work item on MIMO transceivers in downlink has been active for several years within the third generation partnership project (3GPP), and the current target is set to Release 7. Several concept proposals have been presented in standardization meetings and added into the technical report [10]. At the time of writing, the gains of different MIMO proposals are being evaluated against two-antenna receive diversity. Based on the performance evaluation, Work Group 1 will select a single MIMO scheme (two transmit antennas) in meeting No. 44 in February 2006 for further analysis. The decision whether to include the MIMO feature in Release 7 specifications will be made in June 2006, being based on system performance and overall complexity evaluation of the selected scheme.

Introduction of MIMO techniques to UTRA FDD downlink has not proceeded smoothly. First, adding multiple antennas and radio transceiver chains to user equipments is not straightforward due to cost and complexity of implementation. Second, new MIMO techniques need to be backward compatible with earlier releases. In particular, the performance of legacy user equipment, which is able to receive signals from two transmit antennas, should not suffer when new base stations use, say, four antennas for transmission. In addition, 3GPP has outlined several other requirements for MIMO performance. Increasing peak data rates is not considered as the primary target, which does not favor information MIMO (spatial multiplexing) techniques. Instead, the goal is to increase the number of users and coverage of the services. To this end, we introduced a multiuser beamforming technique that improves system throughput without any modifications to user equipments and feedback channels in the current UTRA FDD specification.

References

1. G. Foschini. 1996. Layered space–time architecture for wireless communication in a fading environment when using multi-element antennas, *Bell Labs. Tech. Journal*, pp. 41–59.
2. E. Telatar. 1995. "Capacity of multi-antenna Gaussian channels," *European Trans. Telecomm.*, Vol. 10, No. 6, Nov./Dec. 1999, pp. 585–595, based on AT&T Bell Laboratories, Internal Tech. Memo, June 1995.

3. G. Golden, C. Foschini, R. Valenzuela, and P. Wolniansky. 1999. "Detection algorithm and initial laboratory results using V-BLAST space–time communication architecture," *Electronics Letters*, Vol. 35, No. 1, pp. 14–16.

4. 3GPP. 2004. "Physical layer procedures (FDD)," 3GPP technical specification, TS 25.214, Ver. 6.4.0, December 2004, Available http://www.3gpp.org.

5. ———. 2003. "Tx diversity solutions for multiple antennas (Release 6)," 3GPP TSG-RAN technical report, TR 25.869, Ver. 1.2.0, August 2003.

6. ———. 2001. "Physical layer aspects of UTRA high speed downlink packet access," 3GPP TSG-RAN technical report, TR 25.848, Ver. 4.0.0.

7. ———. 2004. "Multiple-input multiple-output in UTRA," 3GPP TSG-RAN technical report, TR 25.876, Ver. 1.3.0, February 2004.

8. S. Verdú. 1998. *Multiuser Detection*, New York: Cambridge University Press, 1998.

9. X. Wang and H. Poor. 2004. *Wireless Communication Systems. Advanced Techniques for Signal Reception*, Prentice Hall Communications Engineering and Emerging Technologies Series, T. Rappaport, ed., Upper Saddle River, NJ: Prentice Hall, 2004.

10. 3GPP. 2004. "Multiple-input multiple-output in UTRA," 3GPP TSG-RAN technical report, TR 25.876, Ver. 1.5.1, May 2004.

11. R. Knopp. 1997. "Coding and multiple-access over fading channels," Ph.D. dissertation, Ecole Poly-technique Fédérale de Lausanne.

12. D. Chase. 1973. "A combined coding and modulation approach for communication over dispersive channels," *IEEE Trans. Commun.*, Vol. 21, No. 3, pp. 159–174.

13. 3GPP. 2003. "HSDPA enhancements," 3GPP TSG-RAN technical report, TR 25.899 V0.1.0, 2003.

14. K. Pedersen, P. Mogensen, and J. Ramiro-Moreno. 2003. "Application and performance of downlink beamforming techniques in UMTS," *IEEE Commun. Mag.*, Vol. 41, No. 10, pp. 134–143, 2003.

15. E. Tiirola and J. Ylitalo. 2004. "Comparison of beamforming and diversity techniques in terms of UTRA FDD uplink capacity," in *Nordic Radio Symposium*, Aug. 2004.

16. S. Alamouti. 1998. "A simple transmitter diversity technique for wireless communications," *IEEE J. Select. Areas Commun.*, Vol. 16, No. 8, Oct. 1998, pp. 1451–1458.

17. Y. Sanada and Q. Wang. 1996. "A co-channel interference cancellation technique using orthogonal convolutional codes," *IEEE Trans. Commun.*, Vol. 44, No. 5.

18. H.C. Kwon, K.J. Kim, B.H. Park, and K.C. Whang. 1998. "Turbo coded CDMA system with an interference cancellation technique," *IEICE Trans. Commun.*, Vol. E81-B, No. 12.

19. H. Holma and A. Toskala, eds. 2001. *WCDMA for UMTS*, revised ed., Chichester, U.K.: John Wiley & Sons.

20. J. Laiho, A. Wacker, and T. Novosad. 2002. *Radio Network Planning and Optimization for UMTS*. Chichester, U.K.: John Wiley & Sons.

21. ITU. 1997. "Guidelines for evaluation of radio transmission technologies for IMT — 2000," Recommendation ITU-R.M.1225.

22. 3GPP. 2003. "Feasibility study for enhanced uplink for UTRA FDD," 3GPP TSG-RAN technical report, TR 25.896, Ver. 1.0.2, Nov. 2003.

23. S. Ariyavisitakul and L. Chang. 1993. "Signal and interference statistics of a CDMA system with feedback power control," *IEEE Trans. Commun.*, Vol. 41, No. 11, pp. 1626–1634.

24. A. Viterbi. 1995. *CDMA — Principles of Spread Spectrum Communications*, New York: Addison-Wesley.

25. J. Hämäläinen, K. Pajukoski, E. Tiirola, R. Wichman, and J. Ylitalo. 2004. "On the performance of multiuser MIMO in UTRA FDD uplink," *EURASIP Journal on Wireless Communications and Networking*, Dec. 2004, pp. 297–308.

26. GPP. 2003. "BS radio transmission and reception (FDD)," 3GPP TSG-RAN technical specification, TS 25.869, Ver. 6.3.0, Sept. 2003.

27. D. Gesbert, M. Shafi, D. Shiu, P. Smith, and A.A. Naguib. 2003. "From theory to practice: an overview of MIMO space-time coded wireless systems," *IEEE J. Select. Areas Commun.*, Vol. 21, No. 3, April 2003, pp. 281–302.

28. M. Varanasi and T. Guess. 1997. "Optimum decision feedback multiuser equalization with successive decoding achieves the total capacity of the Gaussian multiple-access channel," in *Asilomar Conference on Signals, Systems and Computers*, Vol. 2, pp. 1405–1409.

29. S. Chung, A. Lozano, and H. Huang. 2001. "Approaching eigenmode BLAST channel capacity using V-BLAST with rate and power feedback," in *IEEE Veh. Tech. Conf.*, Fall 2001, Vol. 2, pp. 915–919.

30. C.-N. Chuah, D. Tse, J. Kahn, and R. Valenzuela. 2002. "Capacity scaling in MIMO wireless systems under correlated fading," *IEEE Trans. Inform. Theory*, Vol. 48, No. 3, pp. 637–650.

31. R. Blum and J. Winters. 2002. "On optimum MIMO with antenna selection," *IEEE Comm. Lett.*, Vol. 6, No. 8, pp. 322–324.

32. L. Zheng and D. Tse. 2003. "Diversity and multiplexing: a fundamental tradeoff in multiple-antenna channels," *IEEE Trans. Inform. Theory*, Vol. 49, No. 5, pp. 1073–1096.

33. O. Tirkkonen and A. Hottinen. 2001. "Improved MIMO performance with non-orthogonal space–time block codes," in *Proc. IEEE GLOBECOM*, Vol. 2, Nov. 2001, pp. 1122–1126.

34. S. Bäro, G. Bauch, A. Pavlic, and A. Semmler. 2000. "Improving BLAST performance using space–time block codes and turbo decoding," in *Proc. IEEE GLOBECOM*, Vol. 2, Nov. 2000, pp. 1067–1071.

35. Texas Instruments. 2001. "Double-STTD scheme for HSDPA systems with four transmit antennas: link level simulation results," 3GPP TSG RAN WG1, 21(01)-0701, Release 5 Ad hoc, June 2001.

36. G. Caire and S. Shamai. 2003. "On the achievable throughput of a multiantenna Gaussian broadcast channel," *IEEE Trans. Inform. Theory*, Vol. 49, No. 7, July 2003, pp. 1691–1706.

37. S. Vishwanath, N. Jindal, and A. Goldsmith. 2003. "Duality, achievable rates, and sum-rate capacity of Gaussian MIMO broadcast channels," *IEEE Trans. Inform. Theory*, Vol. 49, No. 10, Oct. 2003, pp. 2658–2668.

38. P. Viswanath and D. Tse. 2003. "Sum capacity of the vector Gaussian broadcast channel and uplink-downlink duality," *IEEE Trans. Inform. Theory*, Vol. 49, No. 8, Aug. 2003, pp. 1912–1921.

39. M. Costa. 1983. "Writing on dirty paper," *IEEE Trans. Inform. Theory*, Vol. 29, No. 3, pp. 439–441.

40. Y. Wei and J. Cio. 2001.Trellis precoding for the broadcast channel," in *Proc. IEEE GLOBECOM*, Vol. 2, pp. 1344–1348.

41. R. Zamir, S. Shamai, and U. Erez. 2002. "Nested linear/lattice codes for structured multiterminal bin-ning," *IEEE Trans. Inform. Theory*, Vol. 48, No. 6, pp. 1250–1276.

42. J. Jing, R. Buehrer, and W. Tranter. 2003. "Spatial T-H precoding for packet data systems with scheduling," in *IEEE Veh. Tech. Conf.*, Fall 2003, Vol. 1, pp. 537–541.

43. Q. Spencer, A. Swindlehurst, and M. Haardt. 2004. "Zero-forcing methods for downlink spatial multiplexing in multiuser MIMO channels," *IEEE Trans. Acoust., Speech, Signal Processing*, Vol. 52, No. 2, pp. 461–471.

44. R. Knopp and P. Humblet. 1995. "Information capacity and power control in single cell multiuser communications," in *IEEE Int. Conf. on Comm.*

45. P. Viswanath, D. Tse, and R. Laroia. 2002. "Opportunistic beamforming using dumb antennas," *IEEE Trans. Inform. Theory*, Vol. 48, No. 6, June 2002, pp. 1277–1294.

46. M. Sharif and B. Hassibi. 2005. "On the capacity of MIMO broadcast channels with partial side information," *IEEE Trans. Inform. Theory*, Vol. 51, No. 2, pp. 506–522.

47. 3GPP. 2003. "Spatial channel model for multiple input multiple output (MIMO) simulations (Release 6)," 3GPP TSG-RAN technical report, TR 25.996, Ver. 6.1.0, Sept. 2003.

48. L. Schumacher, J. Kermoal, K.P.F. Frederiksen, A. Algans, and P. Mogensen. 2002. "MIMO channel characterization," IST, Tech. Rep. IST–1999–11729 ME-TRA, Feb. 2002, Available at http://www.ist-imetra.org/.

49. Lucent, Nokia, Siemens, and Ericsson. 2001. "A standardized set of MIMO radio propagation channels," 3GPP TSG RAN WG1 temporary document, R1–01–1179, Nov. 2001.

50. J. Fonollosa, R. Gaspa, X. Mestre, A. Pages, M. Heikkila, J. Kermoal, L. Schumacher, A. Pollard, and J. Ylitalo. 2002. "The IST METRA project," *IEEE Commun. Mag.*, Vol. 40, No. 7, pp. 78–86.

51. 3GPP. 2001. "Physical layer procedures (FDD)," 3GPP technical specification, TS 25.214, Ver. 4.0.0.

52. J. Hämäläinen and R. Wichman. 2000. "Closed-loop transmit diversity for FDD WCDMA systems," in *Asilomar Conference on Signals, Systems and Computers*, Oct. 2000.

53. G. Corral Briones, A. Dowhuszko, J. Hämäläinen, and R. Wichman. 2005. "Achievable data rates for multiple transmit antenna broadcast channels with closed-loop transmit diversity modes," in *IEEE Int. Conf. on Comm.*, Seoul, Korea, May 2005.

9

Multifunctional Reconfigurable Microelectromechanical Systems Integrated Antennas for Multiple Input Multiple Output Systems

Bedri Artug Cetiner

CONTENTS

9.1   Introduction

A wireless communications system that is capable of performing multiple functions utilizing a single architecture is defined as a multifunctional system. A system consisting of multiple subsystems, each of which performs a separate function, does not fall within this definition. In this context, a true multifunctional system must also be reconfigurable so that a single architecture can reconfigure itself for performing each function. A good example of a multifunctional wireless communications system is a software defined radio that enables a single device to operate over different communications and networking standards. A number of articles on multifunctional radio frequency (RF) systems can be found in the recent special issue in IEEE Transaction on Microwave Theory and Techniques [1].

This chapter focuses on a new class of antenna, henceforth referred to as microelectromechanical systems (MEMS) integrated multifunctional reconfigurable antenna (MRA), which can be employed in many wireless systems applications. In accordance with the definition of a multifunction system given above, an MRA is a single antenna that is capable of performing multiple functions by dynamically reconfiguring its architecture. Dynamic reconfiguration in the antenna architecture is accomplished by MEMS. Progress continues on many existing applications of MEMS technology, such as RF/wireless, biomedical, optical, and microfluidics, while new application areas such as fuel cells and power generators are also opening up. Establishment of MRAs requires a new RF MEMS technology, which has recently been developed by the author and others [2,3]. This technology will be introduced in the next sections. The implementation of MRAs in MIMO systems promises to further improve the system performance and is the topic of this chapter.

The research efforts on MIMO systems with associated transmission algorithms such as space–time codes (STCs) and spatial multiplexing (SM) aim at making the best use of limited and costly wireless bandwidth by exploiting the high spectral efficiencies offered by multiple antennas. The signal processing and space–time coding aspects of MIMO systems are discussed in Chapters 4 and 5. The motivation behind the topic of this chapter is the fact that there is additional room for further exploitation of the theoretical gains of MIMO systems when the antenna/electromagnetic aspects and the associated signal processing and coding aspects are integrated together in a multidisciplinary approach [4]. The performance of MIMO systems depends on several parameters such as the following:

- The physical structure of the channel: scattering density and disposition of the scatterers
- The space–time processing algorithms for MIMO channels: spatial multiplexing (SM), space–time coding (STC), and beamforming
- The antenna array configuration and element properties: radiation pattern, polarization, operation frequency, and input impedance

Due to the time varying nature of the wireless channel, the true benefits of MIMO can be exploited only through a smart design that is able to respond to the channel dynamics. The ability of the network to respond to the channel may be expressed in the following terms.

- **Shaping the Channel Statistics:** The ability to shape the channel statistics in a way that the spatial dimensions can be resolved. This allows multiple data pipes to simultaneously carry information across spatial dimensions and between various users while facing the minimum interference. Key to resolvability is the ability to reconfigure the antenna radiation pattern, polarization, operation frequency.

- **Adaptability:** The ability to adapt to the channel in an opportunistic way where the spatial dimensions associated with good channel conditions are utilized more than the spatial dimensions affected by poor fading conditions. Key to adaptability is the scheduling algorithm as well as the adaptive algorithm that chooses beamforming, or the best space–time coding strategy for the channel condition [5].

While adaptability is important to MIMO systems in general, resolvability is especially important for multiuser MIMO systems where the interference presented by simultaneous communications is the primary bottleneck. It is therefore crucial in a wireless network to first shape the channel statistics and then to use link adaptation algorithms that jointly optimize the modulation level, coding rate, and the transmission signaling schemes such as SM, STC, and beamforming [5]. However, MIMO systems are constrained to employ fixed antenna parameters, which are determined by the initial antenna design, over the varying channel condition. *Thus, non-reconfigurable antenna designs are unable to shape the channel statistics to minimize interference.*

By treating the antenna element properties and array configuration as an additional component in the joint optimization of the adaptive system parameters, an additional degree of freedom is achieved. The goal of joint optimization of antenna array properties and the associated transmission algorithm can only be achieved if the structural geometry of each individual element of the array can be dynamically reconfigured. MRAs can dynamically reconfigure their structural geometry and thus are capable of altering radiation, polarization, and frequency characteristics to adapt to the changes in their operating environments. Therefore, an adaptive MIMO system equipped with MRAs will not be constrained to employ a fixed antenna design over varying channel conditions. This feature will permit the selection of the best antenna properties and configuration in conjunction with the adapted transmission scheme with respect to the channel condition. Thus the gap between theoretical MIMO performance and practice is minimized. Figure 9.1 highlights the performance improvement offered by an adaptive MIMO system equipped with MRAs.

In the next section, first the MRA concept is discussed. Then the interrelationships among transmission signaling schemes, physical channel conditions, and antenna radiation/polarization properties are identified so that the best antenna design for a given transmission-scheme and channel condition is always selected. The following section discusses the RF MEMS

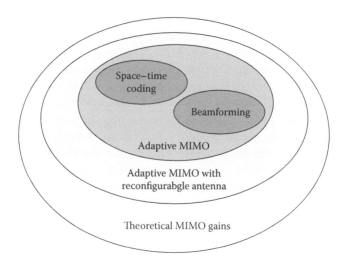

FIGURE 9.1
Potential benefit of adaptive MIMO systems employing MRA arrays.

technology compatible with microwave laminates for constructing MRA systems. Finally, a discussion of two different MEMS integrated antennas is given.

9.2 MRA Concept and Its Links with MIMO Systems

9.2.1 Multifunctional Reconfigurable Antenna

In the literature, a smart, intelligent, or adaptive antenna refers to an antenna array of elements that are typically standard monopoles, dipoles, or patches. The antenna elements themselves do not possess any intelligence. Intelligence is associated with the signal-processing domain where the *time* domain signals from or to the individual antenna elements are weighted and combined such that the resulting radiation pattern (i.e., the *spatial* response of the array) satisfies some conditions. This is the key concept of beamforming through which the electromagnetic energy is focused in the direction of the desired signal, with a null placed in the direction of noise or interference sources. On the other hand, an MRA array consists of antenna elements, each of which has some *intelligence*. This *intelligence* stems from the ability to reconfigure the physical structure of individual elements through which polarization/radiation and frequency properties of the array are changed. The elements in the MRA array have the ability to intelligently process the signals in *spectral* and *angular* domains, adding to the already present *time* domain processing of the system. In the language of phased-array antennas, an MRA array alters not only the array factor but also the element factor. In current phased-array technology, the element factor cannot be modified once individual elements are laid out. The total electric field of a linear array of N identical antenna elements with equal spacing can be given as the sum of the fields due to the N elements at the observation point P (R_0, θ, ϕ):

$$E(R_0, \theta, \phi) = \sum_{i=0}^{N-1} E_i(R_i, \theta, \phi) = f_e(\theta, \phi) \left(\frac{e^{-jkR_0}}{R_0} \right) \times \left[\sum_{i=0}^{N-1} a_i e^{j\psi_i} e^{jikd\cos\theta} \right] \quad (9.1)$$

where R_0 is the distance between the center of the array and the observation point, R_i is the distance between ith antenna element and the observation point, j denotes the square root of -1, a_i and ψ_i represent the amplitude and phase of the excitation, respectively, giving rise to E_i relative to a reference excitation, and $f_e(\theta, \phi)$ represents the element factor. Since the elements are identical, the element factor $f_e(\theta, \phi)$ is the same for all the elements and, hence, this term can be factored out from the summation resulting in the simple expression given in Equation 9.1. As is seen from Equation 9.1, the radiation pattern of a classical phased array antenna is changed only by variation of the amplitude and/or phase of the excitation (a_i and ψ_i). In an MRA array, however, elements are not necessarily the same and do not exhibit the identical directional patterns that result in variable element factor. For example, an MRA array may be reconfigured into a polarization diversity scenario in which the elements would radiate different senses of polarization, such as linear, right-hand circular, and left-hand circular. An MRA that can change its polarization between right-hand and left-hand circular polarization will be introduced in Section 9.4. Block diagrams of a conventional smart antenna array and a reconfigurable antenna array are provided in Figure 9.2a and Figure 9.2b, respectively.

In a reconfigurable array, the antenna element spacing can also be changed allowing efficient selection and application of beamforming and space–time coding schemes. While beamforming requires antennas to be closely spaced (antennas are correlated) to avoid the negative effects of side lobes, space–time coding will perform well if the spacing between antennas is large enough to ensure low correlation. As a result, a MIMO system with reconfigurability in the geometrical domain of antenna will not be constrained to use the same antenna design over varying channel conditions, resulting in better utilization of the available channel capacity.

Finally, a reconfigurable antenna is also advantageous in terms of the physical scale of the antenna systems. In today's miniature, compact, and highly integrated telecommunication devices, the area devoted to antenna elements is typically very limited. This has prompted the antenna community to actively research small efficient antenna design [6,7]. An antenna is said to be an electrically small antenna if it can fit inside a sphere of radius $a = 1/k$, where k is the wave number ($k = 2\pi/\lambda$, λ is the wavelength) associated with the radiated electromagnetic field. A rule-of-thumb formula that relates

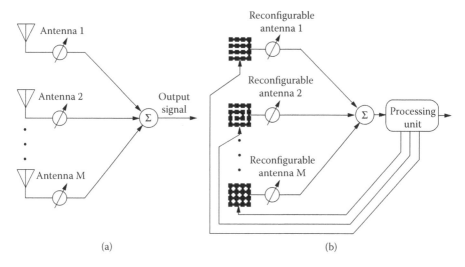

FIGURE 9.2
Block diagrams of (a) conventional and (b) MRA arrays.

the bandwidth (*BW*), efficiency (η), and the volume of small size antennas is given by

$$BW \times \eta \cong \left(k \times a\right)^3 \cong \frac{1}{Q} \tag{9.2}$$

where Q is the minimum radiation quality factor of the antenna, which increases by the reduction in antenna volume. From this formula, one can conclude that there is a compromise between the maximum realizable performances, i.e., bandwidth and efficiency, and the given electrical dimensions of small size antenna. The concept of RF MEMS integrated MRA is a revolutionary approach to circumventing the performance limitations of small size antennas by taking advantage of combined multiple functions in one single antenna. An MRA combines multiple functions in one single antenna architecture with the capability of altering its radiation, polarization, and frequency characteristics. Thus, for given antenna performance characteristics an MRA will occupy only a small fraction of the space required by single-function multiple antenna elements.

9.2.2 Links among Transmission Algorithms, Antenna Properties, and Propagation Environment

In this section, we identify relationships among the transmission algorithms, the radiation/polarization characteristics, and the configuration of the reconfigurable antenna, and the propagation environment. These relationships enable the joint adjustment of the characteristics of the reconfigurable antenna array and the coding schemes over varying channel conditions to optimize performances at all times.

1. *Number of antenna elements:* The most basic relationship, which does not require reconfigurability in the geometrical domain of the multiple antennas, relates the number of antenna elements to a specific transmission algorithm. For multiple antennas, if the number of transmit antennas is larger than two, then it is not possible to design orthogonal space–time block codes (STBCs) [7]. In case of more than two antennas, recently developed quasi-orthogonal STBCs are used to achieve full rate and full diversity at the expense of slight increase in decoding complexity [8]. For high SNRs and very large number of antennas SM is favorable over STBCs, since the data rate of SM increases linearly with an increasing number of antennas, while the diversity gain of STBC will saturate.

2. *Array configuration and polarization:* Besides the number of antenna elements, the subset of the elements selected in an array configuration is an important factor to achieve the majority of the capacity available in the channel [9]. This does not only improve performance but also results in a less complex MIMO system as the number of the transmit and receive RF chains are reduced. The performance can be further enhanced if the polarizations of the elements are also taken into account as the propagation of the electric field for different polarizations differs depending on the environment. It has been shown experimentally that for a line-of-sight (LOS) indoor environment, vertically polarized systems achieve higher capacity than horizontally polarized ones [10]. Moreover, an antenna array with hybrid polarization (i.e., some elements are vertically polarized while others are horizontally polarized) performs better than single polarization systems for both LOS and non-LOS (NLOS) conditions.

In practical communication scenarios, degenerate channel phenomena called the *keyhole channel effect* may arise where the antenna elements both at the receiver and the transmitter have very low correlation due to rich scattering, but the channel matrix has a very low rank, resulting in a single mode of communication [11]. This shows that low correlation itself is not a guarantee for achieving high capacity. It is shown in [11] that in an outdoor propagation scenario, the keyhole problem may be avoided by using a horizontally oriented transmitter array instead of a vertically oriented array. As a consequence, both the array configuration and the polarization of each individual element need to be adaptive in order to maintain the channel performance over varying characteristics of the propagation environment. A reconfigurable antenna array that can change its configuration and polarization has the characteristics necessary to adapt variable transmission/receiving environments.

1. *Spatial and polarization antenna diversity:* A compromise between data rate maximization and diversity maximization (i.e., choosing between SM and STC) is critical in realizing MIMO gains, since the performance of these signaling strategies is strongly dependent on time-varying channel characteristics. As is known, SM performs particularly well in high SNR regions, while STBC has better performance in a low SNR region. It is shown in [12] that, while having multiple linear polarization diversity antennas at both ends of the link degrades the performance of STBC from that of spatial diversity, significant improvements in the symbol error rate for a SM scheme are achieved in certain channel conditions such as in environments with high scattering density and with a high K-factor. This leads to an important conclusion: A reconfigurable antenna array that can readily switch between polarization and spatial diversity schemes is needed to optimize an antenna performance for a given coding scheme (i.e., SM or STBC) in a given channel environment.

2. *Beamforming, MIMO with STC:* When only the receiver knows the channel, STC achieves the maximum diversity in a system with multiple transmit antennas. On the other hand, if the transmitter knows the channel perfectly, beamforming is the optimal solution. In some practical cases, the transmitter has some information about the channel (e.g., the mean or variance) instead of a perfect knowledge. When side information is available at the transmitter, it can be exploited to enhance the performance. Even when the channel information is based on poor channel estimation, its use improves the performance of the system in combating fading. The improvement can be achieved by combining STC and beamforming. Typically, when the quality of the channel feedback is high, the diversity rank is less critical and the transmitter should lay most energy on the "good" beam. On the other hand, when the feedback is unreliable, we should rely more on diversity and distribute energy evenly among different beams. In the extreme case when the channel feedback quality is so poor it is entirely independent of the actual situation, the system becomes an open loop system and the beamforming scheme should gradually fall back to nonbeamformed traditional space–time coding. Therefore, the performance of the scheme should be similar to that of the original space–time code. This requires the design of an adaptive system that can utilize the partial available channel information to change its behavior and provide the optimal performance in all cases [13,14]. Such an adaptive system should converge to space–time coding when the transmitter does not know the channel at all and to beamforming when the transmitter knows the channel perfectly. Performance can be improved further if the optimal array design is employed simultaneously. While antenna elements are closely spaced and correlated in beamforming arrays, MIMO

systems employing STCs require large antenna spacing for uncorrelated antennas. It is also important to note that beamforming is more
effective if the propagation environment has low scattering density
(i.e., LOS or near LOS) resulting in fewer multipaths. Weight selection algorithms can be more easily optimized for a few multipaths
than for many. In contrast, STCs take advantage of the multipath
richness by maximizing data rate or diversity. Variation in antenna
element separation is achieved by employing MRA instead of antennas with fixed properties.

9.3   RF MEMS Technology Compatible with Microwave Laminates for Fabricating MRAs

As indicated earlier, MEMS technology is gaining a greater foothold in many
existing applications, such as RF wireless communications, biomedical, and
optical, to name a few; at the same time, it is finding new applications such
as fuel cells and power generators. In particular, RF MEMS have had a
significant impact due to their potential to revolutionize RF and microwave
system implementation for the next generation of communication applications. One of the first impacts of RF MEMS was single MEMS switches [15]
with their excellent switching characteristics — very low insertion loss, very
low power requirements, and high isolation —which cannot be attained by
semiconductor switches. The true potential of RF MEMS goes beyond a
single switch. The key advantage of RF MEMS can be realized by system-
level implementation through monolithic integration capability with other
circuit components. This capability is key, in particular, for creating multifunctional reconfigurable antenna systems employing a large number of RF
MEMS components and/or operating at high frequencies. Although RF MEMS
devices on semiconductor substrates (tunable RF matching circuits/filters,
variable capacitors, phase shifters, switches) have been demonstrated
[16–19], realization of monolithically integrated RF MEMS antenna systems
requires common substrate housing of both MEMS and antennas. Microwave
laminate printed circuit boards (PCBs) are commonly used for the design of
RF circuits and antennas and represent an ideal substrate for RF MEMS
device integration. However, PCB processing has been historically difficult
due to the process limitations imposed by PCBs, such as low temperatures
and non-planar surfaces. In order to overcome these problems, we recently
developed an RF MEMS technology compatible with microwave laminate
PCBs that overcomes the drawbacks of the silicon-based MEMS technology
in establishing reconfigurable antennas with low cost and high performance.
The details of PCB compatible RF MEMS technology and associated fabrication processes are beyond the scope of this chapter and can be found in

references [2,3,20]. The main advantage of this technology lies in allowing the monolithic integration of the antenna and the switches on the same substrate as part of the same lithographic process. This approach eliminates the need for wire bonds to be interconnected to the RF MEMS switches, thus simplifying the matching circuit as well, while reducing the switch parasitics. Figures 9.3 and 9.4 compare two MRA systems realized by silicon-based and microwave laminate compatible RF MEMS technologies in terms of complexity, cost, and performance. It is obvious from these figures that the latter technology provides higher performance, less complexity, and low cost compared to the former technology.

9.4   RF MEMS Integrated Antennas

In this section, the design, microfabrication, and results of two different RF MEMS integrated antennas will be presented. The first design is a multi-element selection diversity antenna (i.e., cactus antenna) in which RF MEMS actuators are monolithically integrated with antenna feed lines to selectively route the RF feed signal. In the second design, RF MEMS actuators are integrated within the geometrical structure of the antenna to construct a multifunctional reconfigurable spiral antenna capable of dynamically changing its polarization between right-hand and left-hand circular polarizations.

9.4.1   Three-Element Selection Diversity Antenna

We first designed a single antenna, which will henceforth be referred to as the cactus antenna. The geometry of the cactus is depicted in Figure 9.5. It consists of two half-wavelength inductively coupled coplanar waveguide (CPW) slots combined with the half-wavelength capacitively coupled slot antenna, which is fed through a triangular-shaped half-wavelength CPW strip. The inductively coupled slots are bent upward for compactness, necessary in multi-element antenna systems. This structure produces three different resonant frequencies, f_{r1}, f_{r2}, and f_{r3}, defined by the lengths of the two slots (L_1, L_2) and the triangular-shaped CPW line (L_3), respectively (see Figure 9.5). These lengths are determined based on the half-wavelength resonance criterion and are approximately calculated using Equation 9.3. By optimizing L_1, L_2, and L_3 in conjunction with other design parameters (see Figure 9.5), resonant frequencies of f_{r1}, f_{r2}, and f_{r3} are located such that the cactus antenna possesses 42% impedance bandwidth for a voltage standing wave ratio (VSWR) of 2.

$$L_i = \frac{\kappa_i c}{2\sqrt{\varepsilon_{reff}}\, f_{ri}} ; \, 0.8 \leq \kappa_i \leq 1, \, i = 1, 2, 3 \qquad (9.3)$$

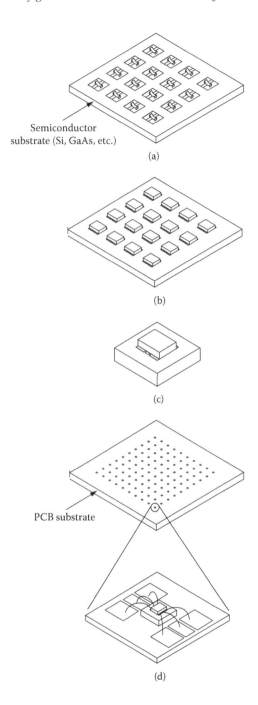

Semiconductor
substrate (Si, GaAs, etc.)

(a)

(b)

(c)

PCB substrate

(d)

FIGURE 9.3

Process flows for silicon-based RF MEMS technology for realizing MRA systems, (a) Fabrication of MEMS switches on a semiconductor, (b) Packaging individual MEMS switches, (c) Dicing, (d) Assembling MEMS and antenna elements through wire bonding (or flip-chip) on PCB to construct the MRA.

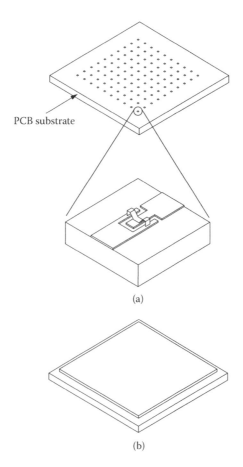

PCB substrate

(a)

(b)

FIGURE 9.4
Process flows for microwave laminate compatible RF MEMS technology for realizing MRA systems, (a) Fabrication of monolithically integrated MEMS antenna on PCB, (b) System level packaging.

In Equation 9.3, c is the speed of light in free space and ε_{reff} represents the effective dielectric constant of the dielectric supporting material. κ_i is a coefficient and ranges from 0.8 to 1 [21].

Next, a three-element selection diversity cactus antenna with RF MEMS switches monolithically integrated on CPW feed network is fabricated on RT/Duroid 5870 (see Figure 9.6). This material is particularly suitable for antenna applications due to its low dielectric constant and low loss property in the desired frequency range ($\varepsilon_r = 2.33$, $\tan\delta = 0.0005$ @ 5 GHz) [22]. The switches used are capacitive and have a structure similar to those previously published [15].

Three RF MEMS switches located on the CPW feed, single-pole three-throw switch (SP3T), route the input to one of the branches in order to maximize the signal-to-noise-ratio (SNR) of the diversity signal at the receiver

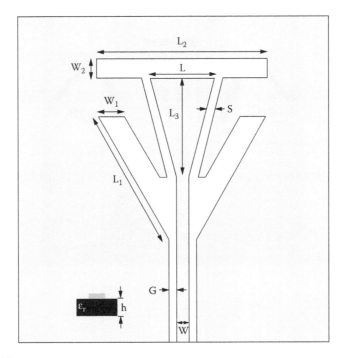

FIGURE 9.5
Top view of the CPW-fed antenna geometry and dimensions in mm. $L_2 = 25$, $W_2 = 3.5$, $L_1 = 28.6$, $W_1 = 4.8$, $L_3 = 18$, $L = 12$, $S = 0.15$, $G = 0.25$, $W = 2.6$, $D = 15$, $h = 1.52$, α: orientation angle.

(see Figure 9.6b). The distances from the centers of switches to the cross-junction are designed to be quarter-wavelength in the substrate at the center design frequency of 5 GHz. Thus, at the center frequency, the diversity branch with switch in the down position is transformed into an open circuit at the cross-junction due to the quarter-wave transmission line. This mechanism slightly narrows the impedance bandwidth of the antenna since the signal propagating on the selected diversity branch sees two quarter-wavelength open transmission lines of the two disconnected diversity branches at the junction. High impedance quarter-wavelength lines using the same concept are utilized to actuate the switches. In this case, the bandwidth degradation is very small due to the combined effect of the quarter-wave line and high impedance of the line. In the current design, a bias voltage of 40V is required to actuate the switch. Three metal-insulator-metal (MIM) capacitors located on the CPW line are employed to decouple the RF signal path from the bias voltage. Air bridges are used to ensure the continuity in the CPW ground planes and to suppress possible slotline modes excited at the junctions. The fabrication of air bridges, as well as MIM capacitors, is compatible with the monolithic process.

The fabrication procedures of the cactus antenna integrated with RF MEMS switches and that of the MRA that will be presented in the next section are very similar. Therefore, only a brief summary of the fabrication process is

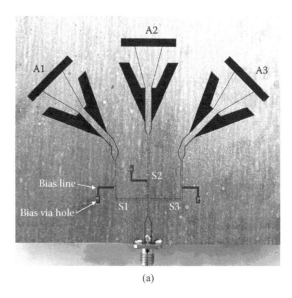

(a)

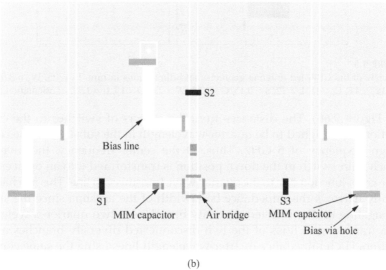

(b)

FIGURE 9.6
(a) Photograph of the selection diversity cactus antenna integrated with RF MEMS switches; S1, S2, S3 are RF MEMS switches; A1, A2, A3 are cactus antennas; (b) Schematic of the feeding structure, single-pole-three-throw switch (SP3T).

given here. The process starts with standard via hole formation for switch biasing. Ease of via layout is an advantage of this process over existing technologies, where deep vertical vias are normally difficult to form. Via holes through the substrate connect the central electrode of each switch to the bias line placed on the backside of the antenna. This separation does not only create a space for multiple antennas but also helps increase the isolation

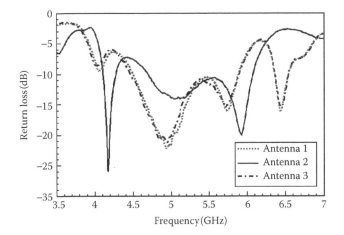

FIGURE 9.7
Measured return loss of individual selected cactus antenna.

between antenna elements and biasing circuit. This, in particular, is an important advantage as the circuit complexity and the number of switches increase. Next, antennas and all feed lines are formed by a simple wet etch process. Finally, MEMS switches, MIM capacitors, and air bridges are monolithically fabricated.

Results are presented in the following figures. Figure 9.7 shows the return losses of each switched antenna. As mentioned earlier, the two quarter-wavelength lines narrow the input impedance bandwidth of the cactus antenna from its original 42% to 30%. However, the covered bandwidth is still very large for most wireless local area network (WLAN) applications. Figure 9.8 illustrates the measured x-y plane co-polar radiation patterns at the center design frequency of 5 GHz, normalized with respect to the same reference, corresponding to the sequential and individual activation of each antenna. As a result of the inter-element 45° rotation angle, radiation patterns with 45° rotation are clearly distinguished in the figure. Appropriate activation leads to angular or polarization discrimination of the signal of interest in multipath scenarios. For example, the axis of the active element can be oriented to the direction of strong interference.

9.4.2 Multifunctional Reconfigurable Spiral Antenna

9.4.2.1 *Antenna Structure and Operational Mechanism*

In the three-element cactus antenna design the role of RF MEMS switches, which are located on antenna feed lines, are limited to selectively routing the RF feed signal. In this design, a number of RF MEMS actuators are monolithically integrated within the geometrical structure of the antenna to construct a multifunctional reconfigurable spiral antenna (MRSA). In other

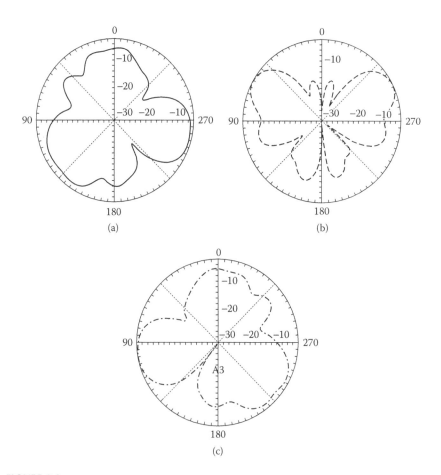

FIGURE 9.8
Measured co-pol radiation patterns at 5GHz of each antenna element in the x-y plane when
the three switches are sequentially activated.

words, RF MEMS actuators are used as part of the physical structure of the
antenna, owing to the monolithic integration capability of the processing
technique, providing a large degree of structural reconfigurability.

MRSA architecture is built on a number of printed rectangular-shaped
metal strips interconnected by RF MEMS actuators on a microwave laminate
PCB substrate, RO4003-FR4 (ε_r = 3.38, tanδ = 0.002). Shown in Figure 9.9a
are two adjacent strips interconnected by a RF-MEMS actuator, which is
made of a metallic movable membrane, suspended over a metal stub protruding from an adjacent strip, fixed to both ends of the strip through metallic
posts. The optimized height of these posts was found to be 8 µm for a good
tradeoff between up position switch coupling and actuation voltage. Metal
stubs are covered by silicon-nitride (SiN$_x$) film to prevent metallic membrane
from sticking onto the stub upon contact. This film also provides a capacitive
contact for the actuator down state isolating RF signal from DC. A DC bias

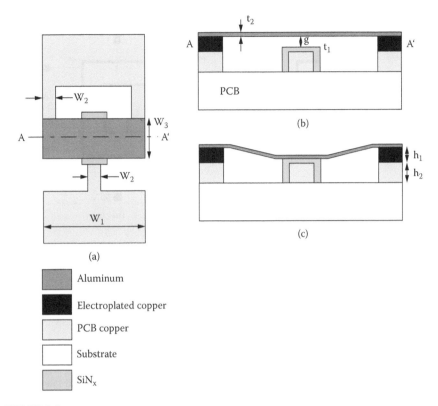

(a)

(b)

(c)

Aluminum

Electroplated copper

PCB copper

Substrate

SiN$_x$

FIGURE 9.9
RF MEMS actuator interconnecting two adjacent metallic strips (a) top view; width of metal strip, w_1 = 800 μm; width of stub, w_2 = 100 μm; width of membrane, w_3 = 150 μm; (b) side view (up position); thickness of nitride, t_1 = 0.2 μm; thickness of membrane, t_2 = 0.5 μm; air gap, g = 7.8 μm; (c) side view (down position); thickness of electroplated copper, h_1 = 8 μm; thickness of PCB copper, h_2 = 16 μm.

voltage of approximately 40 V applied between the membrane and the stub causes an electrostatic force that pulls the suspended membrane on top of the stub (actuator down state or actuator on, see Figure 9.9c), and the actuator connects the strips; otherwise the strips are disconnected (actuator up state or actuator off; see Figure 9.9b). Judicious activation of interconnecting actuators, i.e., by keeping some of the actuators in the up position (zero bias) while activating the rest of them by applying DC bias voltages, allows the reconfigurable spiral to configure its architecture into single arm rectangular spirals with opposite winding sense of the spiral, left or right senses (see Figures 9.10a and b). Accordingly, right- and left-hand circularly polarized (RHCP and LHCP) radiation is achieved. In Figures 9.10a and b, for the clarity of illustration, each configured geometry is depicted separately and actuators in the up state are shown without metallic membrane. The antenna is fed by a single coaxial probe, as shown in Figure 9.10c. The supply voltage is connected to the proper locations on the antenna segments through resistive

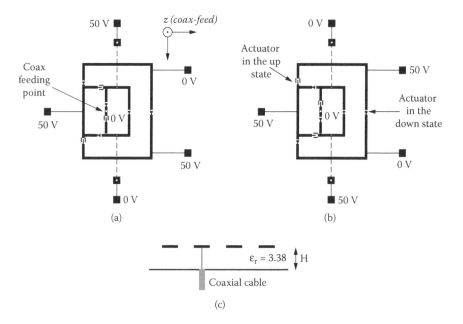

FIGURE 9.10
Schematics of the single arm rectangular spiral antennas, which are reconfigured from the MRSA architecture by judicious activation of the interconnecting RF MEMS actuators, for (a) left-hand circular polarization, (b) right-hand circular polarization, (c) side view of the antenna. The outermost dimensions of the antenna are 9 ∞ 12 (mm), the spiral line width is 0.8 mm.

bias lines so as to prevent the RF signal from being shorted by the DC power supply.

MSRA is designed to radiate an axial beam of RHCP and LHCP fields. It is known that a single-arm rectangular spiral antenna with outermost arm peripheral length (circumference) of C,

$$1\lambda_{eff} < C < 2\lambda_{eff} \tag{9.4}$$

excites only the first radiation mode, giving rise to an axial beam of circular polarization [23], where $\lambda_{eff} = \lambda_0/[(\varepsilon_r + 1)/2]^{1/2}$ is the effective wavelength of the current traveling on the spiral. The number and size of the strips are optimized so that circumference of the antenna, $C = 42$ mm $= 1.04 \lambda_{eff}$, satisfies Equation 9.4 and a minimum number of actuators with associated bias circuitries are needed.

9.4.2.2   *Microfabrication and Results*

The MRSA is microfabricated on a microwave laminate substrate RO4003-FR4, which is conductor backed to ensure that the antenna radiates broadside

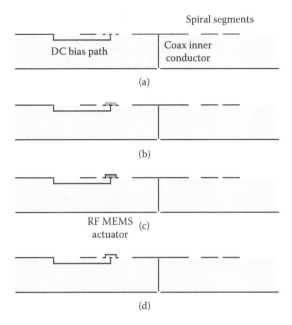

FIGURE 9.11

Fabrication sequence for monolithic integration of RF MEMS actuators with rectangular-shaped strip segments in constructing MRSA (a) Antenna pattern, DC bias path and via formation (b) Dielectric layer deposition and sacrificial layer planarization, (c) Aluminum membrane deposition, (d) Final release.

to the printed spiral surface. Substrate thickness is chosen to be 7.6 mm, which is one of the standard thicknesses for PCB family substrates, the closest one to the quarter-wavelength at a center design frequency of 5 GHz. Theoretical characterization of the antenna structure is conducted by a full-wave analysis tool based on a finite element method, which takes into account the edge effects due to finite size dielectric and conducting plane of the antenna.

A brief fabrication sequence for monolithic integration of RF MEMS actuators with rectangular-shaped strip segments of the MRSA is given in Figure 9.11. Details of the fabrication process can be found in references [2,3,20]. The fabrication begins with RO4003 laminate with copper layers of 16 μm on both sides. First, the segments of the antenna and planar part of the bias circuitry are formed by wet etching copper layer. Vertical vias for bias circuitry and coax feed are created by standard PCB processes. After this step, a thin layer of high-density inductively coupled chemical vapor deposition (HDICP CVD) SiN_x [20] is deposited and etched by reactive ion etching such that the SiN_x covers only the tips of the metal stubs protruding from the antenna segments (see Figure 9.9a). We continue fabricating RF MEMS actuators following the process flow shown in Figures 9.11b–d without affecting the antenna structure.

Figure 9.12 shows the return loss of the MRSA with counterclockwise sense of winding corresponding to the RHCP radiation. The simulated result is

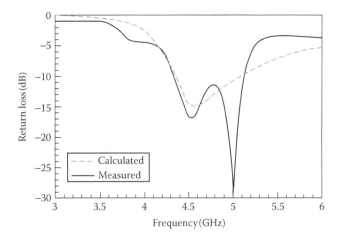

FIGURE 9.12
Return loss of the MRSA for RHCP radiation.

also validated by comparison with experimental data in this figure. Due to the symmetry between two antenna configurations, the RHCP and LHCP spirals exhibited almost identical return loss with a VSWR of less than two covering the frequency band of 4.3–5.4 GHz. Measured and calculated radiation patterns at 5 GHz in two different planes of $\phi = 0°$ and $90°$ are shown in Figures 9.13a–d for RHCP and LHCP configurations. As seen from these figures, the MRSA radiates circularly polarized wave slightly off broadside to the plane of spiral, forming an almost axial beam pattern. This slight tilt from the z-axis is due to the asymmetry of the antenna structure with respect to the z-axis. The measured average half-power beamwidth (HPBW) is approximately 105°. The antenna radiates almost entirely circular wave in the z-axis with an axial ratio value of 0.9 dB. The gain at this direction is 5.3 dB. Variations of axial ratio and gain in the z-direction with respect to frequency are shown in Figure 9.14. The circular polarization bandwidth over which the axial ratio is less than 3 dB is approximately 11%. Gain of the antenna with average value of 4.9 dB shows small variation over this bandwidth. The difference in performance characteristics between the RF MEMS integrated spiral antenna and conventional single-arm rectangular spiral antenna was observed to be negligible.

9.5   Concluding Remarks

Wireless applications that are increasingly bandwidth- and mobility-intensive have driven MIMO research to challenge the physical limits of coding and signaling. The multifunctional reconfigurable antenna technology presented in this chapter greatly impacts adaptive MIMO performance through

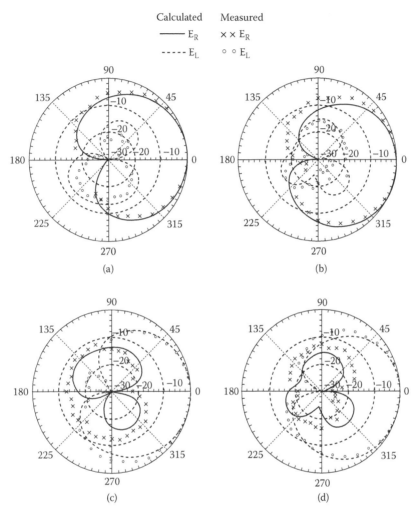

FIGURE 9.13
Radiation patterns for the RHCP configuration in (a) $\phi = 0$ plane and (b) $\phi = 90$ plane; radiation patterns for the LHCP configuration (c) $\phi = 0$ plane and (d) $\phi = 90$ plane at 5 GHz.

the capability to change its performance properties (i.e., radiation and impedance characteristics). The reconfigurable antenna properties integrated with signaling schemes (i.e., spatial multiplexing, space–time coding, beam-forming) to the propagation environment provides an additional degree of freedom in adaptive optimization; thus the gap between theoretical MIMO performance and practice is closed. RF MEMS technology compatible with microwave laminates enables very large-scale monolithic integration of antenna and circuit components on a substrate that best meets the antenna performance characteristics. MEMS integrated antennas; a diversity antenna (three-element cactus) suitable for selection diversity; and a multifunction reconfigurable spiral antenna have been fabricated by using this technology.

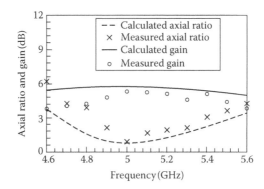

FIGURE 9.14
Frequency responses of axial ratio and gain in the z-direction for MRSA.

A long-awaited design space where an interplay between reconfigurable antenna and adaptive signaling feed back to each other is likely to revolutionize broadband MIMO system design methodology.

References

1. Special Issue on Multifunctional RF Systems. 2005. *IEEE Trans. Microwave Theory and Techniques*, Vol. 53, No. 3.
2. H.P. Chang et al. 2003. "Low cost RF MEMS switches fabricated on microwave laminate PCBs," *IEEE Electron Device Lett.*, Vol. 24, No. 4, pp. 227–229.
3. B.A. Cetiner et al. 2003. "Monolithic integration of RF MEMS switches with a diversity antenna on PCB substrate." *IEEE Trans. Microwave Theory and Techniques*, Vol. 51, No.1, pp. 332–335.
4. B.A. Cetiner et al. 2004. "Multifunctional reconfigurable MEMS integrated antennas for adaptive MIMO systems," *IEEE Communications Magazine*, Vol. 42, No. 12, pp. 62–70.
5. S. Catreux et al. 2002. "Adaptive modulation and MIMO coding for broadband wireless data networks," *IEEE Communications Magazine*, Vol. 40, No. 6, pp. 108–115.
6. L. Jofre, B.A. Cetiner, and F. De Flaviis. 2002. "Miniature multi-element antenna for wireless communications," *IEEE Trans. on Antennas and Propagat.*, Vol. 50, No. 5, pp. 658–669.
7. V. Tarokh, H. Jafarkhani, and A.R. Calderbank. 0999. "Space-time block coding from orthogonal design," *IEEE Trans. on Inform. Theory*, Vol. 48, pp. 611–627.
8. H. Jafarkhani. 2001. "A quasi-orthogonal space-time block code," *IEEE Trans. Commun.*, Vol. 49, pp. 1–4.
9. M.A. Jensen and J.W. Wallace. 2003. "Antenna selection for MIMO systems based on information theoretic considerations," *2003 IEEE AP-S International Symposium Digest*, Vol. 2, pp. 515–518.

10. P. Kyritsi et al. 2003. "Effect of antenna polarization on the capacity of a multiple element system in an indoor environment," *IEEE JSAC*, Vol. 20, No. 6, pp. 1227–1239.

11. D. Chizhik, et al. 2002. "Keyhole, correlations, and capacities of multielement transmit and receive antennas," *IEEE Trans. on Wireless Comm.*, Vol. 1, No. 2, pp. 361–368.

12. R.U. Nabar et al. 2002. "Performance of multiantenna signaling techniques in the presence of polarization diversity," *IEEE Trans. on Signal Processing*, Vol. 50, No. 10, pp. 2553–2562.

13. L. Liu and H. Jafarkhani. 2005. "Application of quasi-orthogonal space-time block codes in beamforming," *IEEE Trans. on Signal Processing*, Vol. 53, No. 1, pp. 54–63.

14. G. Jongren, M. Skoglund, and B. Ottersten. 2002. "Combining beamforming and orthogonal space-time block coding," *IEEE Trans. on Inform. Theory*, Vol. 48, pp. 611–627.

15. Z.J. Yao et al. 1999. "Micromachined low-loss microwave switches," *J. Microelectromech. Syst.*, Vol. 8, No. 2, pp. 129–134.

16. J. Papapolymerou et al. 2003. "Reconfigurable double-stub tuners using MEMS switches for intelligent RF front-ends," *IEEE Trans. Microwave Theory and Techniques*, Vol. 51, No. 1, pp. 271–278.

17. N.S. Barker and G.M. Rebeiz. 1998. "Distributed MEMS true-time delay phase shifters and wide-band switches," *IEEE Trans. Microwave Theory and Techniques*, Vol. MTT-46, No. 11, pp. 1881–1890.

18. C.L. Goldsmith et al. 1999. "RF MEMS variable capacitors for tunable filters," *Int. J. of RF and Microwave Computer-Aided Eng.*, Vol. 9, No. 4, pp. 362–374.

19. E.R. Brown. 1998. "RF-MEMS switches for reconfigurable integrated circuits," *IEEE Trans. Microwave Theory and Techniques*, Vol. MTT-46, No. 11, pp. 1868–1880.

20. C.H. Chang et al. 2002. "RF MEMS capacitive switches fabricated with HDICP CVD SiN_x," *IEEE MTT-S Dig.*, pp. 231–234.

21. R. Garg et al. 2001. *Microstrip Antenna Design Handbook*, Norwood, MA: Artech House, chap. 7.

22. Rogers Inc., Chandler, AZ, Microwave Products Tech. Information, 2001.

23. H. Nakano et al. 2002. "Tilted- and axial-beam formation by a single-arm rectangular spiral antenna with compact dielectric substrate and conducting plane," *IEEE Trans. Antennas and Propagation*, Vol. 50, No. 1, pp. 17–23.

10

Multi-Antenna Testbeds for Wireless Communications

Raghu Rao, Christian Oberli, Stephan Lang, David Browne, Weijun Zhu, Mike Fitz, and Babak Daneshrad

CONTENTS

10.1 Introduction

For some time researchers around the world have developed testbeds to further experimental wireless communications research. The work on multi-antenna systems was initiated at UCLA in 1998 with the high-speed QAM testbed, which incorporated smart antenna processing to deliver 30 Mbps in a 5 MHz band [1]. The MIMO narrowband testbed described in [2] was the first MIMO testbed reported in the literature. This was followed by a fast frequency hopping spread spectrum testbed in 2001 [3] and, finally, two broadband MIMO-OFDM testbeds were recently completed [4,32].

The investment of time, money, and resources required to see a testbed development through is enormous, and it often confronts research teams with the following questions:

- Why is a testbed needed and how can the associated expenditure of time, money, and resources be justified?
- Should it support real-time or non-real-time operation?
- What elements and components are needed to make the testbed and the ensuing research successful?

Perhaps the first question is the most important one to answer. Invariably, in any organization, one finds the "simulation-only" camp that advocates a simulation-only approach to system development. In the past, many semi-conductor vendors have gone straight to silicon after exhaustive simulations and have had successful products. So is a testbed really necessary? The answer resides in the maturity of the technology, market, and communication paradigms that one faces. If the characteristics of a medium are well understood, or the worst-case channel conditions are specified in a standard, and the imperfections of the analog circuits have been thoroughly documented, then a simulation-only approach will suffice. This might be the case for traditional wireline communications or even narrowband cellular communications. However, it is most certainly not the case with MIMO systems, which are ushering in a paradigm shift in wireless data communications, namely the exploitation of the spatial dimension in addition to time and frequency for the transmission of signals.

More generally, our experience has shown that a testbed is justified if one or more of the following conditions are met:

- There is no accurate model of the channel.
- The RF impairments are not known, or if they are known, their impact on the performance of the wireless link is ambiguous.
- Long-term behavior (continuous operation over many hours) of the algorithms and/or hardware are not known.
- Accurate modeling of the interference seen by the unit due to net-worked operation is unknown.

Given the current state of research and commercial activity in the area of MIMO communication systems, it is safely stated that all of the above four conditions are met, either in part or in full. Channel models exist, but are rudimentary and do not properly model the angle of arrival of the rays and the correlation between signals coming into each receive antenna. The RF impairments are not new for MIMO systems; however, the magnitude of their impact on the underlying performance of the system is not known. For example, the impact of an imbalance between the in-phase and quadrature rails (I/Q Imbalance) in RF architectures with zero-IF is substantially more detrimental in MIMO than in traditional SISO systems. Similarly, the effect of any coupling of signals from different RF chains on the resulting performance is unknown. The decoding algorithms needed for MIMO systems are also new and untested for long-term operation. Drifts of adaptive algorithms due to fixed precision implementation and/or bounds on performance for long-lasting links are all unknown. Finally, the performance of MIMO-enabled nodes in the presence of random network interference is unknown. All of this helps motivate and justify testbed development and experimental research in the area of multi-antenna systems.

This chapter provides insight into the development process of wireless communications testbeds, starting with the classification of testbeds to deployment and field measurements with them. To serve as examples, three particular testbeds, all MIMO, will often be referred to. The first one is a mature, narrowband, DSP-based system. It operates in real-time with 4 kHz of bandwidth in the 220 MHz frequency band. A few interesting test setups include 3×4 MIMO system, infrastructure-based networking with multi-antenna support at the base station, and ad-hoc networking with multiple mobile radios. The other two testbeds are both broadband MIMO-OFDM systems, built by two different research groups with entirely different research goals. For this reason, the testbed's architectures are also fundamentally different. One team focuses on the design of high-performance digital VLSI circuits for broadband wireless communications. Accordingly, their testbed's RF section was implemented with a zero-IF architecture for a carrier frequency at 5.25 GHz and a bandwidth of 25 MHz. That choice revealed the problem of I/Q imbalance in MIMO systems, opening up a rich field for applied research that produced valuable new knowledge. The results from that testbed shown here correspond to measurements taken when the testbed had 2×2 capabilities with non-real-time baseband processing. The other group's research is motivated by the goal of furthering fundamental understanding of MIMO communications. Their testbed is, therefore, an instrument for closing the loop of the scientific method through actual field experiments and channel sounding. It operates with a bandwidth of 20 MHz at a carrier frequency of 2.4 GHz. The baseband signals are digitally up and down-converted from to an IF at 70 MHz.

This chapter is organized as follows. Section 10.2 provides a discussion of testbed classifications, followed by the identification of the necessary elements for developing a successful testbed in Section 10.3. Section 10.4 is devoted to lessons learned in the course of developing and calibrating the aforementioned multi-antenna testbeds. Automation of field measurement procedures is treated in Section 10.5, and Section 10.6 provides a summary of the results obtained using three different MIMO testbeds.

10.2 Testbed Classification

The design and development of a testbed is guided by a number of parameters derived from the specific research goal and available funding. While the research goal itself may range from plain Bit Error Rate (BER) measurements to entire networking experiments, other parameters such as desired throughput, form factor, configurability, testbed mobility, development time, and cost are equally relevant. As a whole, all these aspects are tightly coupled with two fundamental properties of a testbed, namely:

1. The technology of choice for implementing the testbed's baseband processing engine
2. Whether the testbed can operate in real-time or not, i.e., whether the processing power at baseband is required to match the throughput of the testbed's RF section or not

The above two qualities are, of course, not independent of each other: they are coupled through the bandwidth (or throughput) of the system. The situation is shown in Figure 10.1.

Considering the above two categories, perhaps the simplest type of testbed is a software-based, non-real-time system. This approach is often used as a

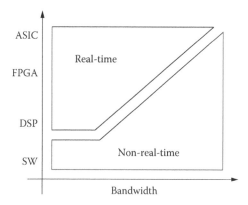

FIGURE 10.1
Real-time operation of a testbed as a function of system bandwidth and baseband processing technology. The boundary between real-time and non-real-time operation moves right as the technology of integrated circuits progresses.

starting point in the implementation of more hardware intensive and/or real-time testbeds, although it may serve as a final goal as well. It typically involves designing and fabricating (or purchasing) the RF front-end modules, plus buffering interfaces and a software-based, PC-hosted baseband processing engine wrapped around them. Overall, such a system has a short development time and low cost.

Although unable to provide real-time functionality, a software-based formulation is very valuable during the initial phase of development because it provides great flexibility for configuring and blueprinting a more complex platform. For instance, it allows for developing the transmission format (e.g., a packet structure) that best fits the research goals, and for testing algorithms for synchronization, channel estimation, equalization, etc. These software tools also serve as an important reference point for calibrating fixed-precision implementations in a later stage of development.

This kind of testbed is not a "one-box" solution; they are difficult to transport for demonstration purposes and do not lend themselves well to experiments with mobility, although it is possible to use them for collecting BER statistics in stationary conditions, while measuring packet or network-level performance would be cumbersome and slow.

When real-time operation is necessary (e.g., for communications with feedback), several options are available depending on the targeted bandwidth and desired throughput of the system. If it is relatively low, then real-time implementations are feasible at low cost with commercially available programmable Digital Signal Processors (DSPs), which replace the PCs and memory boards of the software-based testbed described earlier. On the other hand, when sustained high throughput is required, dedicated Field Programmable Gate Array (FPGA)-based solutions, and even developing Application-Specific Integrated Circuits (ASICs) become appropriate — the latter at much increased cost (Table 10.1).

A DSP-based, real-time system requires significantly less development time than an FPGA or ASIC-based solution (Table 10.1). Its capability enables one to collect performance statistics at a packet level and over extended periods of time, making it well suited for networking experiments. In addition, the real-time nature of this kind of testbed allows for the system to be used as a simulation accelerator.

Dedicated ASICs take time to design and are not reconfigurable, and even though FPGAs provide the flexibility, the hardware design process itself can be time-consuming. Nevertheless, these alternatives have a small form factor

TABLE 10.1

Cost and Development Time for Various Baseband
Processing Technologies

	SW	DSP	FPGA	ASIC
Hardware Cost	Low	Low-Medium	Low-Medium	High
Development Time	Short	Medium	Medium-Long	Long

and are ideal for field trials that involve mobility and networking experiments. ASICs for communication applications also provide graduate level research opportunities in a variety of topics involving efficient signal processing architectures for VLSI implementation.

10.3 Elements of a Successful Testbed

A testbed has a progressively greater chance for success as it increasingly captures the following five characteristics:

1. The testbed is thoroughly calibrated,
2. The testbed interfaces easily to a realistic simulation or emulation platform,
3. The testbed is easily and quickly configurable,
4. The testbed has mechanisms to highly automate field testing,
5. The testbed reflects design tradeoffs that will be present in the final system implementation.

These five characteristics ensure that the testbed experience is fruitful and efficient. Much of the work in a testbed development is spent explaining experimental data that behave in unexpected ways. Most often (in the authors' experience) the unusual behavior is a function of the experimental hardware used in the testbed, while less often the unusual behavior is due to interesting characteristics of wireless propagation and communication algorithms. The above five characteristics allow the engineering team to quickly isolate the source of the unusual behavior and quickly address the issue by (1) attempting to change the experimental hardware or (2) collecting enough data to write a paper of archival quality on the unusual characteristic. The remainder of this section will highlight the important aspects of the five elements of a successful testbed.

Calibration: Thorough quantification of the testbed's performance is essential for understanding the experimental results. Testbed development should only be attempted by teams that are willing to engage in a slow, methodical development effort. The best approach is to have integration of the testbed take place by slowly adding in components of the testbed and completely calibrating the performance. As a minimum, it is important to integrate a communication system (1) at the baseband algorithm level, (2) at the baseband plus radios in a cabled environment, (3) with a full system using channel emulation, and finally (4) a full over the air system. The testbed must be calibrated at each step of the way and compared to any theoretical performance bounds that might be available. In a testbed development there will be performance anomalies, hardware issues, and component failures.

These issues will be much easier to debug if calibration has been completed and documented in a variety of configurations. In addition, the calibration often makes it apparent that hardware imperfections need to be compensated to achieve the desired performance. The sequel will discuss several of these hardware imperfections, e.g., I/Q imbalances and phase noise, and how compensation was implemented to get better performance in the testbed.

Interfacing easily to a realistic simulation or emulation platform: Having an associated emulation and simulation facility for the testbed allows one to hypothesize about performance characteristics and troubleshoot system issues. A critical aspect of a testbed development is the ability to model performance anomalies in a controlled simulation environment. A testbed radio system is an imperfect system, and some of these imperfections can cause noticeable degradations. Having a simulation and emulation system enables the development team to discern which characteristics are important and which are not.

Configurability: Being able to rapidly and reliably configure the testbed produces a system that is more useful in experimental research. The primary (and perhaps only) users of a testbed are its developers. They spend time developing applications, troubleshooting, running calibration tests, running field tests, and doing demonstrations. Similarly, all of the work to be done on the testbed for both development and testing will require compiling and running software (C or DSP code) or firmware (embedded software or FPGA code) on the target system on a frequent basis. Each of these uses of the testbed is greatly facilitated by having the testbed configuration automated, remotely controlled and by having initialization scripts centralizing the definition of all the important parameters in one configuration file.

A major use of the testbed is also in marketing the research. People tend to understand and appreciate work to a much greater degree when it can be seen in action, so being able to quickly and efficiently prepare a demonstration of a testbed is a very important aspect of making it successful.

Automated test mechanics: Automated field testing improves the efficiency of use of engineering resources. While members of the testbed development team enjoy getting out to field test their algorithms and systems and seeing the fruits of their labor, their joy is short-lived when an engineer begins to realize that statistically significant data collection will be time-consuming. The manager also does not enjoy seeing high-priced talent reduced to manual labor (moving antennas and pushing buttons to start tests). Consequently, field tests should be automated to as high a degree as possible.

A wireless modem's performance is strongly dependent on the channels over which the transmission takes place. Therefore, to get a good understanding of a modem performance, a statistically significant number of channel realizations are needed. These channel realizations can be obtained by moving either the transmit or the receive antennas to different positions. In this respect, there are two types of performance averages that are of interest: (1) local averaging and (2) macroscopic averaging. In a local average the antenna deployment will be moved around in an area with dimensions on the order of tens of wavelengths. This motion of the antennas will give statistical averages corresponding to the local fading caused by the multi-path. Section 10.5.1 discusses an automated testing technique that gives local averaging for wireless local area networking applications. Macroscopic aver-aging, on the other hand, is also necessary to understand the performance of a wireless modem in a wide variety of link geometries, e.g., indoor vs. outdoor or line-of-sight vs. rich scattering conditions. Macroscopic automa-tion is a bit harder to achieve, but it can be attained in certain situations (e.g., by deploying on buses or taxis). The automation of these testing environments to as large a degree as possible is very important to gathering statistically significant amounts of data while efficiently using engineering resources.

Representation of design tradeoffs: In general, there are two purposes for a testbed: (1) being a *research platform* for understanding wireless channels and modulations, and (2) being a *technology prototype* for understanding the issues of building a particular wireless application. The testbed's architecture should match the goal. A research platform should have high performance components so that the resulting performance is a function of the channel and of the proposed algorithms and not a function of the hardware imple-mentation. For instance, the authors' experience recommends that a research testbed should use a digital IF, Nyquist sampling, and a digital down-converter chip. While this architecture uses significant power, it does not have I/Q mismatch as in a direct down-conversion receiver (as explained later in Section 10.4). Alternatively a technology prototype for a low cost commercial application should have a much less capable radio system, so that the algorithm development and testing can be done with realistic impairments.

In summary, testbed development should be carefully planned. To achieve the goals of a testbed, the architecture must match the application. The most efficient use of engineering time is achieved with automated configuration and testing. The productivity and output of the testbed will be maximized by a careful calibration and by having a companion software simulation and emu-lation environment. While each of these issues might seem to be over-engineer-ing for some applications, experience has shown that all significant testbed applications benefit from a design flow that uses all of these characteristics.

10.4 Hardware Calibration

Calibration is about quantifying the performance of a testbed under all the operating conditions that will be found during the field measurement cam-paign, and about satisfactorily explaining any performance loss with respect to an ideal system. The performance loss, often called *implementation loss*, is usually a result of hardware imperfections, but can also be related to imple-mentation issues such as fixed-point processing at baseband. Thus, before the testbed is ready for field deployment, it must be calibrated (1) under controlled conditions in the laboratory ("benchtop calibration") and (2) under field conditions ("field calibration"), in that order. While benchtop calibration usually includes adjustable attenuated, wired links between transmitter and receiver branches recreating an additive white Gaussian noise channel (AWGN), field calibration may be attained in the lab by emulating the channel properties that will be found in the field, and by comparing the observed performance with simulations.

10.4.1 Design Tradeoffs

The main calibration issues arise from three specific design tradeoffs. Each one is described next.

Zero-IF or digital-IF architecture: Two main approaches may be taken for up- and down-converting the baseband signals to and from RF. The first one ("digital-IF") considers a digital up-conversion to an intermediate frequency (IF), followed by a D/A conversion and finally the up-conversion to the carrier frequency (and vice versa for the down-conversion process). The 2.4 GHz MIMO-OFDM testbed uses this approach. The second option ("zero-IF") considers the direct D/A conversion of the baseband in-phase (I) and quadrature (Q) rails separately, and then up-converting them to the carrier frequency. The 5.25 GHz MIMO-OFDM testbed uses this architecture. The main disadvantage of a zero-IF approach is I/Q mismatch, i.e., an imbalance of gain, phase and/or delay between the I and Q rails that occurs due to independent analog circuitry at baseband [8]. The problem can be entirely avoided by using digital-IF architectures. However, this solution requires higher sampling rates and higher power consumption. Most of the popular transceiver ICs, especially in the 2.4GHz and 5GHz bands, are zero-IF systems and, hence, these architectures are common in commercial wireless devices. Therefore, when building prototype testbeds, I/Q mismatch will typically be an issue that needs to be addressed.

Signal distribution: A major issue in multiple antenna testbeds is the distribution of signals such as clock, Local Oscillator (LO), and control signals [22]. The problem is two-fold: first, the cabling between the many circuit boards in a MIMO system can be confusing and often leads to a much larger, complex setup. Second, the performance of the MIMO system is degraded when these signals are distributed over long cables.

The three main approaches for the distribution of signals in MIMO systems are discussed next, each one with its own strengths and weaknesses. The first approach is to centralize the distribution of the clock, control, and LO signals on a single board. A single clock source, based on a crystal oscillator, generates the basic timing signal. It could be the sample rate of the D/A converter at the transmitter side or the sampling time of the A/D converter at the receiver. This clock signal is also used on the centralized circuit board to generate the LO signals by means of PLLs. The clock, control, and LO signals are then distributed to all branches with appropriate cables and cable drivers. One caveat is that cable drivers for the distribution of the clock signal add jitter to the clock and can affect the performance of A/D and D/A converters. On the other hand, the advantage is that all the branches run synchronized off a common clock signal, which eases frequency offset compensation at the receiver. A power splitter is needed for the distribution of the LO signals to all the braches. Because of the inherent loss of power splitters, the LO signal needs to be generated with a higher power, which in turn increases its phase noise. Nevertheless, the phase noise properties of all the branches are correlated, and therefore, its effect can be cancelled easily. The cabling effort in this first approach is the most involved, as all the signals are centralized.

In the second approach, the clock and the LO generation are implemented on each of the transmitter and receiver branches' circuit boards, whereas the control signals are centralized. This architecture reduces jitter on the clock and lowers phase noise of the LO signals, but the drawback is that all the transmitters and receivers run off different clock sources and are therefore not synchronized. This issue must be addressed with baseband signal processing. In addition, with this alternative, the LO signals in each branch have uncorrelated phase noise properties. The cabling effort in the second approach is the least complex, as most of the signals are generated in a distributed fashion.

The third approach is a mix of the first two. The clock and control signals are centralized, but the LO signals are generated on each of the transmitter and receiver circuit boards. This guarantees frequency lock among all the branches, but the LO signals have uncorrelated phase noise properties. The cabling effort in this approach is similar to the second approach.

Build or buy: A third important design tradeoff in testbed development is whether to develop the testbed's circuit boards in-house or to outsource them or use commercially available products. The question includes both RF as well as baseband circuitry. Developing circuits in-house requires time and know-how but is the most flexible solution as it is custom made. Outsourcing the design of radios or baseband processing boards (to name a few) helps overcome time and know-how limitations but tends to make it harder to isolate problems and to correct them with a better design. Finally, off-the-shelf solutions may provide the quickest way to set up some modules of a testbed, but are more often than not immature, poorly documented, and weakly supported products due to their recent entry into the marketplace. This is also the least flexible solution.

The sequel illustrates and exemplifies the authors' experience with the above tradeoffs during the calibration process of the three MIMO testbeds introduced in Section 10.1.

10.4.2   I/Q Mismatch

During the calibration of the 5.25GHz MIMO-OFDM testbed, an error floor was encountered at 22 dB of SNR. This was identified to be caused by I/Q mismatch. It was corrected on the testbed at the receiver using baseband signal processing techniques that removed the error floor [8,9].

I/Q mismatch is associated with a link (from transmitter to receiver), i.e., I/Q mismatch at the transmitter can combine either constructively or destructively with the I/Q mismatch at the receiver, causing a certain amount of performance degradation. For SISO systems, I/Q mismatch is usually modeled as being lumped at the receiver, where baseband correction algorithms are employed to remove the mismatch. In MIMO systems, however, there are $M \times N$ links and I/Q mismatch of each of the M transmitters combines differently with the I/Q mismatch of each of the N receivers. Therefore, it cannot be modeled by lumping at the receiver.

I/Q mismatch is an issue in both single-carrier and multi-carrier communication systems. In single-carrier systems, I/Q mismatch causes a distortion of the signal constellation. In SISO-OFDM systems, I/Q mismatch causes interference from the data transmitted on the frequency mirror subcarrier (Figure 10.2). In MIMO-OFDM systems, the interference is more severe because data transmitted on the frequency mirror subcarriers of all transmitter branches contribute to the interference. This is illustrated in the constellation plots of Figure 10.3. Notice in Figure 10.3b that the receive diversity of 1×2 SIMO mitigates the impact of I/Q mismatch completely. This, however, is a suboptimal solution [9].

There are many baseband signal processing techniques for mitigating I/Q mismatch in both single-carrier and multi-carrier SISO systems [5–9]. These techniques involve estimating the I/Q mismatch using training sequences and applying a correction to the I and Q rails at the receiver. Pun et al. [5] use an adaptive filter on the baseband I and Q rails to correct frequency dependent I/Q mismatch. Schuchert and Hasholzner [6] describe an LMS-based adaptive algorithm to cancel I/Q mismatch in the frequency domain. In the research related to the development of the 5.25 GHz MIMO-OFDM testbed, [6] was extended for MIMO-OFDM systems and the new solution was shown to be the optimal joint MIMO decoder-I/Q mismatch canceller [9,31].

The performance of the optimal I/Q mismatch cancellation algorithm was tested in real environments. Figure 10.4 plots the Cumulative Distribution Functions (CDFs) of indoor wireless measurements conducted at UCLA,

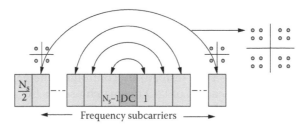

FIGURE 10.2
Illustrating the effect of I/Q mismatch on OFDM systems.

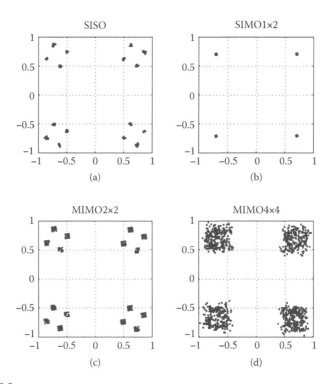

FIGURE 10.3
4QAM at 40dB SNR in the presence of I/Q mismatch in (a) SISO, (b) SIMO1×2, (c) MIMO 2x2, (d) MIMO4 × 4.

with and without I/Q mismatch cancellation for MIMO2×2, SIMO1×2, and SISO systems. These figures show that there is a 5dB improvement due to I/Q mismatch cancellation in a 2 × 2 system, and about 2.5dB in the SISO and SIMO1×2 cases. This highlights the importance and efficacy of I/Q mismatch cancellation in MIMO-OFDM systems. It can also be observed from Figure 10.3(b) and Figure 10.4 that receive diversity ($N \geq 2M$) cancels I/Q mismatch, albeit being a suboptimal solution [9].

10.4.2.1　A Simulation Model for I/Q Mismatch

The models for frequency-dependent I/Q mismatch reported in the literature are simplistic and consider a linearly varying gain and phase on each of the subcarriers [5]. I/Q mismatch can be modeled much more realistically (similar to what is observed on the testbed) by controlling the gain, phase, and delay mismatches. The measure of I/Q mismatch used is the average image suppression (or a plot of the per subcarrier image suppression). On the 2 × 2 MIMO testbed the average I/Q mismatch is approximately –20dB. This can be modeled in the simulator by choosing an I/Q gain mismatch of 0.91 dB, delay mismatch of 5% of Ts, and a phase mismatch of 2.8°. This is illustrated in Figure 10.5.

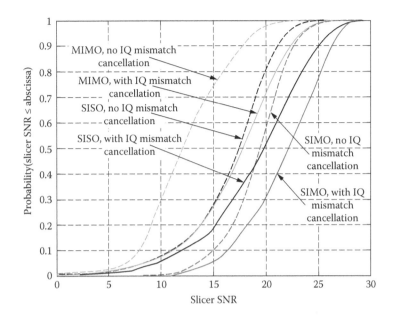

FIGURE 10.4
CDF of slicer SNR with and without I/Q mismatch cancellation on the testbed.

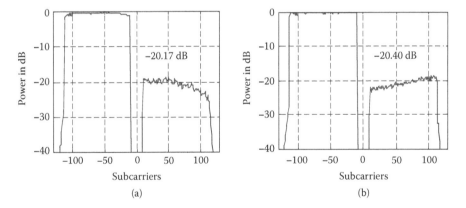

FIGURE 10.5
Image suppression (a) on the testbed and (b) in the simulator.

10.4.3 Phase Noise

An important analog impairment in any wireless transceiver is phase noise. It degrades the performance of the transceiver in the up- and down-conversion process of the RF chain. The Local Oscillator (LO) signals are not perfectly sinusoidal, but rather show random fluctuations of the phase/frequency around the desired LO frequency, which results in skirts in the power spectral density of the LO signal. The wider the power spectrum skirt, the poorer

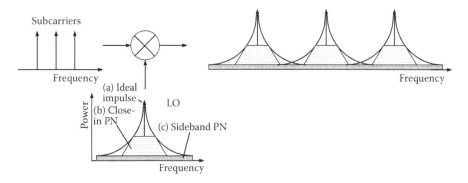

FIGURE 10.6
Phase noise in multi-carrier transceivers.

the quality of the LO. A simple model for phase noise in a multi-carrier system is illustrated in Figure 10.6.

This model [10] represents the three components of the power spectrum skirt: (a) the ideal and desired pulse, (b) close-in phase noise and (c) sideband phase noise. The close-in phase noise has a relatively high power and is band-limited by the loop bandwidth of the Phase Locked Loop (PLL) generating the LO signal. The PLL magnifies the close-in phase noise, which mainly determines the noise performance of the receiver. It can be mitigated with a proper design of the PLL circuit, as described later. The amount of the sideband phase noise is mainly given by the phase noise characteristic of the Voltage Controlled Oscillator (VCO) used in the PLL circuit. In a multi-carrier system (as shown in Figure 10.6), the close-in phase noise is common to all the carriers after an up- or down-conversion and can be cancelled in software. The sideband phase noise overlaps and causes Inter-Carrier Interference (ICI). The interference due to the superposition of the skirts from all the other subcarriers is random in nature.

The effects of the phase noise can be reduced by a proper hardware design or can be cancelled in the demodulator. Both approaches are discussed in the following sections.

10.4.3.1 Design of Low Phase Noise Local Oscillators

Figure 10.7 shows a block diagram of the PLL circuit that was used to generate the low phase noise LOs in the 5.25 GHz MIMO-OFDM testbed.

The basic PLL consists of a phase frequency detector (PFD), a charge pump (CP), a loop lowpass filter (LPF) and a voltage-controlled oscillator (VCO) [9,11]. In the feedback loop is a divide-by-N block to generate the high frequency LOs from the crystal (Temperature Compensated Crystal Oscillator, TCXO). At the input there is also an input reference divider R. Each block contributes to the total phase noise at the output [12]. Analysis of the PLL

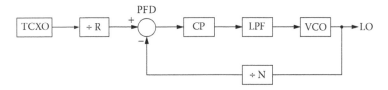

FIGURE 10.7
Basic PLL block diagram

shows that the close-in phase noise can be reduced by minimizing the contributions from the reference and feedback dividers as well as the phase
noise of the TCXO. In addition, choosing the feedback divider N small and
the reference divider R large reduces the close-in phase noise further. For
frequency offsets outside the loop bandwidth, the phase noise is dominated
by the VCO. Figure 10.8 shows the frequency spectrum of the TCXO and
the VCO used to generate the 1.75GHz LO at the transmitter in the 5.25GHz
MIMO OFDM testbed. It can be seen that within the loop bandwidth, the
PLL magnifies the phase noise (close-in part). Outside of the loop bandwidth
(sideband part), the phase noise of the PLL converges with the VCO, as
predicted by theory.

For reference, the phase noise measured at 100kHz frequency offset of the
LO frequencies was in the range of –120 dBc/Hz to –105 dBc/Hz for the
5.25 GHz MIMO-OFDM testbed.

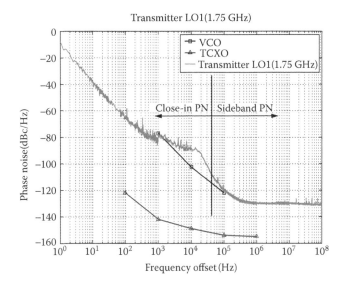

FIGURE 10.8
Phase noise power spectral density.

10.4.3.2   Phase Noise Cancellation in the Demodulator by Signal Processing

The effects of phase noise in SISO-OFDM systems has been extensively analyzed and reported in the literature [13–15]. The results can be extended to MIMO-OFDM systems. The per-subcarrier received signal after the FFT can be represented as

$$Y(k') = H(k')X(k') + \frac{j}{N_S}\sum_{l=0}^{N_S-1}\phi(l)\Big[H(k')X(k')\Big] +$$

$$\frac{j}{N_S}\sum_{l=0}^{N_S-1}\phi(l)\sum_{k=0,k\neq k'}^{N_S-1}H(k)X(k)e^{\frac{j2\pi l(k-k')}{N_S}} + V(k)$$

(10.1)

where $X(k')$, $Y(k')$, and $H(k')$ represent the transmit data vector, receive data vector, and the channel matrix, respectively. N_s is the number of subcarriers and k is the index of the subcarrier under consideration. MIMO decoding involves reversing the effect of the channel and estimating each of the transmitted data streams by left multiplying $Y(k')$ in Equation 10.1 with the weight matrix $W(k')$, which is an estimate of the pseudo-inverse of the channel matrix $H(k')$. The result is

$$\hat{X}_{PN}(k') = W(k')Y(k') = \hat{X}(k') + \frac{j}{N_s}\sum_{l=0}^{N_S-1}\phi(l)\Big[\hat{X}(k')\Big]$$

$$+j\frac{W(k')}{N_s}\sum_{l=0}^{N_S-1}\phi(l)\sum_{k=0,k\neq k'}^{N_S-1}H(k)X(k)e^{\frac{j2\pi l(k-k')}{N_S}} + \tilde{V}(k')$$

(10.2)

The first term represents the estimated data vector. The second and third terms represent the interference due to phase noise. The second term represents the "common phase error" (CPE) and is called this because its effect is an identical phase rotation on all the subcarriers and all the transmit data streams. This term can be estimated and corrected for every MIMO-OFDM symbol. The CPE term is also scaled by the transmit signal on each subcarrier. Therefore, in a MIMO system, as the number of transmit branches increases, the CPE term decreases. The third term is the sum of the projection of the phase noise skirts on all the other subcarriers and is commonly called the inter carrier interference (ICI) due to phase noise. This is random in nature and needs to be minimized by proper choice of the VCO.

The impact of canceling the CPE part of phase noise was evaluated on the 5.25 GHz 2 × 2 MIMO-OFDM testbed. The results are shown in Figure 10.9. It can be observed that canceling CPE has a much bigger impact on SISO systems (approximately 4–4.5dB SNR gain) than on the 2 × 2 MIMO system (1.5–2 dB SNR gain). This is because the testbed is a power constrained

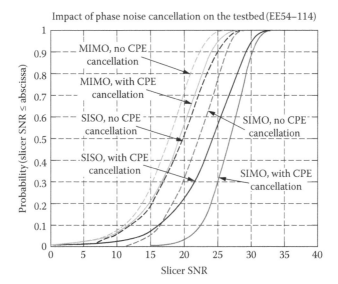

FIGURE 10.9
CDF of slicer SNR for indoor wireless measurements with SISO, SIMO1×2, and MIMO2×2 with and without CPE cancellation.

system (the total transmit power is divided equally among the two transmitters), and the CPE term is scaled by the data transmitted on each subcarrier of an OFDM system. Therefore CPE cancellation yields a smaller improvement in the MIMO system. On the testbed, phase noise caused a performance ceiling at 27dB SNR during the benchtop calibration process, and it was removed after adding the CPE cancellation algorithm to the demodulator.

10.4.3.3 A Discrete-Time Simulation Model for Phase Noise

Although analytical results show that the phase noise process has a Power Spectral Density (PSD) that falls off at $1/f^2$ [16], in reality (on hardware) the phase noise PSD flattens out for large frequency offsets [17]. In [18] Salz mentions (although in the context of laser phase noise) that the spectral density of this phase noise has a $1/f$ to $1/f^2$ characteristic up to around 1 MHz offset and is flat for higher frequencies. Also, at very low frequency offsets, it has a $1/f^3$ characteristic [17]. In discrete time simulations, it is difficult to model phase noise to exactly match with what is measured on the testbed. Approximate models with a $1/f$ characteristic are used. Generally, the phase noise process is modeled with an FIR filter that is generated with a user-defined mask as in [15,21]. The user-defined mask is a rough approximation of the PSD measured using a spectrum analyzer or a phase noise meter. Although this method can model the phase noise behavior to be close to what is observed on the hardware, the FIR filter can be fairly large, and simulations are slow. An alternative method proposed by Kasdin [19,20] in the early 1990s models phase noise using an ARMA process with an IIR filter. This is an elegant method that can model any kind of power law noise (ex. $1/f$, $1/f^2$, etc.). Although not widely discussed and used in the wireless communication literature in the context of phase noise, it is a popular method in industry (an example: MATLAB Simulink). This method was used to model phase noise in the simulator that was used to analyze and compare the performance of the 5.25 GHz MIMO-OFDM testbed.

10.4.4 Benchtop Calibration of the 5.25 GHz 2 × 2 MIMO Testbed

This section briefly discusses the results of the entire benchtop calibration process of the 5.25 GHz 2 × 2 MIMO broadband testbed. The discussion that follows considers uncoded SISO links, i.e., the wireless channel was replaced with RF cables and variable attenuators between individual (paired) radio units, resembling SISO AWGN channels.

The combined effect of CPE phase noise removal and I/Q mismatch cancellation is illustrated in Figure 10.10 for the case with 64-QAM modulation and approximately 20 dB SNR.

Figure 10.11 plots the symbol error rate curves for various signal constellations and compares the testbed with theory and simulation when the testbed is in perfect-timing mode. This means that the clock signal, which also drives the LOs, is common to both the transmitter and the receiver units. The results lie within 1–2 dB of the theoretical curves. The simulator, with all the analog impairments modeled to match the testbed's, is within 1 dB of the testbed's performance. The slight difference between the simulations and the testbed is mainly due to the non-white nature of the noise on the testbed, caused by non-linearities in the RF sections. These were not modeled in the simulator.

An important metric to measure testbed performance is the implementation loss, which compares the average SNR at the input of the decision device

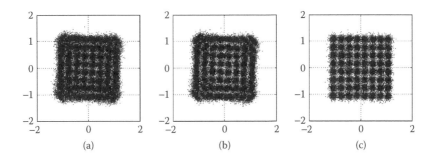

FIGURE 10.10
A 64 QAM constellation measured on the testbed at an input SNR of 20dB (a) without I/Q mismatch cancellation and phase noise (CPE) cancellation, (b) with I/Q mismatch cancellation but no phase noise (CPE) cancellation, (c) with I/Q mismatch cancellation and phase noise (CPE) cancellation.

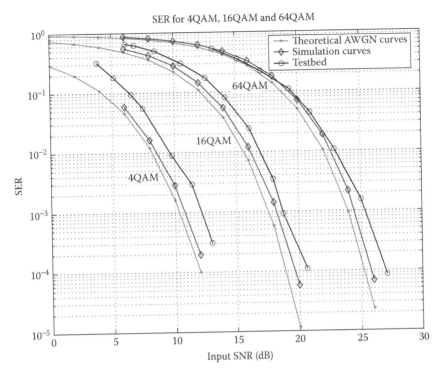

FIGURE 10.11
SER calibration curves for the 5.25 GHz testbed over AWGN channel and in perfect timing mode.

of the receiver ("slicer SNR") vs. the SNR at the input of the receiver. The difference in dB between the input SNR and the slicer SNR is the implementation loss. Figure 10.12 plots the implementation loss of the testbed in perfect and non-perfect timing modes and compares the results with the simulator. Also plotted is the implementation loss when the carrier frequency synchronization algorithm is turned on and the testbed is in perfect timing mode. This quantifies the impact of the frequency offset estimation errors and allows for incorporating it into the system's performance data sheet. In perfect timing mode, when the I/Q mismatch canceller and phase noise canceller are turned on, the implementation loss is within 2 dB in the entire SNR range up to 30 dB. However, the loss is higher in the non-perfect timing mode due to estimation errors in the synchronization algorithms. In this case, it increases to about 3 dB at an input SNR of 25 dB. It can also be observed that the carrier frequency offset estimation errors start having an impact on the implementation loss at about 25 dB, increasing to about 3 dB at 30 dB. In all these cases, it can be observed that with all the analog impairments modeled, the simulator matches the testbed very closely. Other key parameters of the RF and analog frontend (AFE) were measured as well and are listed in Table 10.2.

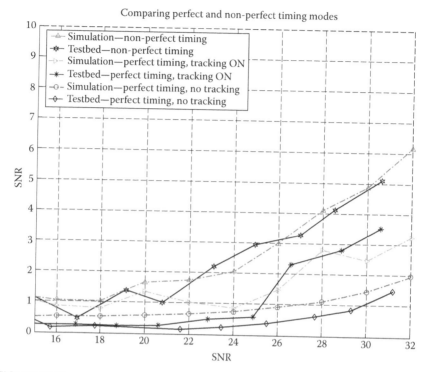

FIGURE 10.12

Implementation loss vs. input SNR curves for the 5.25 GHz 2 × 2 MIMO testbed in perfect timing and non-perfect timing modes.

TABLE 10.2

System Parameters of the 5.25 GHz 2 × 2 MIMO Testbed

Radio architecture	Heterodyne
System setup	2 × 2 (extendible to 4 × 4 antennas)
Carrier frequency	5.25 GHz
Signal bandwidth	25 MHz
Effective bandwidth	21.5 MHz
TCXO temperature stability	<1 ppm (–10°C to +60°C)
Transmitter power	–13 dBm to +18 dBm
Transmitter D/A converter	16 Bits/100 MHz
Receiver noise figure	7.93 dB
Receiver sensitivity	–99 dBm
Receiver dynamic range	70 dB
Receiver IIP3	–25.4 dBm
Receiver A/D converter	14 Bits/25 MHz or 50 MHz
Image suppression	Approx –20 dB
Local oscillator phase noise @100 KHz offset	–120 dBc/Hz to –105 dBc/Hz

10.4.5   Radio Receiver Distortion

An example from the development of the narrowband, DSP-based MIMO testbed might be illustrative of both the impact of radio distortion and for the need of emulation and simulation. The narrowband wireless modem was built and tested in 2002 using the powerful concept of a super-orthogonal trellis code [39,40]. Simulations of the whole baseband system were completed and performance looked quite promising (see lower curves on Figure 10.13a). During the calibration stage the system was subject to a real radio and no fading (benchtop and cabled) and the system had some implementation loss but otherwise performed well. The next stage of calibration introduced rapid time varying fading with a channel emulator and surprisingly a significant error floor was produced (see upper curves on Figure 10.13a). The cause of this degradation was not at first apparent.

The source of the degradation was a non-linearity in the radio receiver for the testbed. The radio was implemented using a single chip radio frequency transceiver that was designed for GSM applications (Analog Devices AD607). This radio system caused some significant distortions to the received linear modulated signal because of a significant non-linearity in the IF amplifier. This distortion destroyed the Nyquist characteristics of the pulses in the linear modulation and resulted in significant inter-symbol interference (ISI) (see the scatter diagram in Figure 10.13b). These distortions in a time flat fading environment did not cause significant degradation as the benchtop calibration tests demonstrated, but for reasons that were not clear at first, these distortions caused significant degradations in time varying fading.

After significant effort [43] that iterated back and forth between the simulation environment and the channel emulator, the cause of the error floor was identified. An error floor is caused by a situation where the distortion in the radio causes an error in the absence of thermal noise. It is pretty clear from the scatter plot that in the static channel an error would not be caused by this ISI. In a rapidly varying channel the situation is significantly different. Two issues impact time varying fading: ISI being relatively larger due to deep fades and ISI causing biases in pilot symbol-based channel estimation during a deep fade.

ISI has a bigger impact during a deep fade. In rapid fading the number of symbols that are impacted by a deep fade is often a very small number. Consequently, the channel gain for the symbol being detected during a deep fading can be small, but the ISI, due to the other interfering symbols, can be large since the channel gain for these symbols is also large. Consequently, the scatter plot does not directly scale but has a form as seen on the right side of Figure 10.14. Simulation showed that the error floor was partially due to this interaction with fading and ISI.

ISI causes biases in the channel estimation, and this bias is another source of errors in the absence of noise. This idea can best be explained by examining the vector diagram on the left side of Figure 10.14. A deep fade occurs when

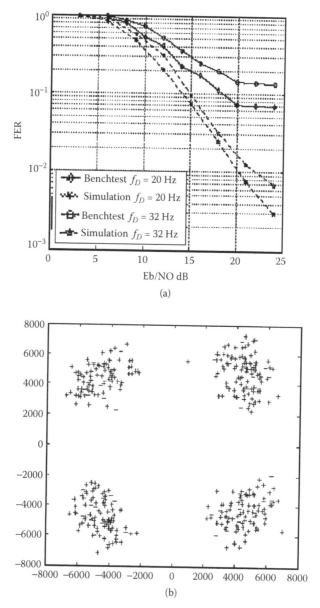

FIGURE 10.13

An example of an error floor in a space–time modem. (a) Frame error rate, (b) Scatter plot.

a channel makes a trajectory in the complex plane that goes near the origin. If the pilot symbols are distorted by ISI, the reconstructed channel trajectory will be in error. When the channel gain is relatively large, a small bias in channel estimates will not have a big impact. When channel gain is near zero in amplitude, a small bias in the channel value can actually result in a deterministic error. This is observed on the left side of Figure 10.14 where

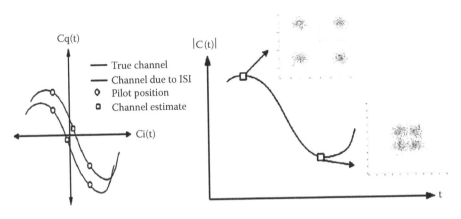

FIGURE 10.14
Causes of performance degradation in rapid fading.

the actual channel estimation error is relatively small in magnitude but will cause a very high error rate since the phase channel estimate is roughly 180° out of phase with the true channel estimate. Simulation showed that this bias in channel estimation leads to an error floor. Without companion simulation and emulation systems, this error floor would have been tough to diagnose.

10.4.6 Frequency Offset Calibration

Accurate frequency offset calibration (i.e., compensation) is fundamental in channel sounding applications of testbeds. In these cases, the transmissions rarely have provisions for running synchronization algorithms at the receiver. Therefore, the frequency offset must be measured between all transmitter-receiver branch pairs prior to the field measurement campaign. This section describes the corresponding procedure.

The sequel derives the mathematical model for the effect of frequency offset among transceiver branches in the most general case, which corresponds to the second approach discussed in Section 10.4.1 on distribution of signals in MIMO systems. The first and third approaches are special cases of the results that follow.

The baseband signal at the mth transmitter branch is modeled as:

$$\hat{x}_m \triangleq k_m(t) x_m(t) \tag{10.3}$$

where $k_m(t) = e^{j2\pi \psi_m t + \theta_m}$ is the relative frequency and relative initial phase modulation introduced by the transmitter branch m with respect to the ideal carrier frequency and phase. $x_m(t)$ is the nominal transmit signal. Similarly, the baseband signal at each receiver branch is modeled as:

$$\hat{y}_n \triangleq g_n\left(t\right) y_n\left(t\right) \tag{10.4}$$

where $g_n(t) = e^{j2\pi\psi_n t + \theta_n}$ is the relative frequency and relative initial phase modulation introduced by the receiver branch n and $y_n(t)$ is the nominal received signal. The frequency offsets and phase offsets are represented by the variables $\{\psi_m, \psi_n\}$ and $\{\theta_m, \theta_n\}$. The time index, t, is implied in the variables $\{k_m, g_n\}$ for notational simplicity.

The frequency and initial phase offset are incorporated in the MIMO signal models according to:

$$Y \triangleq Y\left(t\right)$$
$$= \tilde{H}\left(t\right) X\left(t\right) + N\left(t\right) \tag{10.5}$$
$$\triangleq \tilde{H} X + N$$

where $\tilde{H}$ the distorted channel matrix is given by:

$$\tilde{H} \triangleq \tilde{H}\left(t\right)$$
$$= KHG \tag{10.6}$$
$$= \begin{bmatrix} k_1 h_{11} g_1 & k_1 h_{12} g_2 & \cdots & k_1 h_{1N} g_N \\ k_2 h_{21} g_1 & k_2 h_{22} g_2 & & \\ \vdots & & \ddots & \\ k_M h_{11} g_1 & & & k_M h_{MN} g_N \end{bmatrix}$$

which assumes the channel matrix H to be static during the observation time, and with K and G corresponding to the unitary (time-dependent) distortion matrices due to the frequency offsets at transmitter and receiver branches, respectively. K and G are given by:

$$K = \begin{bmatrix} k_1 & & & 0 \\ & k_2 & & \\ & & \ddots & \\ 0 & & & k_M \end{bmatrix}, G = \begin{bmatrix} g_1 & & & 0 \\ & g_2 & & \\ & & \ddots & \\ 0 & & & g_N \end{bmatrix} \tag{10.7}$$

The model of Equation 10.6 shows that all the frequency offset distortion can be removed at the receiver. This requires estimating all the Δ_ψ $(m, n) = \psi_m - \psi_n$ and all the Δ_θ $(m, n) = \theta_m - \theta_n$. For channel capacity calculations, the initial phase offset between the mth transmit branch and n^{th} receiver branch, Δ_θ $(m, n) = \theta_m - \theta_n$, can be neglected. Note that for an $M \times N$ MIMO system, there are MN frequency offsets between the transmitter and receiver radios that must be determined.

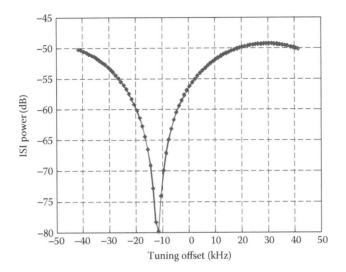

FIGURE 10.15
Power as a function of tuning frequency for a single subcarrier in an OFDM system ($\Delta_f (1,1) =$ 12.5 kHz).

A method for estimating all $\Delta_\psi (m, n)$ and $\Delta_\theta (m, n)$ in MIMO-OFDM systems by measuring Inter-Carrier Interference (ICI) is described next. The effect of a frequency offset in OFDM systems is to cause ICI at the receiver's FFT output. The sensitivity of the ICI to frequency offset is illustrated in Figure 10.15, where the ICI power is plotted for a single subcarrier as a function of a parameter called the tuning offset, $\Delta_T (m, n)$. The tuning offset sweeps all the possible compensation values for $\Delta_\psi (m, n)$. When it takes the value $\Delta_T (m, n) = -\Delta_\psi (m, n)$, it eliminates ICI, revealing $\Delta_\psi (m, n)$.

With this idea, benchtop calibration of the MN frequency offsets in a MIMO-OFDM system is attained using a pilot signaling scheme similar to that described in [33]. This scheme uses pilot tones distributed orthogonally in space and time. The transmitter branches broadcast these pilot signals over the air, and the received signal in the nth receiver is tuned across a range of frequencies as shown in Figure 10.15. The ISI due to the mth transmitter will be minimized when $\Delta_T (m, n) = -\Delta_\psi (m, n)$. Repeating this tuning procedure for each of the MN transmitter/receiver pairs will then fully characterize the frequency offset between all radios. Finally, the frequency offset may vary slowly as a function of time due to changes in temperature in the radio hardware. In this case, the characterization must be repeated regularly.

10.4.7 Frequency Response Calibration

Wideband transceivers almost never have flat frequency responses in their passband. Figure 10.16 shows the measured frequency response for one of the upconverter radios used in the 2.4 GHz MIMO-OFDM testbed. There is

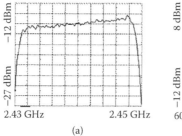
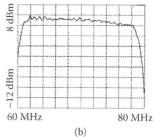

(a)   (b)

FIGURE 10.16
The measured frequency response spectrum of a transceiver radio in (a) transmit mode and (b) receive mode.

a notable 2-dB slope and ripples across the passband. More severe cases with a 5-dB change across the passband were measured for other radio units. The ripple and slope in the transceiver's power spectrum is not usually an important issue for tests involving error rate measurements in data communications because these non-idealities are considered part of the channel response. However, uncharacterized non-ideal power spectra are problematic when the testbed is used for MIMO channel sounding. For such applications, each radio's frequency response must be characterized as part of the calibration procedure. The basic procedure to calibrating a $M \times N$ MIMO transceiver is outlined next.

10.4.7.1 Calibration of Transmitters

The strategy is to generate a known signal that can be used to probe the transceiver's spectrum across the entire transmission band. Two approaches are feasible for generating such a signal: a sweep-frequency tone generator or a wideband signal generator together with a spectrum analyzer.

The swept-frequency tone generator approach is simplest and involves exciting the transmitter with a single tone whose frequency is stepped at regular frequency intervals across the input band. An attenuator is placed between the transmitter and spectrum analyzer to reduce the signal power to a level that is safe for the analyzer. The spectrum analyzer then records the (attenuated) transmitter output signal in peak-hold mode. Figure 10.16 is an example of the resulting spectrum using this technique. The disadvantage of this method is that the peak-hold function records the larger peak noise, instead of an average noise.

The second approach is to excite the transmitter with a set of evenly spaced tones across its input band. This approach requires a wideband signal generator that can synthesize a user-defined signal. A suitable signal design is proposed in [33]. The spectrum analyzer then records the transmitter's output in averaging mode. The benefit of this method is that zero mean AWGN is cancelled out by averaging. The drawback is that the equipment required to generate the wideband excitation signal can be more sophisticated and,

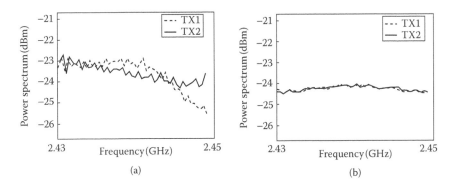

FIGURE 10.17
(a) Measured power spectrum at the transmitter antenna port before pre-equalization by the transmit DSP and (b) after pre-equalization.

hence, more expensive than the signal generator used to generate a swept-frequency tone. Another benefit of using the wideband signal generator in [33] is that the multiple transmitters in a MIMO system can be calibrated simultaneously using a M-to-1 RF combiner and a single spectrum analyzer. This is possible because spatial orthogonality in the excitation signal allows the combined TX spectra to be separated in post-processing. An example of the spectrum generated by this approach for a case with four radios is shown in Figure 10.17. The four transmitter radios used are the same design as the one shown in Figure 10.16a.

10.4.7.2 Calibration of Receivers
The procedure for calibrating the receivers is the same as that for calibrating the transmitters except that connections to the inputs and outputs are swapped. In the case that the signal generator cannot generate an excitation signal at the radio frequency required by the receiver radio's input, a calibrated transmitter radio can be used to up-convert the excitation signal. The transmitter's output can then be connected through an attenuator to the receiver's input.

Once the frequency responses of all transmitter and all receiver branches have been quantified, precise channel sounding measurements can be conducted by pre-compensating the signal transmitted on each transmitter branch, and by post-compensating the received signal of each receiver branch.

10.5 Field Test Administration

Once a testbed is fully calibrated in the laboratory and for its intended environment, it is ready to be deployed for field tests and measurements.

At this point, the logistics of testbed deployment and data collection become an important issue. A well-designed Graphical User Interface (GUI) greatly facilitates setting up the testbed in the field and monitoring its operation. Similarly, the collection of data tends to be cumbersome and labor-intensive when performed manually, while automating the process allows for extensive data collection in a variety of environments in an efficient and reliable way. Both topics are addressed in the following sections.

10.5.1   Automated Field Measurements

The following example illustrates the value of automating measurements in an experiment that tested space–time codec performance. Consider the channel experienced by a receiver as its antenna array moves within a square area of 10 wavelengths on a side. The channel variation in such a small area is due mainly to multipath interference. A typical spatial sampling pattern used for testing the aforementioned space–time codec is specified below in Table 10.3. For comparison, equivalent requirements for testing a SISO codec are provided. For each spatial sample within the area, a full range of SNRs is required to stress the codec in both the low and high capacity regions of the channel realization associated with that spatial sample. This sampling strategy ensures that the codec is subjected to a sufficient range of channels and noise realizations to give a good approximation of its performance at the microscopic (local) level. Several areas have to be measured in this way to subject the codec to a range of macroscopic channel characteristics in the larger environment.

A measurement proceeds according to the pseudo code given below. Lines 2–10 in the Measurement Pseudo Code can be executed automatically by a single component in a MIMO testbed provided that (1) all components of the testbed are able to communicate and (2) the spatial sampling can be performed robotically. The 5 seconds allocated to each measurement (cf. Table 10.3) include the time to execute lines 3–6 in the pseudo code. It is clear from this example that acquiring a large data set requires a significant

TABLE 10.3

Measurement Budget

	SISO Case	MIMO Case
Sampling grid size	$10\lambda \times 10\lambda$ square	$10\lambda \times 10\lambda$ square
Spacing between grid points	0.5λ	0.5λ
Antenna array geometry	Not applicable	Linear
Antenna spacings	Not applicable	0.5λ, 1λ, 1.5λ, 2λ
Total number of spatial samples	21*21 = 441	21*21*4 = 1764
Measurement interval	5 seconds	5 seconds
Measurement time per grid	441*5 = 38 minutes	1764*5 = 2.4 hours
Number of locations in environment	10	10
Total Experiment Time	38*10 = 6 hours	2.4*10 = 24 hours

time investment. It would be impossible to manually place the antenna array and to trigger the radio hardware with the allotted time per measurement (and, at any rate, it would be a very inefficient use of qualified personnel). Measurement repeatability and system reliability are other essential characteristics that would be challenging for the test engineer to achieve but are easily achieved with automation.

Measurement Pseudo Code

```
1  For each location in the environment
2    For each antenna separation
3      For each point in the sample grid
4           Encode data and transmit signal
5           Receive signal decode data
6           Tabulate information error rate
7           Move to the next point in sample grid
8      end
9      Change array geometry
10   end
11 move to the next location in the environment
```

A testbed configuration with the above capabilities is illustrated in Figure 10.18. Robotic antenna arrays are responsible for spatial sampling by moving the transmitter and receiver antenna arrays according to the spatial pattern defined in Table 10.3. TCP/IP connectivity allows any terminal on the Internet to control both radios and robots remotely.

An effective and affordable solution to automating spatial sampling can be realized from LEGO™ Mindstorms components and open source software [32]. Two such robots are shown in Figure 10.19 and Figure 10.20. These robots have a 1.59 mm positioning resolution and a 70 × 70 cm positioning range. These dimensions are well suited to communications within the 1–10 GHz band where wavelengths (30–3 cm) are on the same order of magnitude that can be achieved by Lego structures. The Lego system allows the robots to be configured modularly so that as many as four antennas can be move independently in azimuth. The software that controls the robots is comprised entirely of Java-based code and allows the end user to control the robot with high level commands (move, reset, calibrate, etc.). These commands can be issued via TCP/IP as text strings, thereby enabling control from any application that supports TCP socket communications (C++, MATLAB, web browser, etc.).

Remote control of the robots is enhanced by using a web camera to stream live video of the robot's activity over the Internet. In the case of a mechanical failure, the robots can alert the remote user of the problem by e-mail. This affords the user the option of checking the experiment's progress intermittently,

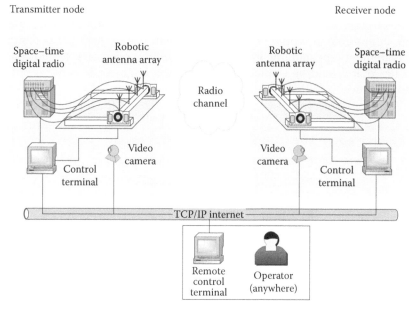

Transmitter node

Receiver node

FIGURE 10.18
A highly automated MIMO testbed.

FIGURE 10.19
Four-antenna transmitter and robotic antenna array.

instead of requiring constant vigil over long and tedious experiments. Another great convenience of the automation system is that it allows a single user to simultaneously control and monitor both the transmitter and receiver

FIGURE 10.20
Four-antenna receiver radio and robotic antenna array.

units even though they may be located in different parts of a building or neighborhood. This would otherwise require a minimum of two users.

10.5.2 Graphical User Interface Design

One vital feature of a testbed is to have well-designed Graphical User Interface (GUI) for controlling its operation. In the authors' experience, the testbed's capabilities can be exploited at a maximum if the GUI provides the following features:

1. Immediate (quasi real-time) feedback of test results and environmental conditions
2. High flexibility for configuring and controlling system parameters "on the fly"
3. A clear and easy-to-understand interface for demonstrating the testbed's capabilities and features to third parties

Each of these three points is described in detail next.

Quasi real-time feedback: A GUI's capability for reporting transmission results and testbed status in real-time or immediately following transmission simplifies the data collection and speeds up the measurement process. It maximizes the efficiency of field measurement campaigns by allowing for early detection of system failure, quick troubleshooting, and on-site assessment of the environment's validity for the desired test (e.g., channel condition, SNR, etc.). Useful information to display includes error statistics, signal and noise power, scatter plots, instantaneous channel state information, channel capacity, etc.

On-site configurability: A testbed is often very flexible in terms of system parameters and algorithm components. However, this flexibility is of limited use if it cannot be exercised quickly and with little effort through a GUI that centralizes the configuration of system parameters. Similarly, in some cases a transmission may require handling the data flow in sequential steps, and more often than not, it is useful and even necessary to repeat measurements in a same setting with different parameters. A GUI can greatly facilitate all these tasks by providing dialog boxes for manual parameter input as well as push buttons for automated script execution of commonly repeated transmission and/or configuration sequences.

Adequate display for demonstration purposes: Exhibiting the testbed to sponsors and visitors is very important for the continuous development of the testbed. Unfortunately, people who see a demonstration of the testbed often do not have a strong background in theory or implementation of wireless communication systems. This suggests that a GUI designed solely for monitoring and controlling the testbed at a technical level may not be adequate for demonstration purposes. Therefore, the GUI should have a mode in which transmissions can be controlled and results can be observed at a reduced technical level. A good GUI design and development takes significant time; however, it pays off in the long run.

Figure 10.21 illustrates the GUI developed for the narrowband MIMO testbed. The GUI displays a variety of information at different technical depths, and it is used for both demonstration and field test monitoring. Some of the components in the GUI are very helpful for troubleshooting. For instance, the bar plots of the channel powers can help identify which part of the hardware chain has failed when something is not working as expected. Tabbed windows are used to extend the display size, and it is especially useful to switch between information at different technical levels.

10.6 Field Test Results

A variety of tests and measurements can be performed with the testbed based on its capabilities. This section has a discussion on the metrics used to measure MIMO system performance followed by some field test results using two different testbeds. Some of these metrics measure the quality of the wireless channel to sustain MIMO communications, and others measure the performance of the communication system.

10.6.1 Measuring the Quality of the MIMO Channel

The information theoretic capacity of a MIMO communications scheme is a function of the average received SNR and the channel matrix. For a given

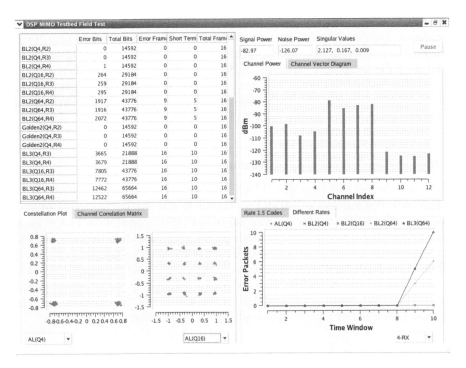

FIGURE 10.21
The graphical user interface of the narrowband testbed.

SNR, the "quality" of the channel matrix determines the achievable capacity. This can be analyzed by decomposing the channel matrix using the singular value decomposition (SVD) [23–27].

$$H = U\Sigma V^H$$

$$C = E_H \left[\log_2 \left[\det \left(I_M + \frac{\rho}{M} V\Sigma^H \Sigma V^H \right) \right] \right] \text{bits/sec/Hz} \qquad (10.8)$$

U and V are unitary matrices and $UU^H = VV^H = I$. This can be simplified to

$$C = E_H \left[\sum_{i=1}^{k} \log_2 \left(1 + \frac{\rho}{M} \sigma_i^2 \right) \right] \text{bits/sec/Hz} \qquad (10.9)$$

σ_i represents the singular values of channel matrix H. $k \leq \min(M, N)$ is the rank of the channel matrix H. From Equation 10.9 it can be observed that the total capacity of a MIMO channel is the sum of parallel AWGN SISO sub-channels, each of which has a gain given by their corresponding singular values (or power given by their corresponding eigenvalues). When the SNR is very low, the roughly linear increase in capacity expected with MIMO systems may not be achieved. Similarly, when there is severe fading correlation, the minimum singular value could be very small thereby significantly reducing the achievable capacity [28,29]. Since the capacity of a given channel can be expressed in terms of its singular values, these can in turn be used to measure the "quality" of the channel matrix. One popular metric is the effective degrees of freedom (EDOF) [28–30].

10.6.2 Effective Degrees of Freedom (EDOF)

Effective degrees of freedom (EDOF), introduced in [28], is a metric that represents the number of MIMO subchannels actively participating in conveying information under a given set of operating conditions. EDOF gives an indication of the capacity gain achievable in a MIMO channel. For a SISO channel, a G-fold increase in the transmitter power results in an increase in the channel capacity of $\log_2 G$ bits/sec/Hz. If a MIMO channel is equal to EDOF SISO channels in parallel, a G times increase in the transmit power should increase the capacity of the system by (EDOF*$\log_2 G$) bits/sec/Hz. EDOF is defined at a given transmit power and outage probability.

$$EDOF = \frac{d}{d\delta} C_q (2^\delta \rho) \Big|_{\delta=0} \qquad (10.10)$$

The capacity at outage q is evaluated for a δ increase in SNR. EDOF is a real number in $[0, k]$ where k is the rank of the channel matrix H. Equation 10.10 was analytically solved to get another convenient means to compute EDOF in [30]. This is given in Equation 10.11.

$$EDOF = \sum_{i=1}^{k} \left[\frac{1}{1 + \dfrac{M}{\rho \sigma_i^2}} \right] \qquad (10.11)$$

Although EDOF was defined at an outage probability, it can also be defined as the average EDOF (EDOF averaged over all the channel realizations) to indicate the average capacity gain achievable in the MIMO system [31]. The EDOF decreases as fading correlation increases. Similarly EDOF decreases in a LOS channel compared to a rich scattering NLOS channel. In fact, in a LOS channel with a dominant propagation path with very few scatterers it could be close to 1.

10.6.3 Analysis of Field Test Measurements

The 5.25 GHz 2 × 2 MIMO system was deployed in an indoor quasi-static environment, and the performance gain of the communication system as a function of the number of antennas was measured. The same set of measurements was used to measure the capacity gain with the 2×2 MIMO system compared to the SISO system and the EDOF of the environment. The wireless measurements were carried out in a laboratory environment. The transmitter and receiver were located in the same room of the ICSL and, thus, a line of sight existed between them. The distance between them was kept constant at about 6–7 m. The total transmit power was fixed at –3 dBm (0.5 mW) for all the experiments. 3dB attenuators were placed before the antennas for MIMO 2×2 to keep the total transmit power constant at –3 dBm. A second set of measurements was conducted with the total transmit power fixed at –10 dBm. The transmitter was moved in steps larger than the wavelength in order to get independent channel realizations. 200 distinct measurements for each of SISO, 1×2 and 2×2 setups were conducted.

Figure 10.22 plots the CDF of slicer SNR comparing 1×1, 1×2, and 2×2 systems. The same plot also has simulation results with identical amounts of Carrier frequency offset, I/Q mismatch, and phase noise added to mimic the real system. It can be observed that the simulation results closely match with the real measurements. The Slicer SNR for a 2×2 system is 3dB worse than a SISO system, since the transmit power is uniformly distributed among both the transmit antennas. Figure 10.23 plots the theoretical capacity using the estimated channels along with the capacity using the post-processing SNR (slicer SNR), which indicates the capacity that can be achieved with this communication system. The difference between the two indicates the capacity loss due to the implementation loss of the system.

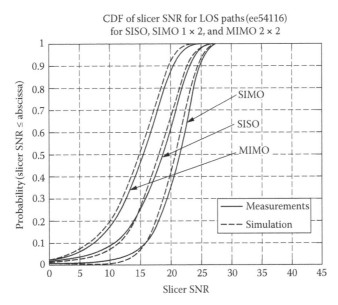

FIGURE 10.22
CDF of slicer SNR for LOS channels with excess I/Q mismatch at Tx 1 fixed.

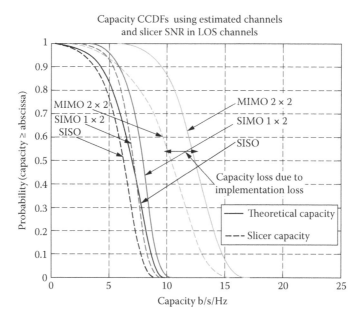

FIGURE 10.23
CCDFs of capacity for MIMO, SIMO, and SISO in LOS channels.

The average capacity gain observed with the communication system is plotted in Figure 10.24. The capacity computed using the estimated normalized channels is plotted along with the average achievable capacity (capacity computed using slicer SNR). This shows that there is a capacity loss of about 2.5 bps/Hz in this communication system using the measurements made at a received SNR of about 22 dB (when the transmit power was fixed at –10 dBm), and the capacity loss grew to 4 bps/Hz when the received SNR increased to approx 28 dB (when the transmit power was fixed at –3 dBm). This was mainly due to the non-linearities in the RF frontend at the higher power. The capacity loss in a SISO system is much less than the MIMO system because the impact of analog impairments such as I/Q mismatch is much less in a SISO system than in a MIMO system. We also observe a linear increase in achievable capacity with the MIMO system.

Table 10.4 lists the capacity and capacity gain observed along with the computed EDOF in the LOS channels. The theoretical MMSE capacity gain was also computed to be 1.73. The average EDOF using the estimated channels was measured to be 1.9, and the observed capacity gain was 1.67.

10.6.4 Channel Impact on Space–Time Coding

The pioneering work by Foschini and Gans [23] and Telatar [24] showed that multiple antennas in a wireless communication system could greatly

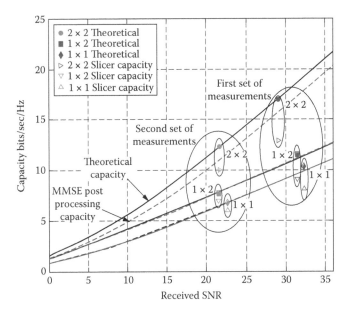

FIGURE 10.24
Capacity vs. SNR curves overlaid with measured average capacity values for LOS measurements.

TABLE 10.4

Capacity Gain in LOS Channels at 20 dB SNR

Average capacity	11.30 bps/Hz
Capacity at 10% outage	8.64 bps/Hz
Average EDOF	1.9
EDOF at 10% outage	1.77
Observed capacity gain	1.67
Observed capacity gain at 10% outage	1.57
Theoretical slicer capacity gain for MMSE MIMO decoder at 20 dB SNR	1.73

improve performance. For M transmit antennas and N receive antennas in Rayleigh fading, it was shown that with spatial independence, there are essentially $M*N$ levels of diversity available and there are $\min(M, N)$ independent parallel channels that could be established. These information theoretic results spawned two lines of work: one to increase throughput by exploiting spatial multiplexing [34] and one to improve error performance by exploiting spatial diversity [35,36]. The code design and performance analysis paradigm in the area typically assume a spatially white Rayleigh fading channel. However, a real channel could neither be Rayleigh nor be spatially white.

Hence, it is of great interest to test and compare the performance of the space–time codes in a realistic environment. This section reports on a series of experiments whose goal is to evaluate the impact of antenna separation on code performance for each of these lines of work. Two types of signaling are considered in this section: (1) multiplexing type schemes (space–time constellations and precoding) and (2) space–time codes designed to harvest diversity or performance.

The experimentation was done on the UCLA campus and the surrounding West Los Angeles area using the UCLA Narrowband real-time multi-antenna DSP testbed. The testing reported in this section was limited to the scenario where the transmitter was deployed on the top of a five-story building and the receiver on a vehicle (a van). The test consisted of the receiver radio being driven around the campus area. The speed of the driving was maintained at a rate of less than 5 miles per hour as the space–time trellis codes that were tested were all designed for quasi-static fading. The UCLA campus area is heavily urbanized, and a line of sight was not achieved in any significant portion of the testing. An example of the testbed deployment is shown in Figure 10.25.

Three antenna configurations are considered: (1) 2 λ spacing at the transmitter with 0.5 λ spacing at the receiver, (2) 2 λ spacing at the transmitter with 0.25 λ spacing at the receiver, and (3) 1 λ spacing at the transmitter with 0.25 λ spacing at the receiver. For the case of precoding and space–time constellations the Alamouti [37] coding scheme has the most robustness to different antenna geometries. All constellations considered here use $R = 4$ bits per symbol and the comparison for frame error rate is in Figure 10.26 for $M = 2$ and $N = 2$. The Alamouti code must use a 16 QAM constellation to achieve this rate while spatial multiplexing and the threaded architectures can use QPSK constellations. When the spacing of the array is larger (closer

TX deployment					RX deployment

(a)						(b)

FIGURE 10.25
Test deployment.

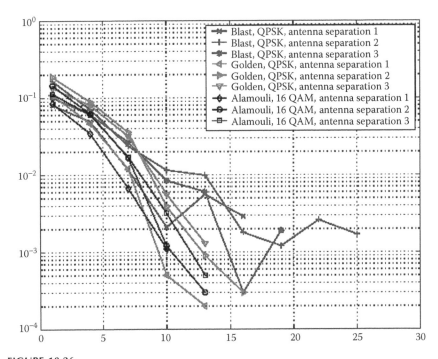

FIGURE 10.26
FER of 2×2 spatial multiplexing and Alamouti schemes.

to spatial independence) the Golden Code [38] has the advantage due to the smaller constellation point differences. As the spacing become closer and the channels become correlated, the Alamouti code has the advantage. Direct spatial multiplexing with a BLAST-like architecture has clearly lower performance than either of the two considered architectures and does not achieve full diversity. For the case of trellis codes, the universal code has the most robustness to different antenna geometries. All constellations considered here use a $R = 2$ bits per symbol, and the comparison for frame error rate is in Figure 10.27 for $M = 2$ and $N = 2$. The universal code has good performance in most cases with moderate degradation due to spatial correlation. The super-orthogonal space–time trellis codes [39,40] and CYV [41] codes show more significant degradation due to spatial correlation. For more detailed results, refer to [42].

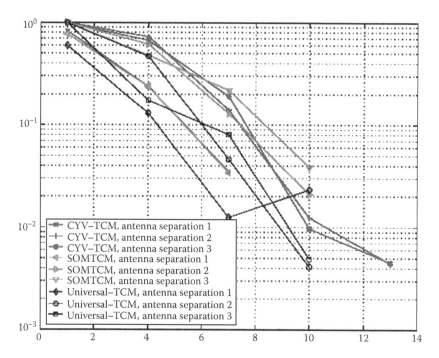

FIGURE 10.27
PER of 2 × 2 space–time trellis codes.

References

1. J.F. Frigon and B. Daneshrad. 2002. "Field measurements of an indoor high-speed QAM wireless system using decision feedback equalization and smart antenna array," *IEEE Transactions on Wireless Communications,* Vol. 1, No. 1, Jan. 2002, pp. 134–144.
2. M.P. Fitz et al. 1997. "ITS wireless narrowband land mobile data communications," Vehicular Technology Conference, IEEE 47th, Vol. 2, pp. 909–913.
3. H. Zou, S. Mohan, D. Cabric, and B. Daneshrad. 2001. "Implementation and field trial results of a fast frequency hopped FSK testbed for wireless communications," ICC 2001.
4. S. Lang, R.M. Rao, and B. Daneshrad. 2004. "Development of a software configurable broadband wireless communication platform," IST Mobile Communication Summit, June 27–30, Lyon, France.
5. K.P. Pun et al. 2001. "Correction of frequency dependent I/Q mismatches in quadrature receivers," *Electronics Letters,* Nov. 8, 2001.
6. A. Schuchert and R. Hasholzner. 2001. "A novel I/Q imbalance compensation scheme for the reception of OFDM signals," *IEEE Trans. On Consumer Elect.,* Aug. 2001.
7. T. Ylamurto. 2003. "Frequency domain I/Q imbalance correction scheme for OFDM systems," WCNC 2003.

8. R. Rao and B. Daneshrad. 2004. "Analysis of I/Q mismatch and a cancellation scheme for OFDM systems," IST Mobile Communication Summit, June 27–30, Lyon, France.

9. R. Rao and B. Daneshrad. 2004. "I/Q mismatch cancellation for MIMO-OFDM systems," Personal Indoor Mobile Radio Communications Conf. (PIMRC), Barcelona, Spain, Sept. 5–8, 2004.

10. D.B Leeson. 1996. "A simple model of feedback oscillator noise spectrum," *Proceedings of the IEEE*, Vol. 54, Feb. 1996, pp. 329–330.

11. F.M. Gardener. 1980. "Charge-pump phase-lock loops," *IEEE Transactions on Communications*, Vol. COM-28, Num. 6, Nov. 1980, pp. 1849–1858.

12. M. Curtin and P. O'Brien. 1999. "Phase-locked loops for high-frequency receivers and transmitters — part 1–part 3," *Analog Dialogue* 33-3, 33-5, 33-7, 1999.

13. E. Costa and S. Pupolin. 2002. "M-QAM-OFDM system performance in the presence of a nonlinear amplifier and phase noise," *IEEE Transactions on Communications*, Vol. 50, No. 3, March 2002.

14. A. Armada and M. Calvo. 1998. "Phase noise and subcarrier spacing effects on the performance of an OFDM communication system," *IEEE Communication Letters*, Vol. 2, No. 1, Jan. 1998.

15. M. El-Tanany, Y. Wu, and L. Hazy. 2001. "Analytical modeling and simulation of phase noise interference in OFDM-based digital television terrestrial broadcasting systems," *IEEE Transactions on Broadcasting*, Vol. 47, No. 1, March 2001.

16. A. Papoulis. 1984. *Probability, Random Variables, and Stochastic Processes*, New York: McGraw-Hill.

17. T.H. Lee and A. Hajimiri. 2000. "Oscillator phase noise: a tutorial," IEEE *Journal of Solid-State Circuits*, Vol. 35, No. 3, March 2000.

18. J. Salz. 1986. "Modulation and detection for coherent lightwave communications," *IEEE Communications Magazine*, Vol. 24, No. 6, June 1986, pp. 38–49.

19. N.J. Kasdin. 1995. "Discrete simulation of colored noise and stochastic processes and $1/f^{\alpha}$ power law noise generation," *Proceedings of the IEEE*, Vol. 83, No. 5, pp. 802–827, May 1995.

20. N.J. Kasdin and T. Walter. 1992. "Discrete simulation of power law noise," *Proc. 1992 IEEE Frequency Control Symposium*, Hershey, PA, May 1992, pp. 274–283.

21. P. Robertson and S. Kaiser. 1995. "Analysis of the effects of phase noise in Orthogonal Frequency Division Multiplex (OFDM) systems," *ICC 1995*, Seattle, Vol. 3, June 1995, pp. 1652–1657.

22. S. Lang and B. Daneshrad. 2005. "From architecture to implementation of a wireless, multiple antenna testbed," *2005 IEEE International Symposium on Circuits and Systems*, ISCAS, Kobe, Japan, May 23–25, 2005.

23. G.J. Foschini and M.J. Gans. 1998. "On limits of wireless communications in a fading environment when using multiple antennas," *Wireless Pers. Commun.*, Vol. 6, March 1998, pp. 311–335.

24. E. Telatar. 1995. "Capacity of multiantenna Gaussian channels," AT&T Bell Laboratories, Tech. Memo., June 1995.

25. D. Gesbert, M. Shafi, D. Shiu, P.J. Smith, and A. Naquib. 2003. "From theory to practice: an overview of MIMO space-time coded wireless systems," *IEEE Journal on Selec. Areas in Comm*, Vol. 21, No. 3, April 2003, pp. 281–302.

26. A.J. Paulraj, D.A. Gore, R.U. Nabar, and H. Bolcskei. 2004. "An overview of MIMO communications — a key to gigabit wireless," *Proc. of the IEEE*, Vol. 92, No. 2, Feb. 2004, pp. 198–218.

27. A. Paulraj, R. Nabar, and D. Gore. 2003. *Introduction to Space-Time Wireless Communications,* Cambridge, U.K.: Cambridge University Press.

28. D. Shiu, G.J. Foschini, M.J. Gans, and J.M. Kahn. 2000. "Fading correlation and its effects on the capacity of multielement antenna systems, " *IEEE Trans. on Commun.,* Vol. 48, No. 3, March 2000, pp. 502–513.

29. C.-N. Chuah, G.J. Foschini, R.A. Valenzuela, D. Chizhik, J. Ling, and J.M. Kahn, "Capacity growth of multi-element arrays in indoor and outdoor wireless channels," *IEEE WCNC 2000,* Chicago, Sept. 23–28, 2000, Vol. 3, pp. 1340–1344.

30. T. Svantesson and J. Wallace. 2003. "On signal strength and multipath richness in multi-input multi-output systems," *ICC 2003,* Vol. 4, May 11–15, 2003, pp. 2683–2687.

31. R. Rao. 2004. "Performance Analysis of MIMO-OFDM systems," Ph.D. Thesis, University of California, Los Angeles, Oct. 1, 2004.

32. W. Zhu, D.W. Browne, and M.P. Fitz. 2005. "An open access wideband multi-antenna wireless testbed with remote control capability," IEEE TridentCom, Feb. 2005.

33. D. Browne, W. Zhu, and M.P. Fitz. 2005. "A signaling scheme and estimation algorithm for characterizing frequency selective MIMO channels," *IEEE VTC,* May 2005.

34. G.J. Foschini. 1996. "Layered space-time architecture for wireless communications in a fading environment when using multiple antennas," *Bell Labs Technical Journal,* Autumn 1996, pp. 41–59.

35. V. Tarokh, N. Seshadri, and A.R. Calderbank. 1998. "Space-time codes for high data rate wireless communication: Performance criterion and code construction," *IEEE Info. Theory,* Vol. 44, March 1998, pp. 744–765.

36. J.-C. Guey, M.P. Fitz, M.R. Bell, and W.-Y. Kuo. 1996. "Signal design for transmitter diversity wireless communication systems over Rayleigh fading channels," *1996 IEEE Vehicular Technology Conference,* Atlanta, GA, pp. 136–140.

37. S.M. Alamouti. 1998. "A simple transmit diversity technique for wireless communications," *IEEE J. Select Areas Commun.,* Vol. 16, Oct. 1998, pp. 1451–1458.

38. J.-C. Belfiore, G. Rekaya, and E. Viterbo. 2004. "The golden code: A2 × 2 full-rate space-time code with non-vanishing determinants," *IEEE International Symposium on Information Theory,* Chicago, IL, June 2004, p. 308.

39. S. Siwamogsatam and M.P. Fitz, "Improved high rate space–time TCM via orthogonality and set partitioning," *International Symposium on Wireless Personal Multimedia Communications,* Alborg, Denmark, September 2001.

40. S. Siwamogsatam and M.P. Fitz. 2002. "Improved high rate space–time TCM via concatenation of expanded orthogonal block codes and MTCM," *IEEE International Conference on Communication,* New York, April 2002.

41. Z. Chen, J. Yuan, and B. Vucetic. 2001. "Improved space-time trellis coded modulation scheme on slow Rayleigh fading channels," *Electronics Letters,* Vol. 37, March 2001, pp. 440–441.

42. W. Zhu, H. Lee, D.N. Liu, and M.P. Fitz. "Antenna array geometry and coding performance," submitted to IEEE International Symposium on Information Theory, 2005.

43. L. Wei, O.Y. Takeshita, P. Gupta, W. Zhu, and M.P. Fitz. 2003. "Implementation and testing of super-orthogonal space time trellis codes," *IEEE Vehicular Technology Conference,* Volume 3, Fall 2003, pp. 1748–1752.

11

Gigabit Mobile Communications Using Real-Time MIMO-OFDM Signal Processing

Volker Jungnickel, Andreas Forck, Thomas Haustein, Christoph Juchems, and Wolfgang Zirwas

CONTENTS

11.1   Introduction

There is an ever-increasing demand for higher bandwidths in mobile communication systems. While the target of the first generations was mobile telephony, currently deployed systems aim to enhance people's lives by enabling high-speed mobile access to the Internet anywhere and at any time. While the integration of heterogeneous mobile and wireless access techniques may be one important aspect of the next generation (4G), a second goal is the support of even higher data rates, similar to fixed networks, as well as to provide high-quality multimedia applications at reasonable costs for multiple mobile users. These applications need a dramatically increased sum capacity to be realized in parallel to the existing mobile services. Services are expected to become available between 2012 and 2015, most likely in an extended, but definitely limited spectrum with a scalable bandwidth between 5 and 100 MHz [1].

Recent work on wireless data transmission for indoor applications at Gbit/s data rates is merely based on classical microwave [2–4] or infrared [5] techniques. There are commercial products combining both approaches, which replace, in principle, an optical fiber for wireless bridge applications. But these systems generally use directional transceivers to extend the range and to suppress unwanted multipath components: By tracking the direction of a principal path with a narrow-beam antenna, the multipath fading becomes negligible. Since the power of the other paths is so reduced, the link becomes almost "transparent," i.e., the coherence bandwidth is no longer limited by the multipath propagation. Simple single-carrier modulation and demodulation techniques can then be used with no equalization. On the other hand, this concept limits the mobility of the users, since transmitter (Tx) and/or receiver (Rx) must be aligned to each other and a free line-of-sight (LOS) or a strong reflection is required as the principal path.

In a mobile communications environment, the base station often has sector antennas and knows the direction of the terminal equipped with omni-directional antennas only coarsely. True mobility requires such widely opened antenna apertures, and with them it is difficult to select and track the direction of a principal path out of the potentially large number of received signals coming from many different directions. As a consequence, the data signals are corrupted by the statistical fading in time, frequency, and space. Very robust signaling schemes need to be used for reliable communications.

A modern approach for wireless fading channels is to use multiple antennas both at the Tx and Rx in combination with sophisticated space–time (ST) signal processing techniques, as space–division multiple access (SDMA) and multiple-input multiple-output (MIMO). The implementation of advanced signal processing becomes more and more attractive even at mobile terminals. But the complexity of the ST algorithms is considerable, in particular when they are combined with the WCDMA technology used in the third generation of mobile communication systems [6,7]. To reduce this complexity, one may want to simplify the transmission, detection, and coding techniques in future air interfaces.

Almost a decade ago it was noticed [8] that the complexity of the ST processing for multipath MIMO channels can be substantially reduced with the discrete matrix multitone technique operating in the space–frequency domain, which became popular under the term MIMO-OFDM. By performing an inverse fast Fourier transform (IFFT) and inserting a cyclic prefix (CP) between consecutive OFDM symbols at each transmitter, and by removing the CP and performing a fast Fourier transform (FFT) at each receiver, the equivalent space–frequency channel matrix becomes block-diagonal. The well-known SDMA and MIMO algorithms for frequency-flat fading channels can therefore be used also in frequency-selective channels, by applying them individually on each OFDM subcarrier. The complexity for the adaptation of the signal processing to the time-variant channel scales then linearly with N, which is the length of the OFDM symbol without CP (and equal to the number of subcarriers), rather than cubic with N for the space–time approach. Also the complexity of the data reconstruction scales linearly with N rather than quadratic. Using a joint water-filling across the spatial eigen-modes and subcarriers (see Section 11.2), the channel capacity is asymptotically achieved for infinite symbol length with this approach.

But despite this fundamental complexity reduction, it is still challenging to implement MIMO-OFDM in real-time with current digital signal processors (DSPs). Most MIMO-OFDM prototypes in the recent literature transmit the signals over the air and perform the processing off-line, as in [9,10]. But the joint water-filling requires real-time adaptation to the time-variant channel along the entire transmission chain, and the real-time implementation of the signal processing is an essential step toward the application of these new techniques.

The algorithmic part of new signal processing schemes is highly developed in the literature on wireless communications. But there are surprisingly few detailed reports about hardware implementations, and how state-of-the-art real-time signal processing components, such as digital signal processors (DSPs), field-programmable gate arrays (FPGAs), and application-specific integrated circuits (ASICs) can be used to implement new algorithms in order to bring them closer to application.

It is well known that the step from the algorithmic to the register-transfer level is a highly critical process in any implementation where teamwork between algorithm and hardware specialists and a good portion of creative intelligence are needed to get an efficient and reliable system at the end. So it might be helpful to describe such a hardware implementation by using MIMO-OFDM as a timely example.

The first real-time MIMO-OFDM prototype was implemented using a DSP farm [11]. It was intended for the evolution of existing 3G systems and allowed a peak data rate of 37 Mbit/s with two transmit and up to four receive antennas in a bandwidth of 5 MHz. However, one would have to increase the DSP processing power by more than an order of magnitude to cover the projected 4G cell capacity. The ITU-R Recommendation M.1645 states that "potential new radio interface(s) will need to support data rates of up to approximately 100 Mbit/s for high mobility such as mobile access and up to approximately 1 Gbit/s for low mobility such as nomadic/local wireless access."

In order to meet this challenge, we develop a dedicated implementation concept for MIMO-OFDM. As a proof-of-concept, the system is implemented on a reconfigurable experimental prototype realizing the projected 1 Gbit/s data rate in real-time in a bandwidth of 100 MHz. For this, we needed to

- Perform a basic complexity study of the MIMO-OFDM algorithm concerning real-time implementation
- Find a suitable partitioning of the signal processing using state-of-the-art reconfigurable signal processing components (DSP, FPGA)
- Define an efficient preamble for the channel estimation and to realize the corresponding channel estimator
- Analyze the complexity for the adaptation of the MIMO-OFDM signal processing to the time variation of the channel
- Develop an efficient MIMO-OFDM detector that is capable of processing the received data continuously
- Integrate the MIMO-OFDM signal processing with forward error correction and data source and sink
- Perform transmission experiments and measure the performance

This chapter is organized as follows: Section 11.2 describes the implementation concept. Section 11.3 develops the structure of a new preamble for the channel estimation. Section 11.4 introduces the channel estimator and the implementation using a wrapped pipeline structure. Section 11.5 considers methods to speed up the adaptation to the time-variant channel and reports benchmark results concerning the achieved adaptation rate in the experimental system. The data reconstruction unit is described in Section 11.6. Section 11.7 sketches the simplified channel coding, data mapping, and integration of the data source and sink. Section 11.8 gives insights into the signal processing platform and the integration of the system. Section 11.9 describes the measurement set-up and reports results from over-the-air transmission experiments.

11.2 Implementation Concept

Basic algorithm: The implementation concept is derived from the optimal information-theoretic structure of the MIMO-OFDM system [8], which is first briefly reviewed. The transmission is conveniently formulated in the frequency domain as

$$\mathbf{y}_n = \mathbf{H}_n \mathbf{x}_n + \mathbf{v}_n \tag{11.1}$$

where the $(n_{Tx} \times 1)$ vector $\mathbf{x}_n$ contains the signals transmitted from all Tx antennas at the OFDM subcarrier with index $n = 1 \ldots N$ where N denotes the number of OFDM subcarriers. The $(n_{Rx} \times 1)$ vectors $\mathbf{y}_n$ and $\mathbf{v}_n$ contain the received signals and the noise, respectively, at all Rx antennas. The integers n_{Tx} and n_{Rx} denote the numbers of antennas at the transmitter and receiver, respectively. The $(n_{Rx} \times n_{Tx})$ matrix $\mathbf{H}_n$ denotes the channel matrix for sub-carrier n with the complex-valued channel coefficient between each transmit and each receive antenna, and it can be obtained from the time-domain channel impulse response matrices $\mathbf{H}_l$ as

$$\mathbf{H}_n = \sum_{l=0}^{L-1} \mathbf{H}_l exp\left(-j2\pi \frac{nl}{N}\right) \qquad (11.2)$$

where L denotes the number of resolved multipath components. In the optimal way, based on full channel state information (CSI) both at the Tx and Rx, the channel capacity is approached asymptotically by performing a singular value decomposition (SVD) of $\mathbf{H}_n$ on each subcarrier,

$$\mathbf{H}_n = \mathbf{U}_n \mathbf{D}_n \mathbf{V}_n^H \qquad (11.3)$$

which gives the matrices $\mathbf{V}_n$ and $\mathbf{U}_n$ containing the eigenvectors of the channel matrix in the Tx and Rx spaces, respectively, and the diagonal matrix $\mathbf{D}_n$, which contains $i = 1 \ldots \min(n_{Tx}, n_{Rx})$ singular values λ_i^n, referred to as the amplitude gains of the spatial eigenmodes. The superscript H denotes the conjugate transpose of a matrix. The capacity is asymptotically approached for infinite N by a joint water-filling across all spatial eigenmodes i and all subcarriers n [8].

Realization: In a practical system, the Tx signal $\mathbf{x}_n = \mathbf{V}_n \mathbf{d}_n$ is obtained from the transmitted data vector $\mathbf{d}_n$ and the spatially multiplexed data signals are reconstructed at the Rx as $\hat{\mathbf{d}}_n = \mathbf{D}^{-1} \cdot \mathbf{U}_n^H \mathbf{y}_n$. The noise in each stream is then boosted differently, according to the actual singular values at the given subcarrier index. Unlike in the information theory, in practice we employ discrete instead of continuous modulation alphabets. A joint bit-loading and power allocation algorithm is used with individual modulation on each eigenmode and each subcarrier, according to the current channel state, so that important optimization criteria (throughput, fairness, stability of queues) can be fulfilled.

Depending on the availability of the CSI, there are modifications. When CSI is available at the receiver only, no pre-processing can be applied. Assuming additionally linear detection, which requires a simple matrix-vector multiplication, the transmitted signals on each subcarrier may be reconstructed using the minimum mean-square error detector given by the formula

$$\hat{\mathbf{x}}_n = \left(\mathbf{H}_n \mathbf{H}_n^H + \sigma^2 \mathbf{I}\right)^{-1} \mathbf{H}_n^H \mathbf{y}_n \qquad (11.4)$$

where $\mathbf{I}$ and σ^2 are the $(n_{Tx} \times n_{Tx})$ identity matrix and the noise variance at one Rx antenna, respectively.

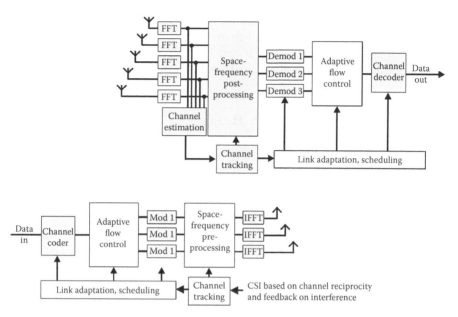

FIGURE 11.1
Implementation concept for the experimental MIMO-OFDM system.

The two proposed concepts are jointly shown in Figure 11.1. The first concept based on the eigenmode signaling requires all components, while the second concept based on the linear MMSE needs only the shaded ones. In what follows, we describe the first concept. The input data are jointly interleaved and encoded across the spatial channels and subcarriers before passing them to the adaptive flow control unit. The consecutive link adaptation uses an individual modulation for each stream and each subcarrier.[1] The next step is a joint pre-processing of the data signals on all streams, which is performed individually for each subcarrier. The resulting signals are then passed through the IFFT, the CP is inserted, and the signals are transmitted over the air. The CSI for the spatio-temporal pre-processing can be obtained by exploiting the reciprocity of the radio channel (i.e., transmit pilot signals in the reverse link and reuse CSI for the forward link), which is possible in the time-division duplex mode or by a suitable feedback technique in frequency-division duplex mode. Additional information on the interference is needed for proper pre-processing and link adaptation. It can be provided via feedback from the Rx over the reverse link.

At the receiver, the CP is removed and the signals are passed through the FFT. The channel estimation is performed in the frequency domain during a dedicated time slot in each frame. Based on the estimates for each subcarrier, for each frame a new set of SVD or MMSE matrices is calculated. The data signals are then reconstructed using spatial post-processing individually for each subcarrier. Thereafter, the signals are demodulated and decoded so the binary data stream can be reconstructed.

The second concept based on the MMSE detector is actually implemented, since the experimental link is unidirectional, due to effort limitations. Due to the lack of CSI at the transmitter, pre-processing and link adaptation cannot be used. Also, the adaptive flow control unit in Figure 11.1 is not yet implemented. Note that the linear MMSE exploits a fraction of the available spatial diversity only. An asymmetric antenna configuration with 3 Tx and 5 Rx is therefore used to add more diversity and to stabilize the link.

Structural analysis and functional mapping: At first, Figure 11.1 reveals that two basic groups of tasks must be realized in real time: In the first group (sample rate processing), we have the IFFTs and FFTs with their complex-valued serial inputs and outputs and a number of tasks that must be performed consecutively and individually for each subcarrier, as modulation, demodulation, channel estimation, and space–frequency pre- and post-processing. This kind of processing includes elementary operations that must be performed at a very high processing clock, which is 100 MHz in the experimental system. The flow of data signals is naturally arranged in a chain. It can be realized so that the input of one operation is the output of the previous operation. Signals are then fed through different steps in the processing chain and signals for consecutive subcarriers are processed simultaneously while being in a different pass. Effectively, each signal needs then a single clock cycle to be processed, which is called "pipelining" in the literature. Pipelining is the first basic principle of how complex computations can be broken down into elementary operations which can be executed in real time so that the effective number of operations performed in each clock cycle is much larger than unity. The second basic principle is parallel computing, which is heavily used in the MMSE MIMO-OFDM detector described below.

The algorithm assessment shows that the sample rate processing for MIMO-OFDM can be organized so that only primitive look-up, multiply, add, or combined multiply-and-accumulate operations are needed with a high potential of parallel processing. This workload is ideally suited for implementation in FPGA.

This is in great contrast to the second group of operations called channel rate processing, which is shown in parallel below the signal path in Figure 11.1. This group includes the adaptation of the entire signal processing chain to the time-variant channel, i.e., the renewal of pre- and post-processing matrices, the link adaptation and the scheduling units. These algorithms operate on a totally different time scale, which must not be confused with the requirements for the sample rate processing. Besides other factors, the required update interval scales with the coherence time of the time-variant wireless channel, which is on the order of 10 ms for 1 m/s velocity in an indoor scenario at 5 GHz carrier frequency.

Some of these channel rate algorithms operate on matrices that sometimes become singular. The results may then have a high dynamic, which is most easily handled in floating point arithmetic. Furthermore, these algorithms depend on the available type of the CSI, and we want to maintain and exchange them easily. For testbed purposes, the channel rate processing is best implemented in a floating point DSP operating in a loop, which renews all settings in a fraction of the channel coherence time.

11.3 A Code-Multiplexed Preamble

For MIMO and SDMA, a precise CSI must be available. In [9], the simplest time-multiplex approach is used for the training signals, i.e., the same preamble is transmitted consecutively by all antennas. But the experimental results in Figure 11.12 in [9] indicate that there is a significant penalty when this kind of channel estimation is used. An obvious drawback of the time-multiplexed preamble is that only one out of n_{Tx} antennas is active at once, and so the potentially valuable contribution of the other antennas to the signal-to-noise ratio is lost.

Here we use a code-multiplexed preamble being orthogonal over multiple OFDM symbols. All antennas and subcarriers are active during the whole training phase, i.e., each antenna transmits at the maximal energy. The use of all subcarriers enables a rapid carrier-wise raw estimation, which may be of advantage in space–frequency scheduled multi-user systems where it enables the ultimate granularity of the resource allocation — on each spatial mode and on each carrier.* In real time and at very high data rates, there are additional constraints. Normally it takes some time until the weight matrices are obtained from the channel estimates. Moreover, there is no buffer memory in the experimental system to delay the sampled and Fourier-transformed signals for one or more frames.** Only in this way can they be processed/time aligned with the corresponding weight matrices. Hence the weights are applied with some delay in our system, which is minimized when the channel estimates on all subcarriers and for all antennas are available instantly after the last training symbol.

The preamble structure is most easily explained in the frequency-time grid: On the time axis (i.e., OFDM symbol by OFDM symbol), the pilots for different antennas shall be mutually orthogonal. This is achieved by assigning a characteristic code $H_j(k)$ to the *j-th* Tx antenna out of the well-known Hadamard sequence set. To reduce the hardware effort at the receiver, it is essential to reuse the same sequence $H_j(k)$ for all subcarriers. But this would yield an undesired sharp pulse after the IFFT. So the subcarrier signals are scrambled in the frequency domain using the same binary sequence S_n on all Tx antennas. This way the binary signal structure in the frequency-time grid is maintained, and we can use a simpler correlation circuit for the

---

* In particular, we need no multiplication in the estimator.
** For future implementations, it is highly recommended to foresee such first-in first-out (FIFO) buffer memory with an interface of considerable bandwidth to transfer the 12 bit I and Q signals for multiple antennas from and to the FPGA.

First antenna

Third antenna

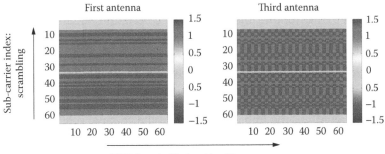

OFDM symbol index: identify antennas

FIGURE 11.2
Graphical representation of the preamble according to Equation 11.1: red: 1, blue: –1, green: 0.

channel estimator.* In the experimental system, we have used the well-known scrambling sequence $S_{-26...26}$ from the 802.11a standard. For other numbers of subcarriers, a suitable scrambling sequence may be found by extensive computer search among random sequences having a small peak-to-average power ratio of the Tx signal after the IFFT. The elementary form of the preamble is given as

$$p_n^j(k) = \cdot S_n \cdot H_j(k) \qquad (11.5)$$

where $j = 1...n_{Tx}$. Note that this form of the preamble is also suggested by others [12,13]. In Figure 11.2, the training signals at the first and third antenna according to Equation 11.1 are shown in the subcarrier-index vs. OFDM training-symbol-index grid.

In Figure 11.2, one can observe that the same scrambling is used for all antennas along the frequency axis. The first antenna is marked by the all-ones Hadamard sequence, while the third Hadamard sequence used for the third antenna alters the sign after each second symbol.

Some modifications have been added to this elementary preamble form for practical reasons.

- We have observed that it is good to avoid the all-ones sequence since it regularly causes slightly biased estimates not present for other sequences, most likely due to an imperfect DC offset compensation in our analogue IQ demodulators. So the original Hadamard matrix (which is taken over from MATLAB) is flipped, i.e., the last sequence is now assigned first, and so the critical sequence is not used.
- We have assigned different sequences to the I and Q branches of the complex-valued Tx signal, similar to our recent flat fading real-time

---

* A time-domain estimation would need a larger bandwidth to predict the channel on a single subcarrier.

MIMO implementation [14], to estimate I and Q branches independently. This doubles the number of correlation circuits, but it fits
better to the real-valued signal processing used in the experimental
system.

• OFDM systems are often subject to a residual IQ imbalance in the
RF units. The resulting cross-talk between equally indexed image
carriers in the upper and lower side-band depends, in addition to
the strength of the imbalance, also on the channel coefficient of the
image carrier [15]. When we use the same sequence $H_j(k)$ in both
sidebands, the image carrier may leak signal energy into the desired
carrier, which is likely to cause a certain channel estimation error.
We have occasionally observed ill-reconstructed signal constellations, accordingly, if the channel of the desired subcarrier is faded.
Visibly, the constellations on respective subcarriers become more
stable when the upper and lower side-bands are marked with a
different sequence, allowing also a base-band correction of the
imbalance in the future when the increased complexity becomes
affordable.

Note that these modifications increase the preamble length. The minimal
length is $P = n_{Tx}$ OFDM symbols without and $P = 4^* \, n_{Tx}$ training symbols
including all modifications.

11.4 Channel Estimation

The number of channel coefficients to be estimated for MIMO-OFDM is
enormous. In the experimental system, we have explicitly estimated all coefficients between each I and Q input and output with 3 Tx and 5 Rx antennas
and 48 subcarriers, which results in a number of $4 \cdot n_{Tx} \cdot n_{Rx} \cdot N = 2880$ coefficients. The above-described preamble has the inherent advantage that we
can get a raw estimate for the coefficients on all subcarriers almost instantaneously after the end of the last training symbol. In principle, additional
interpolation between adjacent carriers (see below) is not required.

Raw estimation: Channel estimation is performed in the frequency domain
using a correlation over multiple OFDM training symbols. Using a separate
correlation circuit (CC) for each of the 2880 coefficients would be prohibi-
tively complex. But since the frequency-domain scrambling is the same for
all Tx antennas, it can be reversed prior to the channel estimation. Then we
have effectively the same sequence on all subcarriers for a given channel
input, and the same CC can be reused for all carriers. This allows an efficient
implementation of the channel estimator, as explained in the following.

Figure 11.3 shows a frequency-time grid where each column corresponds
to one OFDM training symbol and each row to one subcarrier. After the

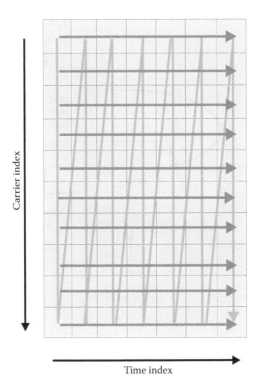

FIGURE 11.3
Frequency-time grid to explain the principle of the channel estimation (see text).

pipelined FFT for each Rx antenna in the FPGA realization, the grid is scanned column-by-column, as indicated by the zig-zag line. A separate correlation must now be performed for each subcarrier from OFDM symbol to OFDM symbol as indicated by the horizontal arrow.

Let us assume that we have a separate memory for each subcarrier and each pair of inputs and outputs where intermediate results can be stored. Now the desired correlation is performed in a piecewise manner: If we arrive at a particular carrier index n and at OFDM symbol index k, we recall the last intermediate result from the memory, then we add or subtract the received signal, according to the current sign of the Hadamard sequence, and finally we store the new intermediate result in the memory. Then we go to the next subcarrier. In this way, one correlation circuit can be reused for all subcarriers, for one pair of inputs and outputs, which reduces the complexity by the factor N.

However, these three steps (read, add/subtract, write) must be performed consecutively, which needs a higher clock in the channel estimator than for other sample rate operations. To avoid this higher clock, the circuit is actually implemented as a wrapped pipeline using a dual-port RAM and an adder as shown in Figure 11.4. At first, the scrambling is reversed and the signal

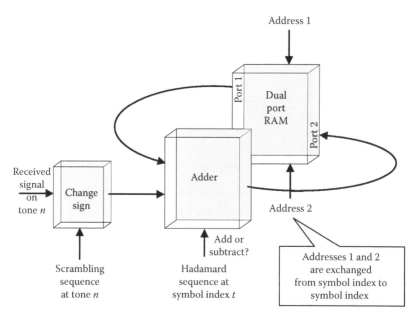

FIGURE 11.4
Wrapped pipeline structure used to operate the channel estimation at sample clock rate.

is either added to or subtracted from the last intermediate result. The latter is recalled from port 1 of a dual-port RAM, and the result is stored in port 2. The idea is then to exchange the two-port address spaces for read and write operations from training symbol to training symbol, where the address of each subcarrier is used once in each address space and counted through by the carrier index n. All processes are organized in this wrapped pipeline so that the read, add or subtract and write tasks are effectively performed in a single cycle. We have operated $4 \cdot n_{Tx} \cdot n_{Rx} = 60$ of these circuits in parallel at 100 MHz in the experimental system. This highly parallel channel estimator operates very reliably, even in complex FPGA designs. We like to point out that the realization of the channel estimator using the wrapped pipeline was one of the key ideas for stable operation of the entire MIMO-OFDM signal processing core at 100 MHz clock.

Raw estimation error: The received signal at the ith Rx antenna and n^{th} subcarrier is given as

$$y_n^i(k) = \sum_{j=1}^{n_{Tx}} H_n^{ij} \cdot p_n^j(k) + n_n^i(k) \tag{11.6}$$

where H_n^{ij} is the channel coefficient to be estimated and $n_n^i(k)$ the noise at the receiver. The pilot sequence $p_n^j(k)$ identifies the jth Tx antenna and all sequences together are normalized as

$$\sum_{j=1}^{n_{Tx}} \sum_{n=1}^{N} |p_n^j(k)|^2 = 1 \qquad p_n^j = \sqrt{\frac{1}{n_{Tx} \cdot N}} \cdot \{\pm 1\} \qquad (11.7)$$

Now the channel estimation is performed as a correlation over all consecutive OFDM training symbols at the nth subcarrier of interest as

$$\hat{H}_n^{il} = \frac{n_{Tx} \cdot N}{P} \sum_{k=1}^{P} p_n^{*l}(k) \cdot y_n^i(k) =$$

$$\frac{n_{Tx} \cdot N}{P} \sum_{j=1}^{n_{Tx}} H_n^{ij} \cdot \sum_{k=1}^{P} p_n^{l*}(k) \cdot p_n^j(k) + \frac{n_{Tx} \cdot N}{P} \sum_{k=1}^{P} p_n^{l*}(k) \cdot n_n^i(k) \qquad (11.8)$$

Provided that the sequences are orthogonal

$$\sum_{k=1}^{P} p_n^{l*}(k) \cdot p_n^j(k) = \frac{P}{n_{Tx} \cdot N} \cdot \delta^{lj} \qquad (11.9)$$

where δ^{lj} is the Kronecker symbol ($\delta^{lj} = 1$ for $l=i$ and $\delta^{lj} = 0$ elsewhere), one obtains

$$\hat{H}_n^{il} = H_n^{il} + \frac{n_{Tx} \cdot N}{P} \sum_{k=1}^{P} p_n^{l*}(k) \cdot n_n^i(k) \qquad (11.10)$$

Since the pilot tones have just positive or negative sign, the statistics of the Gaussian noise are not modified by the multiplication in Equation 11.10. Now the noise is modeled as a random process as

$$n_n^i(t_k) = \sqrt{\frac{1}{N \cdot SNR}} r_n^i(t_k) \qquad (11.11)$$

where SNR is the signal-to-noise ratio at one Rx antenna and r a complex Gaussian number with zero mean and unit variance, then the sum in Equation 11.5 reduces to

$$\hat{H}_n^{il} = H_n^{il} + \sqrt{\frac{1}{G \cdot SNR}} N_n^{il} \qquad G = \frac{P}{n_{Tx}} \qquad (11.12)$$

where G is called the estimator gain and N_n^{il} is a complex Gaussian random number with zero mean and unit variance.

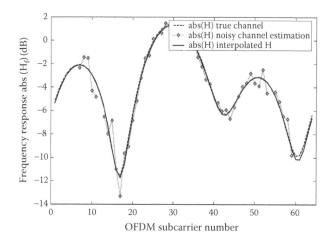

FIGURE 11.5
The OFDM channel estimation can be enhanced by interpolation of raw estimates.

With the above described preamble, the minimum number P_{min} of OFDM training symbols to identify the Tx antennas is n_{Tx}. In addition, P_{min} must be a power of 2. By doubling the length of the preamble, one can double G and correspondingly reduce the raw estimation error.

Interpolation: Actually, Equation 11.12 is the same result as for flat fading MIMO channel estimation, when each antenna is marked with a sequence out of an orthogonal set. In OFDM systems, however, there is inherent redundancy in the channel estimates, which can be used to reduce the channel estimation error. This method is called interpolation in the following.

The mathematical formulation is based on the elementary observation that Equation 11.2 states an over-determined set of N equations for the frequency-domain estimates, with only L variables being the time-domain channel coefficients. The redundancy can be exploited by first solving this set for the time-domain coefficients, from which a filtered version of the frequency-domain estimates is obtained in a second step by discrete or fast Fourier transform. The algorithm is described in Appendix 11A. For MIMO-OFDM, it is applied separately for each pair of Tx and Rx antennas.

Figure 11.5 shows a simulation result for $N = 48$ and $L = 16$, according to the IEEE 802.11a standard where L is set equal to the cyclic prefix length. The dashed line corresponds to the true channel, from which we get noisy estimates on only 48 of the total number of 64 subcarriers (red dots) after the raw channel estimation. After application of the interpolation, the estimation error is reduced and the true channel is almost perfectly reconstructed (solid line).

As shown in Figure 11.6, the variance of the channel estimation (CE) error can be described by an enhanced estimator gain

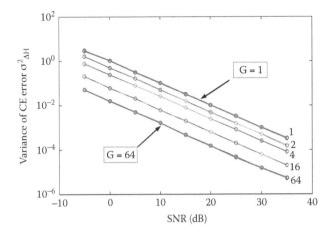

FIGURE 11.6
Simulation for the variance of the channel estimation error in Gaussian noise.

$$\hat{H}_n^{il} = H_n^{il} + \sigma_{\Delta H} \cdot N_n^{il} = H_n^{il} + \sqrt{\frac{1}{G \cdot SNR}} \cdot N_n^{il} \qquad G = \frac{P \cdot N}{n_{Tx} \cdot L} \qquad (11.13)$$

where N is the number of active subcarriers and L the number of multipath components in the channel. It is obvious from Equation 11.13 that there is a well-defined tradeoff between the parameters of the channel estimation. Within these limits, the improved estimator gain can be used to

- Reduce the length of the preamble P
- Reduce the number of subcarriers N on which pilots are transmitted
- Reduce the channel estimation error by using *a priori* knowledge about the multipath channel. When L is known in advance, for instance by exploiting channel data from the most recent frames, L could be much smaller than the length of the cyclic prefix. So the estimator gain can be boosted.

The interpolation must be performed once for each pair of Tx and Rx antennas. In the experimental system, we have not had sufficient resources for performing the interpolation. The implementation on the current DSP consumed a significant part of the processing power, due to the large number of 15 complex-valued channel frequency responses to be interpolated. So we have skipped the interpolation routine in favor of a faster adaptation to the time-variant channel. Equivalently, we have used a longer preamble as described above to improve the raw estimates. According to Equation 11.12, the estimator gain in the experimental system is $G = 64/3 = 13.3$ dB. Further research is concerned with extending this algorithm to a bunch of subcarriers (chunk). Chunkwise interpolation is a key component in future MIMO-OFDM signal processing chains. It is needed for systems with large numbers of subcarriers to distribute the calculation of weights onto multiple DSPs, as explained in the next section, and to reduce the computational complexity accordingly.

11.5   Adaptation to the Time-Variant Channel

Adaptation to the time-variation of the fading channel is challenging with the limited DSP processing power in the experimental system. First results with a single-carrier MIMO system [14] already indicated that substantial effort is needed here to fulfill the mobility requirements even for indoor environments.

For the MMSE detector, an update of the N weights matrices for the space–frequency post-processing is needed in a fraction of the channel coherence time. To get a rough idea of the required processing power, we have implemented the calculation of the pseudo-inverse matrix in MATLAB 5.3 and counted the required number of floating point operations (FLOPs). For a MIMO system with 2 Tx and 2 Rx or 4 Tx and 4 Rx, one needs 419 or 4.800 FLOPs, respectively, to get the matrix $(HH^H)^{-1}H^H$ for a single carrier needed in Equation 11.4 for $\sigma^2 = 0$.* Using these values, the required processing power P_{DSP} is lower bounded by

$$P_{DSP} > \frac{N \cdot FLOPs}{T_{frame}} \tag{11.8}$$

which gives 10 and 115 MFLOPs/s in the two examples above when the number of subcarriers and the frame length are 48 and 2 ms, respectively. So the adaptation to the time variation is, in principle, feasible for moderate numbers of subcarriers and antennas with current DSPs claiming peak processing powers of more than 1 GFLOPs/s.

But DSPs are a special kind of processor. In principle, they may perform multiple add- and multiply-operations as well as memory accesses in a single cycle, which makes them favorable for operations with large matrices and vectors. Only in such applications and with highly optimized code does a DSP approach the performance limit claimed in the data sheet. Our dedicated optimization of the MIMO algorithms on the DSP is described in detail in [17]. Here we report some quantitative results as a benchmark of what is

---

* Note that MATLAB is based on the highly optimized LAPACK matrix-vector routines. The results obtained in this way may be optimistic for a DSP implementation, where, to our knowledge, routines with a truly comparable performance are not yet commercially available.

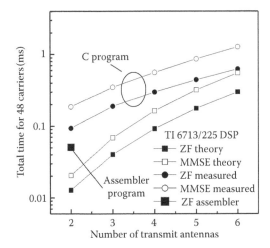

FIGURE 11.7
Measured times needed to track the multipath channel using the TI 6713 DSP at 225 MHz.

feasible with the experimental system. The performance bounds in the two bottom lines in Figure 11.7 have been derived from an analysis of the MMSE and zero forcing algorithms. The data on the time axis have been measured with the floating-point DSP TI 6713 clocked at 225 MHz. We roughly need 150 or 400 µs for the 2 Tx and 2 Rx or 4 Tx and 4 Rx configurations, respectively, to renew the MMSE weight matrices for all 48 subcarriers, which is shorter than the channel coherence time in indoor scenarios, as required. Note that additional time (384 µs) is needed twice in the experimental system: to transfer the channel estimates and MMSE filter weights from and to the FPGA, respectively. The transfer time is due to the slow external memory interface between the DSP and the FPGA. An attempt with the Greville algorithm [18] in assembly language has reduced the value for the 2 Tx and 2 Rx system down to 50 µs, but for this case the numbers of antennas is not variable. Note the increasing gap to the performance bound toward smaller matrix dimensions where the pipelined multiply-and-accumulate operations in the DSP are less effective than for larger matrices. Pipelining is, in general, more effective when large numbers of identical operations can be performed consecutively.

Future mobile systems may have a larger number of subcarriers, and they may need faster adaptation to the channel variation than our experimental system. According to Equation 11.8, the required processing power scales with the ratio of the number of subcarriers and the frame length, which is a favorable characteristic of OFDM. More processing power will consequently be needed.

It is therefore proposed to use multiple DSPs connected to the FPGA or ASIC in a star configuration as sketched in Figure 11.8. Each DSP is responsible for a certain subset of subcarriers, and it has an individual connection to the respective channel and weight matrices, so that the transfer time and

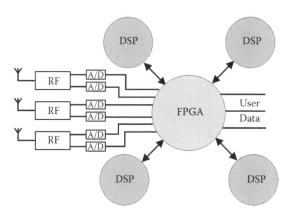

FIGURE 11.8
A star of DSPs can be used around an FPGA to speed up the channel tracking.

computational effort can be shared among the DSPs. With this star configuration, the engineer can trade off hardware costs against mobility.

Intuitively, interpolation of the weight matrices between adjacent subcarriers might be a promising option to speed up the adaptation to the time-
variant channel [19]. However, the gain of currently known interpolation
techniques becomes noticeable only when the delay spread of the wireless
channel is rather small, relative to the CP duration. The fixed base-point
interpolation in [19] causes an error floor, which can be steered using an
adaptive step width between base points [20] or even avoided by using
perfect interpolation as shown in [21,22]. A detailed complexity analysis [23]
confirmed that there are few scenarios where the algorithms proposed in
[21,22] actually realize a reduced computational complexity. As observed in
[20], this is true only when the delay spread is very small. A mathematical
analysis of the reasons, therefore, is given in [21]: The interpolation of the
channel matrix needs L base points but it takes $n_{Tx} \cdot L$ base points for the
channel inverse matrix, which is equal to the total number of 48 subcarriers
used for 3 Tx antennas in the 802.11a standard when L is set equal to the
cyclic prefix length of 16. Based on what is currently known about the inverse
matrix interpolation, we feel that the pragmatic approach used in our experimental system, to optimize the calculation of the MMSE weight matrix for
the DSP processor architecture and applying this routine independently for
all subcarriers, is an efficient solution for most cases.

11.6 Data Reconstruction

In the optimal approach described in Section 11.2, the pre- and post-processing requires linear matrix-vector multiplications to be performed individually for each subcarrier. Also the linear MMSE detector requires this

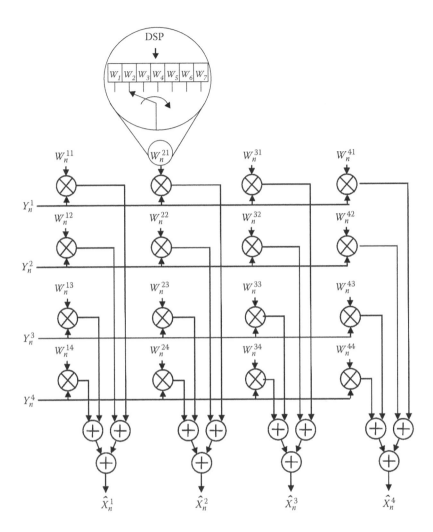

FIGURE 11.9
The MIMO-OFDM data reconstruction uses a pipelined matrix-vector multiplication unit. Inputs are the I and Q signals from each Rx antenna (Y, left), outputs are the reconstructed data signals ($\hat{X}$, bottom).

operation. So it is straightforward to equip the detection unit with this elementary operation. For each data stream, a scalar product with an individual weight vector must be realized after the FFTs for each received signal vector. To realize an instantaneous reconstruction of the streams, a pipelined matrix-vector multiplication unit is implemented as sketched for a 2 Tx and 2 Rx MIMO-OFDM configuration in Figure 11.9. The signals Y_n^i incoming from the left are the received signals for subcarrier index n leaving the pipelined FFT unit subcarrier by subcarrier. Odd indices $i = 2 \cdot k - 1$ denote the real part of the complex-valued signal from the kth receive antenna, and even indices $i = 2 \cdot k$ the imaginary part from that antenna, where k is a natural number. In the example shown in Figure 11.9, each signal branch is distributed to four multipliers for which dedicated hardware building blocks in the FPGA have been used. All multiplications are performed simultaneously (i.e., in one 10 ns cycle), and the results are added pairwise in a subsequent pipeline until a stream is reconstructed. Due to the pipelining, the final result is obtained after a short delay depending on the number of antennas. The entire data reconstruction with 3 Tx and 5 Rx requires 60 simultaneous multiplications (in one cycle) for 10 received and 6 transmitted I and Q signals and four cycles for pairwise executed additions, resulting in only 50 ns delay after the corresponding subcarrier signals leave the pipelined FFT units, which is negligible compared to the OFDM symbol duration of 0.8 µs.* So the data reconstruction unit performs the required 100 million matrix-vector multiplications per second continuously in real time.

The major challenge is that the weight matrices differ from carrier to carrier and need to be exchanged rapidly. This is shown exemplary for the weight W_n^{21} in Figure 11.9. Approximately once per frame, the frequency response of each weight is written as a vector by the DSP into a dual-port RAM assembled from the dedicated hardware memory blocks in the FPGA. The second port of the dual-port RAM is connected to the dedicated multiplier used. Now the address of the weight in the vector is counted through at 100 MHz clock synchronous with the increasing subcarrier index n of the incoming signals (leaving the FFT unit subcarrier by subcarrier). In this way, the matrix vector multiplication pipeline is reused for all subcarriers. The number of multipliers needed is related only to the numbers of antennas used while the memory effort scales in addition with the number of subcarriers.

The rapid exchange of weights is a key idea enabling the implementation of MIMO-OFDM with current FPGAs at 100 MHz bandwidth as shown in the experimental system. Even higher bandwidths may be possible using the same technique in an ASIC clocked at higher speed, accordingly. The spatially multiplexed data streams appear separated from each other after the matrix-vector multiplication unit. The subsequent signal processing can be organized in parallel using conventional pipelined OFDM receiver processing chains. System integration is straightforward (see Figure 11.11).

Therefore, we have organized the weight matrices for all carriers in register pages assembled of 60 dual-port RAM blocks, where each RAM block contains the weights for all subcarriers for one input-output pair and is located next to the corresponding multiplier. The weights are written once per 2 ms frame into the RAM blocks by the DSP using the outer port. Via the inner port, the matrix-vector pipeline reads out the corresponding weight for the current subcarrier and pair of input and output once per sample clock cycle. In the next cycle, it switches to the next weight for the next subcarrier by incrementing the weight address. This switching is performed simultaneously for all 60 weights. Once per OFDM symbol, hence, all weight matrices

---

* The FFT and IFFT units are implemented in parallel for all antennas using an FPGA core module from XILINX. The pipelined units need about 1.5 OFDM symbol durations (1.2 µs) to provide the output.

are consecutively used in the right order. The number of subcarriers that can be handled so is limited by the number of the Block-RAM units in the FPGA. The current FPGA design can handle up to 256 subcarriers without using more Block-RAM units. For more subcarriers, the weight memory blocks must be assembled from the available units in the FPGA.

11.7 Framing, Mapping, Channel Coding, and Real-Time Data Interface

The experimental system uses 48 data subcarriers out of a total number of 64. The sampling clock of the complex base-band signals is 200 MSPS both in the I and Q branches and the processing clock after the digital filtering is 100 MHz (which is called the sample clock). The CP covers 160 ns, which is sufficient for small rooms. The OFDM symbol duration is 0.8 μs.

Data are continuously transmitted and received in frames with a fixed length of 2 ms with no additional pauses between the frames. The structure of the 1 Gbit/s frame is shown in Figure 11.10. It consists of 64 training symbols for the channel estimation, an idle gap of 16 symbols, a data block with 2400 symbols, and another idle gap of 20 symbols. With 64-QAM modulation and three antennas, we transmit 864 bits per symbol (48·18). The 2400 data symbols per 2 ms block, hence, correspond to a payload data rate of 1036.8 Mbit/s.

The channel coding concept is shown in Figure 11.11. The payload is taken over from the data source, scrambled and multiplexed with pseudo-random data. Then the stream is split into four parallel streams individually encoded using a convolutional code with a constraint length of 7 as in the IEEE 802.11a standard. In the experiments, channel coding with rate ½ is used. The encoded data are fed into a pseudo-random interleaver with a block length of 288 bytes of encoded data. With 64-QAM modulation (48·6 bits/symbol), the blocks are mapped onto eight consecutive OFDM symbols. The interleaving covers more symbols when the number of bits per symbol is reduced,

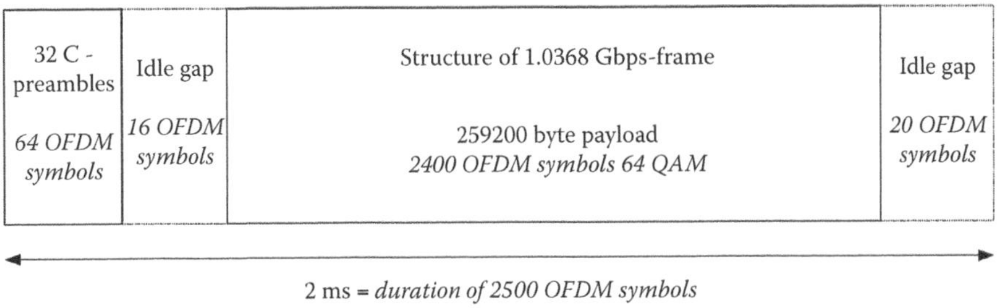

32 C - preambles	Idle gap	Structure of 1.0368 Gbps-frame	Idle gap
64 OFDM symbols	*16 OFDM symbols*	259200 byte payload *2400 OFDM symbols 64 QAM*	*20 OFDM symbols*

←──→

2 ms = *duration of 2500 OFDM symbols*

FIGURE 11.10
Structure of the 1 Gbit/s frame. The structure is the same for all antennas.

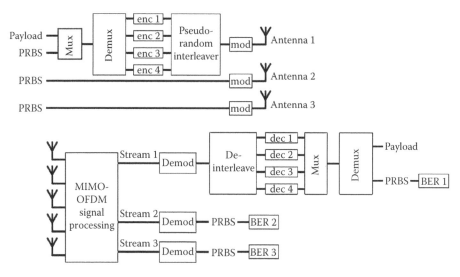

FIGURE 11.11
Parallel decoding is used to overcome the speed limitations in the FPGA.

as for 16-QAM and QPSK modulation. The encoded stream is passed through the modulator and then used at the first antenna only. The other two antennas are loaded with pseudo-random data. At the receiver, the reconstructed stream of the first antenna is at first de-mapped and de-interleaved. Then the original coded stream is reconstructed, de-multiplexed, and fed into four parallel Viterbi decoders (realized in the second FPGA with an FPGA software core module provided by XILINX) with a trace-back length of 96, according to the multiplexing at the transmitter. The parallel coding concept is required to overcome the speed limitations of the Viterbi decoder core in the FPGA, which can handle up to 100 Mbit/s, depending on the code rate.

The parallel decoding concept is obviously scalable to 1 Gbit/s, but the hardware effort scales linearly with the data rate. A third FPGA would be required to fully decode all streams. We feel that encoding the first stream is sufficient to proof feasibility. With the full concept, it would, of course, be helpful to interleave over all antennas, which better exploits the transmit diversity.

The MIMO-OFDM detector has knowledge about the post-detection SINR on each subcarrier and each stream. Once per frame, this information is quantized and delivered from the DSP to the Viterbi decoder to enable soft-decision decoding (one hard plus two soft bits). The soft decoding has an obvious effect on the performance: besides the higher coding gain, it better exploits the multipath diversity. Finally, the decoded payload is de-scrambled and fed into the data sink.

Data source and sink are added to the system for public demonstrations and realized with two Real-time Linux-based PCs running a simplified medium access control (MAC) protocol stack up to the Internet protocol (IP) layer where the payload is bridged to an Ethernet interface. Measurements showed stable operation up to 80 Mbit/s, indicating that the current MAC concept must of course be revised and simplified in the future to enable the envisaged Gbit/s data rates on top of the MAC protocol stack. Two notebook computers are coupled to the data source and sink PCs via Ethernet, respectively, and a conventional IP network is operated, with the unidirectional radio link in between. The reverse link is replaced by an Ethernet cable. In this way, MPEG-encoded HDTV video transmission and data transfer are enabled.

11.8 Implementation, Complexity, and System Integration

The payload data from the source PC are coupled into the base-band processing FPGA via flat ribbon cable connected to a dedicated 64-bit PCI interface card. The Tx signal processing is distributed over two Virtex2-6000 FPGAs placed on the *Chip-it Gold* platform [24]. While the first FPGA contains the data mapping and forward error correcton encoder, the second one contains the three OFDM transmit chains. Each complex base-band signal is connected to two 12-bit AD9753 digital-to-analogue (DA) converters operating at 200 MSPS on a small printed circuit board (see Figure 11.12) having a large number of short parallel connections to the input-output ports at the FPGA. Three such boards are used to realize the three Tx chains.

The operation of OFDM links requires radio front-ends with negligible IQ imbalance. The base-band signals are up-converted to a 900 MHz intermediate frequency (IF) using three AD8349 IQ modulators, which have been carefully adapted to each DA converter, so that the phase error is below 0.1° and base-band responses in I and Q branches are quasi-identical. The IF signal is up-converted in a second stage to the 5.26 GHz radio frequency. The received signals are down-converted to the same IF as used at the transmitter. AD8347

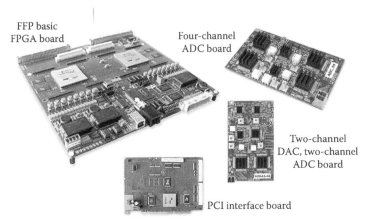

FIGURE 11.12
Commercially available components are used for the reconfigurable signal processing platform.

IQ demodulators are used to down-convert the IF signals to the complex base-band, and the IQ mismatch in each demodulator is carefully adapted to each analogue-to-digital (AD) converter, so that the cross-talk between image carriers is negligible in the entire OFDM transmission chain. For the matching procedure, see [25]. For lower bandwidth designs, one may consider digital IQ modulation, which is not yet available for 100 MHz bandwidth.

At each Rx antenna, the base-band signals are sampled at 200 MSPS with two AD9430 AD converters placed on a small printed circuit board having a large number of connections in parallel to the input-output ports at the FPGA. The ADC board already hosts four converters (see Figure 11.12), and three such boards are used in the experimental system for the 5 Rx chains.

For the receiver base-band processing, a dedicated rapid prototyping platform is used (*FFPbasic*, see Figure 11.12 and [25]) carrying two Virtex-2/PRO-100 FPGAs (speed grade 6). The whole MIMO-OFDM base-band processing, except the adaptation to the time-variant channel, is implemented in the first FPGA, while the other realizes the de-interleaving, decoding, and de-mapping, as well as the interface to the host PC.

The adaptation to the time-variant channel is implemented in a floating-point DSP TI 6713 placed on a standard TI development board clocked at 225 MHz clock and coupled to the FPGA using the external memory interface (EMIF). The DSP is asynchronously coupled, and it can read from and write to dedicated memory blocks in the FPGA (see [14]).

In the table below, the resources used in the FPGA are reported after compiling our MIMO-OFDM software core with Version 6.3 of the XILINX ISE development tool (from the FFT inputs to the separated data streams), for the 3 Tx, 5 Rx configuration. The 1 Tx, 1 Rx configuration is given in brackets as a reference. It can be observed in Table 11.1 that the FPGA is not fully occupied, which allows efficient routing of the design and reliable operation at the 100 MHz sample clock. In the trial implementation described in this chapter, the focus is to demonstrate that the MIMO-OFDM signal processing (being the heart of the new system) is feasible in real time. Synchronization is initially realized by cables. Over-the-air synchronization for MIMO-OFDM is very similar to conventional OFDM, except that the performance can be enhanced by exploiting spatial diversity. It has been added later in the experimental system together with an enhanced number of subcarriers and the hardware effort required for this will be reported elsewhere.

TABLE 11.1

FPGA Resources Used for the Implementation of the 3 Tx, 5 Rx (1 Tx, 1 Rx) Signal Processing

	FFT	Channel Estimation	Data Reconstruction	Total (Integrated)
BlockRAMs	95 (19)	60 (4)	60 (4)	215 (27) of 444
Multipliers	45 (9)	—	60 (4)	105 (13) of 444
Slices	9878 (2401)	1272 (48)	2904 (24)	15.235 (2.473) of 44.036

TABLE 11.2

Computation Times and Latencies of Different Processing Steps in the 3 Tx, 5 Rx Signal Processing

Function	Computation Time	Latency	Comment
IFFT/FFT	No	1.2 μs each	Symbol-wise pipelined XILINX FFT core, Ver. 2.1
QAM mapping/demapping	No	10 ns	Carrier-wise pipelined uses look-up tables
Synchronization	—	—	Not yet implemented
MIMO channel estimation	51.2 μs equal to preamble length	No	Wrapped pipeline
Calculation of MMSE weight matrices	0.5 ms in DSP	No	Application of weights is delayed by one frame
Data reconstruction	Carrier-wise pipelined	50 ns	See text
De-interleaver and decoder	Pipelined	3.4 μs	16-QAM, see text

In lower bandwidth designs, parts of the components can be reused, such as for the FFTs, which can be consecutively used for all antennas at a correspondingly higher clock. Also, the logic in the channel estimator can be reused, when the correlation circuit is implemented once for each Tx antenna and used consecutively for all Rx antennas with correspondingly increased clock speed. In the matrix-vector multiplication engine, one may reconstruct the parallel streams consecutively at a correspondingly higher clock. However, the memory effort does not reduce in this way for both the channel estimator and the data reconstruction unit, since it depends on the number of subcarriers and on the antenna configuration. No components are reused in the experimental system. The maximal clock of the integrated MIMO-OFDM processing core is predicted by the XILINX ISE synthesis tool as 147 MHz.

As already mentioned, the sample rate processing is fully pipelined, in order to allow a continuous signal flow. So one cannot assign a time consumption to a required operation, as this is familiar for block-wise signal processing preferred in DSP-based implementations. In a pipelined FPGA implementation, there is just latency for each operation since the output of one operation is the input of the next operation in the chain, and operations are performed simultaneously on consecutive signals. For the sake of completeness, we report the latency of important system functions in Table 11.2. The latency for the de-interleaving and decoding is measured as the time between the first bit input of a data block at the de-interleaver and the first bit output at the Viterbi decoder for 16-QAM modulation, and it is larger for QPSK and smaller for 64-QAM.* The given value is intended here as a rule-of-thumb for estimating the total latency which varies, accordingly.

---

* De-interleaving and decoding operate on large bit sequences and input registers must first be filled before output can be generated. This is faster at higher data rates.

FIGURE 11.13

Top left: 3-antenna transmitter. Top right: 5-antenna receiver. Bottom left: Received signal spectrum in a 100-MHz frequency span. Bottom right: Local monitoring terminal (LMT) showing the received 64-QAM signal constellations of the three data streams, for all subcarriers. The first stream is zoomed.

Elementary system functions are supervised using the USB 2.0 interface on the Rx FPGA board, which is frequently read out with a local monitoring terminal (LMT) running on a notebook computer. The LMT makes available samples from the reconstructed signal constellations on each stream, the mean powers between each Tx and Rx antenna, the singular values of the channel matrices, as well the good-put, which is the number of correctly transmitted bits per second, as well as the un-coded and coded bit error rates. Long-term statistics for these quantities are recorded in a log-file, for off-line evaluation purposes.

The integrated transmitter and receiver racks are shown in Figure 11.13, as well as a received signal spectrum and an LMT screenshot in the 64-QAM mode. Each rack contains an integrated RF unit (bottom) and a base-band unit (top). The base-band unit contains the PC (left), the FPGA board, and (at the Rx only) the DSP (right). The received signal spectrum and the separated 64-QAM signal constellations have been measured after transmitting the signals over the air. The actual good-put is 1035.8 Mbit/s, and the un-coded bit error rate is 10^{-4}. The experimental system was presented at the 3GSM World Congress, February 14–17, 2005 in Cannes, France.

11.9 Transmission Experiments

Next, we report on over-the-air measurements with the experimental system in a mobile communications scenario. Measurements are conducted in a

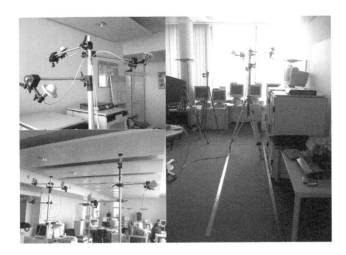

FIGURE 11.14
Measurement scenario. Top left: Tx antenna configuration. Bottom left: Rx antenna configuration. Right: The Tx is moved along a line through the room.

$15 \times 7 \times 2.8$ m^3 office room in a SIEMENS building at Werinher-Strasse in Munich, Germany. At the Tx, three antennas are used addressing different field directions (see Figure 11.14, top left). The omni-directional Tx antennas (360°h, 25°v, 20° down tilt) are placed at 2-m height on a wheeled photo stand. The stand is moved like a cable car using a DC motor and guided by a rail on a 4-m track through the room at 1-m distance from the short front of the room (see Figure 11.14, right). The movement over 70 wavelengths at the 5.26 GHz carrier frequency forms well-reproducible channel statistics. The Rx sector antennas (65°h, 35°v) were fixed at irregular positions approximately looking toward the Tx in a line forming a distributed antenna scenario (Figure 11.14, bottom, left). The distance between the Tx track and the Rx line was about 4 m.

We first used the built-in channel estimator to obtain information about the broadband MIMO channel. The channel is recorded along the whole measurement track. In Figure 11.15, the average power delay profile is shown. The time axis is scaled in samples and one sample corresponds to 10 ns.

It is observed that the peak is much wider than a sample. The 100 MHz bandwidth obviously resolves multiple paths, mostly due to the multipath propagation in the room. When the paths are independently faded, the larger bandwidth is a valuable source of multipath (or frequency-) diversity, which is a new feature of broadband compared to narrow-band MIMO systems. Not only is the capacity a figure of merit in these systems but also the multipath diversity can be exploited to improve the system performance. Diversity effects are traditionally observed in the bit error rate curves at high SNR. But one may observe them also in the cumulative distribution of the channel capacity.

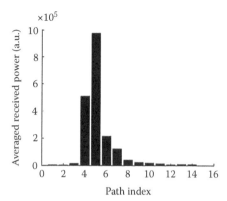

FIGURE 11.15
Average power delay profile along the measurement track.

The capacity distribution is obtained from the measured channel data as follows. The data are at first corrected for the frequency response of the transmitter and receiver chains, which exhibit some ripple due to the analogue and digital base-band filters. Since we have not observed significant slow fading on the relatively short track, the mean received power is averaged over both all carriers and all snapshots, which gives the normalization factor η in Equation 11.9. Finally, the broadband capacity is obtained using the formula

$$C_{broadband} = \frac{1}{N}\sum_{n=0}^{N-1}\log_2 \det\left(1 + \frac{SNR}{n_t \cdot \eta}\mathbf{H}_n\mathbf{H}_n^H\right) \qquad (11.9)$$

The narrowband capacity immediately follows from Equation 11.9 for $N = 1$, i.e., for a single carrier.* Results are shown in Figure 11.16. The statistics for the broadband capacity are based on calculating Equation 11.9 with $N = 48$ for 296 snapshots along the track. For the narrowband case, the capacity of all 48 channels for each subcarrier is individually calculated for all 296 snapshots, which forms a smoother statistic, due to the larger ensemble.

As expected, the cumulative distribution is shifted right when the numbers of antennas are increased. The slope is slightly steeper, and the curve is shifted right with additional Rx antennas (compare the 3 Tx, 3 Rx and the 3 Tx, 5 Rx configurations). The benefit of the multipath diversity becomes obvious since the broadband distributions (solid lines) are significantly steeper than in the narrow-band case (dashed lines). For the 3 Tx, 5 Rx

---

* Actually, we have used the real-valued representation of Equation 11.9, according to our measured channel data, which means that the real-valued channel matrices have the dimensions $(2 \cdot n_r \times 2 \cdot n_t)$ so that the normalization factor η and the capacity are multiplied by two and divided by two, respectively.

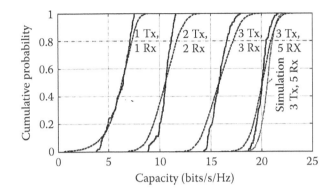

FIGURE 11.16
Statistics of the narrow- (dashed lines) and broad-band capacities (solid lines) in the measured scenario for various antenna configurations at SNR = 20 dB. A simulation curve (dotted line) for the 3 × 5 configuration for the Rayleigh fading channel with four independently and identically distributed paths is given as well.

configuration, we have also plotted a simulation result for the Rayleigh fading channel with four paths (dotted line). Note that the measured capacities are slightly smaller than the numerical results, most likely due to the presence of the LOS signal.

The steeper capacity distribution in multipath fading is well known from simulations [26]. It can quantitatively be described using the standard deviation of the distribution. In Figure 11.17, simulation results are plotted vs. the number of paths L, for the 3 Tx, 3 Rx and 3 Tx, 5 Rx configurations. There is a certain offset between the curves, depending on the number of excess antennas at the Rx. But in general, the width of the capacity distribution is approximately inversely proportional to $\sqrt{L}$, which is not so obvious in

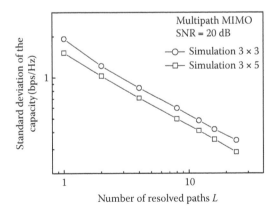

FIGURE 11.17
Standard deviation of the capacity distribution for different numbers of resolved paths.

Equation 11.9, where the capacity is averaged over the number of subcarriers. Effectively, the capacity is averaged over the number of independent paths in the channel, which is intuitive. Large numbers of paths with significant separation in the delay domain are typical for urban outdoor scenarios where the line of sight between the base station on a rooftop of a higher building and the mobile terminal in a street canyon may be frequently blocked or shaded so that the received signal is mostly due to multipath propagation. The more independently faded paths are resolved, the more such a broad-band MIMO system performs as a wired link, despite the fading in the radio channel.

In Figure 11.18 (top), the measured bit error rates (BERs), spatially resolved along the measurement track, are individually plotted for the three spatially multiplexed streams using 64-QAM modulation, which is the 1 Gbit/s mode. At the bottom of Figure 11.18, the coded error rate for the first data stream is plotted for the same measurement. Due to the fading, the un-coded error rates vary along the track between 10^{-3} and $3 \cdot 10^{-2}$. The regions with lower and higher error probability are reproducible also for the 16-QAM and QPSK modes, of course at substantially reduced error numbers. Approximately at 10^{-2}, the coded 64-QAM data are error free. Only at three positions, a minor outage occurs in the first stream, but the coded error rate always remains below 10^{-6}. Note that there are substantial regions along the track where all three data streams have an un-coded error rate below 10^{-2}. In these regions, the good-put (i.e., the number of correctly detected bits per second) is larger than 1 Gbit/s, and all three encoded streams may simultaneously show no outage, if the channel coding is fully implemented.

In Figure 11.19 (top), we have plotted the average un-coded bit error rates for BPSK, 16-QAM, and 64-QAM modulation vs. attenuation at the transmitter. The Tx power at all antennas is set equal (±1 dB) to 0 dBm at first. The automatic gain control (AGC) is then regulated at a central position on the track and then held fixed for all measurements.* The Tx is moved along the whole track for each attenuation at the transmitter.

At no attenuation, the average bit error rate in the 64-QAM mode is about 10^{-2}, and the coded data of the first stream have an average error rate of $1.2 \cdot 10^{-8}$, indicating the low outage probability despite the statistical fading in the channel. For the 3 Tx and 5 Rx antenna configuration, with the MMSE detector, one would expect a spatial diversity order of 3 in the Rayleigh fading channel, which is comparable with the slope of the un-coded bit error rate curve for QPSK in the high SNR region, as expected. The coded bit error rate curves are, of course, much steeper, which is partly attributed to the additional multipath diversity.

---

* The AGC is set jointly for all antennas such that the antenna signal with the largest received power is below a threshold empirical determined by minimizing the bit error rate in all streams due to clipping of peak amplitudes at the AD converters.

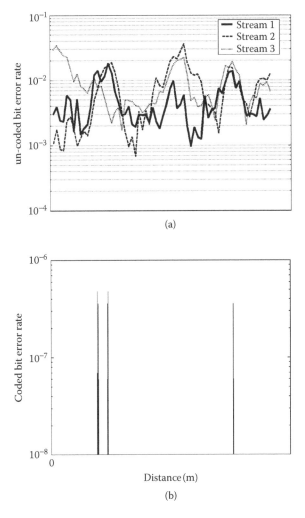

FIGURE 11.18
(a) Un-coded bit error rates of the three data streams along the measurement track in 64-QAM mode. (b) Coded error rate for stream 1 in the same measurement.

11.10 Conclusions

We have presented a real-time concept for mobile communications that is suitable for data rates of 1 Gbit/s, based on MIMO-OFDM. Perhaps the most striking feature of MIMO-OFDM is that it is easily scaled to higher bandwidth, as demonstrated in this chapter, whereby the required signal processing complexity increases just linearly with the bandwidth (in the same channel).

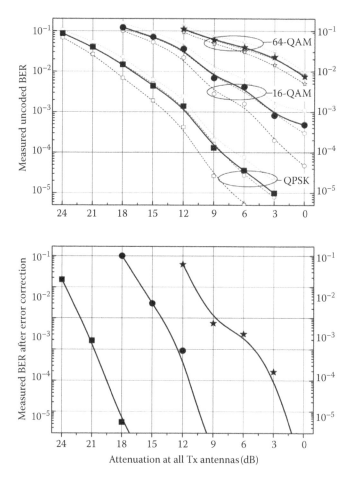

FIGURE 11.19
Measured un-coded and coded bit error rates vs. the attenuation at all Tx antennas.

For this reason, the complexity of the spatio-temporal coding can be handled with currently available signal processing components. Additionally, the same fundamental approach as used for narrow-band can be applied for broad-band applications as well. The concept may, therefore, be a good starting point for the next generation of mobile systems, when it is further equipped with the required multi-user functionality.

We have shown how the elementary base-band signal processing for MIMO-OFDM can be implemented efficiently on a reconfigurable real-time signal processing platform and integrated with the MIMO-OFDM core in a fully functional radio system.

Mobile transmission experiments have been conducted in an indoor environment, and the system allowed reliable transmission of data at a gross data rate of up to 1 Gbit/s. After error correction coding, the link is free of error at almost all locations along the measurement track. Mobile data and video transmission was also demonstrated with the experimental system.

The excellent performance of the experimental system may be attributed to three major reasons. The indoor channel offers sufficiently rich scattering. The influence of channel estimation errors is negligible due to the high estimator gain, and the two excess antennas at the receiver enable a performance similar to a 3 Tx, 3 Rx MIMO system using the optimal maximum-likelihood detection instead of the linear MMSE.

But there are some points noteworthy for future research. The overall MIMO-OFDM physical layer is currently feasible in real time for low mobility with a moderate number of subcarriers. The limiting factor here is the large number of matrix inversions or singular value decompositions to be performed in the DSP in a fraction of the channel coherence time. With the same number of antennas but 20 times the number of subcarriers (as recently discussed in the 3GPP Long Term Evolution study item), about 20 times the DSP power would be needed for the adaptation to the time-variant channel compared to our experimental system. In the FPGA, just the memory effort would be increased by factor 20. Fortunately, a single user is assigned to only a fraction of the bandwidth and so the computational effort at the terminal may still be in a realistic region for the down-link. For the up-link at the base station, however, the full processing power must be available. A practical way out is to perform those matrix computations in a chunkwise manner, i.e., for subgroups of subcarriers, with multiple DSPs operating in parallel in order to meet these challenging requirements. In this way, the transfer times of channel estimates and weights can be reduced as well.

Also the MAC functions must be further simplified and partly implemented in hardware, similar to the sample rate processing for MIMO-OFDM, to allow such high data rates at the interface of the application in the future.

11.11 Summary

We develop a concept and describe the realization of an experimental mobile communication system that can transmit data continuously up to 1 Gbit/s in mobile scenarios by means of real-time base-band signal processing.

To allow for mobility, omni-directional antennas are used. This is in contrast to fixed wireless links, which normally use directional antennas and line-of-sight connections. A coded OFDM scheme is used to overcome the resulting multipath fading effects. The spectral efficiency is boosted with multiple transmit and multiple receive antennas. In order to identify at the receiver the signals from multiple transmit antennas, we apply a novel preamble enabling precise channel estimates on each OFDM subcarrier by means of a highly parallel correlation circuit. Spatial multiplexing is individually applied on each OFDM subcarrier. Techniques to speed up the adaptation of the base-band processing to the time-variant channel are described, and it is shown how the spatially multiplexed streams can be instantaneously reconstructed.

Key functions of the concept have been realized and tested in an experimental system. The real-time base-band processing is implemented on a hybrid FPGA/DSP platform. Using a few elementary processing components, the implementation is easily scalable with respect to the numbers of subcarriers and antennas. For the gross data rate of 1 Gbit/s, the experimental system is configured with three transmit and five receive antennas, and it uses 48 out of 64 OFDM subcarriers. All carriers cover a bandwidth of 100 MHz at an RF carrier frequency of 5.26 GHz.

Finally, the system is used for mobile transmission experiments. Using the built-in channel estimator, we investigate the channel properties of an exemplary mobile indoor scenario. Then we report on over-the-air bit error rate measurements with the experimental system for various modulation formats.

11.12 Acknowledgments

The authors wish to thank Stefan Schiffermueller and Clemens von Helmolt (both from Fraunhofer HHI); Frank Luhn, Marian Pollock (from IAF GmbH); and Matthias Lampe, Joseph Eichinger, and Egon Schulz (from SIEMENS) for their valuable contributions in completing this project. We are grateful to the Bundesministerium für Bildung und Forschung (BMBF) and SIEMENS for financial support in research projects Coverage, HyEff, 3GeT, and WIGWAM.

Appendix 11A

Here we describe an elementary interpolation scheme to improve the frequency-domain channel estimates. It is based on the elementary relation (Equation 11.2) between frequency- and time-domain channel coefficients (H_n and h_l, respectively) in OFDM systems. Equation 11.2 is a set of N equations with L variables, where L is the number of resolved multipaths. To solve for h_l, the values H_n are stacked in a $(1 \times N)$ vector $\mathbf{H}$ where N is the total number of subcarriers. Similar to the discrete Fourier transform, Equation 11.2 is rewritten as a multiplication of a $(N \times L)$ matrix $\mathbf{W}$ with a vector containing the time-domain channel coefficients stacked in the $(1 \times L)$ vector $\mathbf{h}$

$$\mathbf{H} = \mathbf{W} \cdot \mathbf{h} \tag{11A.1}$$

where the elements of the matrix $\mathbf{W}$ are given as

$$W_{nl} = \exp\left(-j2\pi\frac{nl}{N}\right) \tag{11A.2}$$

In practice, some values in $\mathbf{H}$ are not available, due to a spectrum mask or the presence of pilot carriers. The available estimates are, hence, described by a reduced form of Equation 11.2 as

$$\mathbf{H}_{red} = \mathbf{W}_{red} \cdot \mathbf{h} + \mathbf{N}_{red} \tag{11A.3}$$

where the index *red* means that the rows corresponding to the missing estimates in $\mathbf{H}$ and $\mathbf{W}$ are filled with zeros. Also, there is a vector $\mathbf{N}_{red}$ in Equation 11.A.3 describing the estimation error. In a first step, the time-domain estimates are obtained by using the pseudo-inverse $\mathbf{W}_{red}^{+}$ as

$$\hat{\mathbf{h}} = \mathbf{W}_{red}^{+} \cdot \mathbf{H}_{red} \tag{11A.4}$$

when $\mathbf{W}_{red}$ is well conditioned. This is true if $N_{red} \geq L$ (interpolation rule), where N_{red} is the number of subcarriers on which channel estimates are available. In a second step, Equation 11.2 is used to interpolate the channel coefficients.

This way we also get useful results for channel coefficients on those sub-carriers where estimates are not available. Note that the product $\mathbf{W} \cdot \mathbf{W}_{red}^{+}$ has dimension $(N_c \times N_c)$ but rank L, which explains the filtering effect observed in Figure 11.5 and Figure 11.6.

Appendix 11B

Here we like to mention a particular phenomenon when the signal processing and the radio frequency (RF) front ends are combined to operate the OFDM link over the air. Beside the IQ imbalance problems already mentioned in the text, in addition the subcarrier assignment is mixed up at first sight.

The proper assignment for a given RF transmitter can be tested in the lab. Use an arbitrary waveform generator (like the SMIQ- or SMU-ARB from R&S or a flexible transmitter signal processing platform) to create time-domain OFDM test signals from MATLAB, for instance. Load the test signal into the generator, feed the signal into the complex-valued base-band IQ inputs of the transmitter front end, and observe the spectrum of the trans-mitted signal with a spectrum analyzer (as the FSQ form R&S). With our RF front ends, we have observed the following:

- Starting from MATLAB, the subcarrier index n varies from 1 to N.
- The subcarrier with index 1 is centered at the RF carrier frequency. Due to the instable DC offset at the receiver* it is normally not used for communication.
- The subcarriers with indices $2 \leq n \leq N/2$ are mapped "en block" to frequencies larger than the RF carrier.
- The subcarriers with indices $N/2+1 \leq n \leq N$ are mapped "en block" to frequencies smaller than the RF carrier such that the first subcarrier in this group with index $N/2+1$ has the smallest RF frequency and the last subcarrier with index N is close to the RF carrier. Intuitively this assignment is clear from the periodicity of the discrete Fourier transform: if $y(n)$ is the FFT of $Y(k)$, then $y(n) = y(n + N)$. Starting from the natural placement of the subcarrier with $n = 1$ at the RF carrier frequency, the observed carrier placement is explained by continuing the FFT spectrum periodically to low RF frequencies.
- If base-band I and Q inputs are exchanged, the carrier assignment is reversed.
- Once subcarriers are properly addressed at the transmitter such that the rectangular spectrum of OFDM signals is realized, the order of signals at the receiver can be checked accordingly. If the order is not correct, changing the sign of the Q branch in time domain may be helpful.

We have matched the OFDM signal processing according to our RF front ends such that subcarriers are placed properly both in the RF domain at the transmitter and in the base-band domain at the receiver. In general, take care at the base-band-to-RF interface; design the OFDM signals accordingly. Do not confuse base-band I and Q inputs or outputs.

---

* It is related to the "carrier-feedthrough" signal from the IQ modulator at the transmitter, which is more or less faded in the radio channel and then down-converted to a very unstable DC signal at the receiver.

References

1. 3GPP TSG RAN Future Evolution Workshop, November 2–3, Toronto, CA, ftp://ftp.3gpp.org/workshop/2004_11_RAN_Future_Evo/.
2. P.F. Driessen and L.J. Greenstein. 1995. "Modulation techniques for high-speed wireless indoor systems using narrowbeam antennas," *IEEE Transactions on Communications*, Vol. 43, No. 10, Oct. 1995.
3. C. Evci, A. de Hoz, R. Rheinschmitt, M. Araki, M. Umehira, M.A. Beach, P. Hafezi, A. Nix, Y. Sun, S. Barberis, E. Gaiani, B. Melis, G. Romano, V. Palestini, and M. Tolonen. 1999. "AWACS: system description and main project achievements," *Proceedings ACTS Mobile Communications Summit*, Sorrento, Italy, June 8–11, 1999.
4. G. Grosskopf, A. Norrdine, D. Rohde, and M. Schlosser. 2004. "Gigabit Ethernet Transmission Experiments at 60 GHz," *Proceedings 13th International Plastic Optical Fibres Conference*, Nürnberg, Germany, Sept. 2004 (on CD-ROM).
5. D.R. Wisely. 1996. "A 1 Gbit/s optical wireless tracked architecture for ATM delivery," *Proceedings IEE Colloquium on Optical Free Space Communication Links*, London, pp. 14/1–7.
6. C.B. Papadias and H. Huang. 2001. "Linear space-time multiuser detection for multipath CDMA channels," *IEEE Journal of Selected Areas in Communications*, Vol. 19, No. 2, pp. 254–265.
7. V. Jungnickel, H. Chen, and V. Pohl. 2005. "A MIMO RAKE receiver with enhanced interference cancellation," *Proceedings Vehicular Technology Conference (VTC Spring)*, Stockholm, Sweden, May 30–June 2, 2005 (on CD-ROM).
8. G.G. Raleigh and J.M. Cioffi. 1998. "Spatio-temporal coding for wireless communications," *IEEE Transactions on Communications*, Vol. 46, No. 3, March 1998.
9. A. van Zelst and T. C.W. Schenk. 2004. "Implementation of a MIMO OFDM-based wireless LAN system," *IEEE Transactions on Signal Processing*, Vol. 52, No. 2, Feb. 2004, pp. 483–494.
10. W. Xiang, D. Waters, T.G. Pratt, J. Barry, and B. Walkenhorst. 2004. "Implementation and experimental results of a three-transmitter three-receiver OFDM/BLAST testbed," *IEEE Communications Magazine*, Vol. 42, No. 12, Dec. 2004, pp. 88–95.
11. C. Dubuc, D. Starks, T. Creasy, and Y. Hou. 2004. "A MIMO-OFDM prototype for next generation wireless WANs," *IEEE Communications Magazine*, Vol. 42, No. 12, Dec. 2004, pp. 82–87.
12. K. Mizutani, K. Sakaguchi, J. Takada, and K. Araki. 2004. "Development of 4×4 MIMO-OFDM system and test measurements," *Proceedings 12th European Signal Processing Conference (EUSIPCO)*, Vienna, Austria, Sept. 6–10, 2004 (on CD-ROM).
13. S. Häne, D. Perels, D.S. Baum, M. Borgmann, A. Burg, N. Felber, W. Fichtner, and H. Bölcskei. 2004. "Implementation aspects of a real-time multi-terminal MIMO-OFDM testbed," *Proceedings IEEE Radio and Wireless Conference (RAWCON)*, Atlanta, GA, Sept. 2004 (on CD-ROM).
14. V. Jungnickel, T. Haustein, A. Forck, U. Krueger, V. Pohl, and C. von Helmolt. 2004. "Over-the-air demonstration of spatial multiplexing at high data rates using real-time base-band processing," *Advances in Radio Science*, Vol. 2, pp. 135–140, available: http://www.copernicus.org/URSI/ars/ARS_2_1/135.pdf.

15. T. Ylämurto. 2003. "Frequency domain IQ imbalance correction scheme for OFDM Systems," *Proceedings IEEE Wireless Communications and Networking Conference (WCNC)*, New Orleans, LA, March 16–20, 2003 (on CD-ROM).
16. G.L. Stuber, J.R. Barry, S.W. Mclaughlin, Y.G. Li, M.A. Ingram, and T.G. Pratt. 2004. "Broadband MIMO-OFDM wireless communications," *Proceedings of the IEEE*, Vol. 92, Feb. 2004, pp. 271–294.
17. T. Haustein, A. Forck, H. Gäbler, V. Jungnickel, and S. Schiffermüller. "Real time signal processing for multi-antenna systems: algorithms, optimization and implementation on an experimental test-bed," accepted for *EURASIP Special Issue on MIMO Implementation Aspects*, 2005.
18. F. Gantmacher. 1986. *Matrizentheorie*, Berlin: Springer-Verlag.
19. J. Wang and B. Daneshrad. "Performance of linear interpolation-based MIMO detection for MIMO-OFDM systems," *Proceedings IEEE Wireless Communications and Networking Conference*, Atlanta, GA, March 21–25, 2004 (on CD-ROM).
20. V. Jungnickel, T. Haustein, A. Forck, S. Schiffermueller, H. Gaebler, C. von Helmolt, W. Zirwas, J. Eichinger, and E. Schulz. 2004. "Real-time concepts for MIMO-OFDM," *Proceedings CIC/IEEE Global Mobile Congress*, Shanghai, China, October 11–13, 2004.
21. M. Borgmann and H. Bölcskei. 2004. "Interpolation-based efficient matrix inversion for MIMO-OFDM receivers," *Proceedings 38th Asilomar Conf. Signals, Systems, Computers*, Pacific Grove, CA, Nov. 2004 (invited).
22. D. Cescato, M. Borgmann, H. Bölcskei, J. Hansen, and A. Burg. 2005. "Interpolation-based QR decomposition in MIMO-OFDM systems," *Proceedings 6th IEEE Workshop on Signal Processing Advances in Wireless Communications (SPAWC)*, New York, June 2005 (invited).
23. T. Haustein, S. Schiffermueller, V. Jungnickel, M. Schellmann, T. Michel, and G. Wunder. 2005. "Interpolation and noise reduction in MIMO-OFDM — a complexity driven perspective," *Proceedings ISSPA*, Sydney, Australia, Aug. 28–Sept. 1, 2005.
24. http://www.prodesign-europe.com/ce/CHIPitGoldEdition.htm
25. http://www.iaf-bs.de/products/add-on-boards/
26. A.J. Paulraj, D.A. Gore, R.U. Nabar, and H. Bölcskei. 2004. "An overview of MIMO communications — a key to Gigabit wireless," *Proceedings IEEE*, Vol. 92, No. 2, Feb. 2004, pp. 198–218.

12

Network Planning and Deployment Issues for MIMO Systems

Thomas Neubauer, Ernst Bonek, and Christoph Mecklenbräuker

CONTENTS

12.1 Network Planning

Network planning for MIMO systems is still in its infancy while the planning of radio networks with beamforming enhancements is more mature. The possible gain obtained from MIMO transmission and reception is currently being intensely investigated by information theorists and baseband signal processing experts. From a purely physical layer perspective, MIMO systems provide three different types of gain that can be traded off against each other:

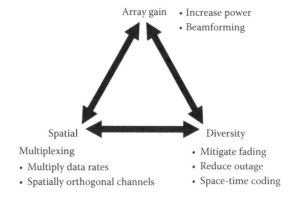

FIGURE 12.1
The magic triangle of Array Gain, Diversity, and Spatial Multiplexing [1].

spatial diversity gain [2], spatial multiplexing gain [3], and beamforming gain. These tradeoffs are illustrated in the triangle in Figure 12.1. In his doctoral thesis, Weichselberger has shown that this "magic triangle" can be broken up into three dichotomies, depending on whether the channel is directive or diverse at either link end [1]. The chart in Figure 12.2 illustrates the possible dichotomies depending on the available diversities or directivities of the MIMO channel. The diversity gain decreases outage probabilities, while the multiplexing gain increases physical layer symbol rate. Decreased outage probabilities translate into increased coverage without increasing transmit power.

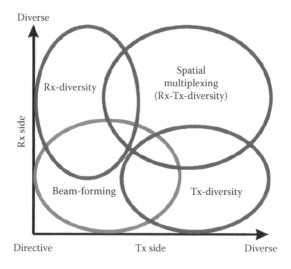

FIGURE 12.2
Diversity vs. directivity dichotomies.

TABLE 12.1

Information-Theoretic Limits on Gains in SIMO, MISO, and MIMO Systems [3]

Configuration	Channel State at Tx	Expected Array Gain (Beamforming Gain)	Diversity Order
SIMO	Unknown	M_R	M_R
SIMO	Known	M_R	M_R
MISO	Unknown	1	M_T
MISO	Known	M_T	M_T
MIMO	Unknown	M_R	$M_R M_T$
MIMO	Known	$E\{\lambda_{max}\}$	$M_R M_T$

Diversity gains have been exploited *on the receiver side* in mobile communications for several decades. Relatively recent is the introduction of *transmit diversity* through space–time coding in third generation cellular networks. According to the 3GPP air interface specification, transmit diversity must be supported by the terminals. This means that network operators have the choice of whether or not to provision for transmit diversity in their radio networks. A significant challenge associated with specifying the performance of downlink transmit diversity in cellular networks is the identification of a representative mobile terminal performance. The mobile terminal implementation influences the performance of transmit diversity schemes. For packet services with automatic repeat request (ARQ), diversity gains at the physical layer translate into a reduced number of required retransmissions. Thus, transmit diversity gains result in increased MAC-layer throughput for packet services by reducing the average retransmission count. Spatial transmit diversity schemes are particularly suited for microcells where the capacity gains are large and beamforming is less appropriate due to large angular spread [4].*

Spatial multiplexing gains increase the symbol rate on the transmitter side and translate into higher peak data rates. Currently, spatial multiplexing is not specified as a mandatory feature in any type of commercial radio access network, although proprietary MIMO systems with spatial multiplexing gains are commercially available. Spatial multiplexing is anticipated as a future enhancement for third generation cellular systems, wireless LANs, and wireless metropolitan networks (MANs). For spatial multiplexing, the identification of a representative mobile terminal performance is even more troublesome than for the case of transmit diversity.

A summary of the information-theoretical limits on the available gains from multiple transmit and receive antennas is given in Table 12.1. The SIMO configuration refers to a single transmit antenna and M_R receiver antennas,

---

* Actually, the use of angular spread as a means for characterizing a MIMO radio channel is rather coarse. It is more appropriate to evaluate angular spreads after *clustering* the propagation paths of the MIMO channel [5].

which correspond to receiver antenna diversity. We note that the diversity gain becomes M_R and the expected array gain is also M_R. For the MISO configuration, multiple transmit antennas are deployed with a single receive antenna. This corresponds to transmit diversity when the transmitter is ignorant of the channel and to transmit beamforming if the transmitter knows the channel, respectively. The limits in this table assume that the receiver always knows the channel. We note that the gains are not just limited by the numbers of antennas in a cellular network configuration, but also by the availability of channel state information at the transmit side. For further details, we refer to [3].

A key problem of network planning for MIMO systems is a lack of available MIMO propagation models, which can be used for predicting the signal field strength jointly with the number of available dimensions in the signal space for specific sites. In contrast, propagation models for single antenna transmission and reception can be considered mature. In modern radio network planning tools, the propagation model consists of a basic pathloss model, line-of-sight checking, and corrections for topography, morphography, and street orientation [4, chap. 3]. Propagation models for MIMO transmission are much less understood, but the recently developed model within COST 273 carries great promise [6].

For MIMO systems in which the transmitter does not have channel state knowledge, the instantaneous capacity can be calculated from

$$C = \log \det \left(I_{M_R} + \frac{\rho}{M_T} HH^H \right)$$

where I_{M_R} denotes the identity matrix of order M_R and H is the MIMO channel matrix, ρ is the mean receive SNR, and X^H is the Hermitian transpose of X. If the distribution of H were known, the distribution of C could be determined. Sadly, the distribution of H parameterized by receiver location, base station antenna array height, and configuration in real environments is little understood. Most current MIMO literature treats the matrix elements of H as (more or less) independent identically distributed, which leads to optimistic values of capacity. It would mean that adjacent receive antenna elements carry fully decorrelated signals, e.g., as a consequence of random multipath from all directions in a non-line-of-sight (NLOS) environment. In [7], it was shown that MIMO capacity can have a large local variation in an indoor scenario, depending on the position of the antenna arrays and on the environment. The only solution currently at hand is to carry out detailed MIMO channel measurements to link topology and morphography with MIMO channel statistics.

Finally, we point out that the gains summarized in Table 12.1 affect the signal to interference plus noise ratio (SINR) of individual links. These gains affect the link quality statistics directly. In interference-limited situations, these gains translate into throughput gains. If the situation is operating near the hard-blocking limit, no such throughput gains can be achieved.

12.1.1 Introduction to Network Planning

In order to understand why effective radio network planning is essential for every operator, we will present some numbers from an investor's perspective [8]: Governments in Europe raised about 120 billion Euro in fees for UMTS licenses. As an example, the government in the U.K. raised 39.3 billion Euro in fees from five Mobile Network Operators (MNOs), which is about 31.2 Euro per head of the population per year, for the duration of the license validity (20 years).

All of a sudden, many national governments realized that UMTS fees could be extremely valuable, and this set the scene for a feeding frenzy among governments trying to maximize the proceeds of access rights to UMTS frequency bands. The German auction resulted in a per capita fee (30.9 Euro per head of the population per year — for 20 years) that was similar to the per capita fee in the U.K. market. This enormous amount of money, which is just an investment for the "admission to further invest money," has to be earned by the operators during the license validity period of typically 15–20 years. For many other governments the outcome of further auctions was disappointing, or in other words, operators had a more realistic view of business cases and license fees.

3G operators are currently making heavy investments in UMTS infrastructure. The analysis in [8] shows that the costs of building 3G networks in Europe will be in excess of 140 billion Euro.

In order to launch these networks, strategies for the deployment must be coupled with realistic business cases in order to determine the demand for future services and applications, as well as investments required for network infrastructure.

The evaluation of network infrastructure requirements can be achieved by system dimensioning, i.e., radio network planning. The network planning must be able to model the system behavior for individual business cases, service profiles, and network loadings. Since radio network planning and optimization are ongoing processes, the network efficiency, and hence the required investments, will heavily depend on them.

The capacity and coverage analyses in 3G CDMA networks can no longer be separated from the actual traffic demand in the system. However, the future 3G traffic demand can be divided into two individual aspects.

- Increased data rates: In GSM the average data rate is on the order of 10 kbit/s. In UMTS we expect an increase in the service data rate of up to 384 kbit/s or even higher.

- Increased number of subscribers: The number of users in UMTS networks is expected to grow substantially over the next few years.

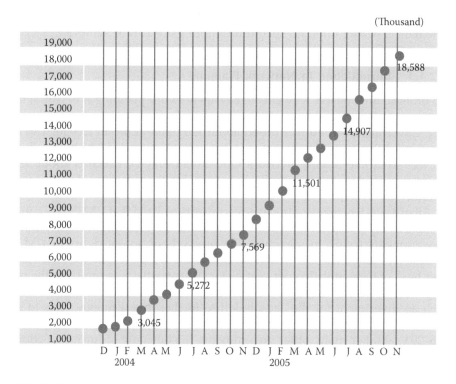

FIGURE 12.3
Subscriber growth from the world's first commercial W-CDMA network, FOMA (NTT DoCoMo, Japan). Source: NTT DoCoMo.

Figure 12.3, as an example from the world's most experienced UMTS network FOMA, shows the increase in the number of users in the W-CDMA network of NTT-DoCoMo in Japan, reaching more than 12 million by April 2005.

Since 3G CDMA networks are interference-limited, higher traffic means higher interference, and hence, a permanent network improvement and optimization based on future traffic and service expectations will be essential. A direct consequence of this is the development of enhanced radio network planning and optimization tools, with the aim of adjusting the UMTS radio network configuration depending on the actual requirements.

12.1.1.1 Planning and Optimizing for Advanced 3G Technologies

While the initial planning and deployment of 3G networks is going on, a number of enhanced and advanced features are expected to be deployed and used over the next few years. In order to maximize the return on investment of these advanced technologies, their usage needs to be comprehensively planned and optimized.

Examples for such technologies are Remote Electrical Tilt (RET) antennas (3GPP [9]), Tower Mounted Amplifiers (TMA), Tower Mounted Boosters (TMB), micro- and picocells, higher sectorization, additional carriers, Multi User Detection (MUD), higher order receive/transmit diversity, smart antennas* and, most recently, MIMO transceivers.

With the introduction of HSDPA (high speed downlink packet access), UMTS will provide data rates of up to 14 Mbit/s.** HSDPA will use a higher-order modulation scheme (namely 16-QAM) which will bring new challenges for the planning and optimization process in UMTS.

12.1.1.2 Advanced Planning and Optimization Methods

While automatic frequency planning was of little interest during the early days of GSM, it is highly important for all GSM networks these days. Basically every single GSM planning tool has an automatic frequency planning functionality included.

In 3G-CDMA networks the use of advanced methods is even more important due to the more complicated nature of CDMA networks. The equivalent to automatic frequency planning tools in GSM are automatic design and optimization tools that support the planning and optimization process, as well as the introduction of advanced radio technologies in 3G and beyond. The basis for decisions on the network quality, key design, and optimization performance indicators for UMTS networks include:

- **Clear Pilot dominance:** Distinct cell boundaries with limited overlapping areas are in high demand.
- **Pilot E_c/I_o:** Pilot E_c/I_o planning and optimization is the most important step in any CDMA design.
- **Pilot Pollution:** A polluter is a received pilot sequence in the downlink that meets all criteria to enter a given mobile's active set but is not admitted due to the active set's size limit or restrictions by the implemented RAKE receiver.
- **Handover Areas:** Desired handover percentages are in the range of 20 and 35%.
- **System load:** Using traffic forecasts, service types, and traffic distribution maps, Monte Carlo simulations can be carried out to identify cell loading and throughput of the network.

---

* Smart antennas become finally available for 3G radio network technologies. In the North American CDMA2000 networks vendors such as Nortel, Lucent, and Samsung have already introduced smart antenna base stations. In the Chinese 3G standard TD-SCDMA smart antennas are used at every single base station. Smart antenna base stations are commercially available, such as the Siemens NB430TS and the NB450TS. Even though they are not mandatory, smart antennas are becoming an industry standard for TD-SCDMA.

** NTT DoCoMo committed to HSDPA and introduced this technology before the end of 2005.

Automated design and optimization tools are currently attracting great interest from 3G operators and vendors. They will be highly important to maximize the coverage, quality, and capacity tradeoff in 3G networks, and at the same time, to maintain a high flexibility for future CDMA needs and requirements.

12.1.2 Coverage and Capacity Enhancement Methods

The key reason for operators to introduce advanced coverage and capacity enhancement methods is to achieve a better return on investment. The return on investment in wireless network infrastructure is given by the income enabled due to the usage of the infrastructure equipment. However, since the availability of the infrastructure is a requirement, rather than a driver for higher revenues, it is more and more considered as bit pipe for wireless services.

Therefore, a general business strategy for wireless service providers is to reduce the infrastructure cost as much as possible, while providing sufficient coverage, quality, and capacity for the network.

12.1.2.1 *Fewer Base Stations*

Automated design optimization tools in combination with advanced radio technologies are currently developed and used for 3G radio network planning and design. These tools compute the minimum resources required to satisfy coverage/quality/capacity requirements. Simultaneously, design optimization finds the best radio configuration for each site/sector as well as the best strategy to introduce advanced radio technologies. By doing so, such tools boost the overall network performance. The more efficient use of the infrastructure results in fewer base stations for the same performance, compared to a manual design.

12.1.2.2 *Savings in CapEx and OpEx*

The design and deployment of a radio network is an ongoing process. The requirements for the network change during the network lifecycle. While the provision of sufficient coverage is of highest priority at rollout, service quality and capacity requirements dominate in the longer term.

To reduce the costs of the initial network, staged deployment plans are developed. Again, automated design optimization tools and advanced radio technologies deliver these staged network plans. They can provide sufficient coverage at low cost within the rollout stage.

Operator costs for the network are often expressed as capital expenditure (CapEx) and operating expenditure (OpEx) [10]. Later on, when the network is more mature, the design objectives will change to sufficient system capacity. The staged network deployment features of design optimization tools ensure that the infrastructure investment is put in the right place at the right time. This results in significant savings in both CapEx and OpEx; see Figure 12.4.

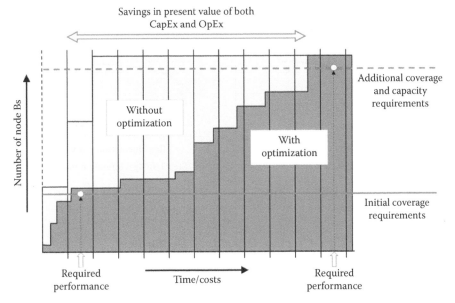

FIGURE 12.4
Savings in the present value of both CapEx and OpEx with optimization techniques.

It also helps reduce the risk of overbuilding networks. This means, for example, that a radio network is deployed to deliver high service capacity. However, with the lack of available handsets on the market, high capacity is not a top priority. It would be sufficient to provide high capacity at a later stage.

By applying staged deployment methods for radio networks, large portions of budget can be saved on the present value of the required CapEx and OpEx to realize these networks; see Figure 12.4.

Examples of UMTS deployments have shown that up to 12% of the present value of both CapEx and OpEx can be saved.

12.1.2.3 Best Deployment of Budget

Another important aspect for a good radio network plan is the efficient use of corporate budgets. This means that a radio network planner should make the most efficient use of an available budget.

At a corporate level, different budgets are applied to different markets or deployment areas. With cost and efficiency analysis available in automated planning, design, and optimization functionalities corporate budgets can be spent in the most efficient way. An example of this is shown in Figure 12.5.

The planning target here is network coverage in both Market A and Market B. Based on the "coverage per cost" analysis, the radio engineer can make the decision to move the available budget to the area where the most positive effect on coverage can be achieved.

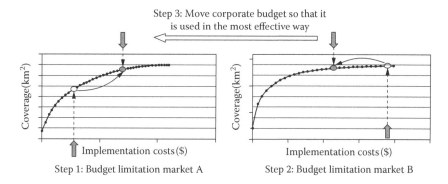

FIGURE 12.5
Efficient use of budget.

12.1.2.4 Most Cost-Effective Technology Deployment

3G radio networks are expected to include a mix of various advanced technologies such as Remote Electrical Tilt (RET) antennas, cf. [9], Tower Mounted Amplifiers (TMA), microcells, transmit diversity, smart antennas, etc. This leads to the question of the best utilization, penetration, and deployment of the more advanced (but also more expensive) technologies. From these deployments, a number of questions arise, such as:

- Which technologies should be considered when?
- How many smart antennas should be deployed based on the costs?
- Where should advanced techniques be deployed in the most effective manner?
- How should advanced techniques be configured?
- How should conventional antennas be configured?

These questions can be answered by using enhanced radio network planning and optimization tools in order to maximize the overall radio network performance, constrained by investment and operational costs.

12.1.3 Base Stations with Downlink Transmit Diversity and Beamforming

The gain in spectral efficiency for downlink beamforming in UMTS Time Division Duplex (TDD) mode was evaluated in simulation studies [11]. System-level simulations were carried out to evaluate the network throughput for the downlink by means of beamforming at the base station transmitter and multiuser detection in the mobile station receiver. The system-level model is based on the sectorized macro-environment specified in [12]. For downlink speech traffic at 8 kbit/s using transmit beamforming at the base stations and multiuser-detection at the mobile stations, the gain in spectral efficiency due to downlink beamforming is 55% higher than the gain resulting from power control and dynamic channel allocation combined. With downlink beamforming and power control together, there is no additional gain from employing dynamic channel allocation. It was found that there is a 2.6-fold spectral efficiency gain if antenna arrays of eight elements are deployed in a sectorized macrocellular environment. For further details on the simulation scenario and additional results, we refer to [11]. Very similar results apply for Time Division Synchronous Code Division Multiple Access (TD-SCDMA), which is also known as the low chiprate TDD-mode of UMTS.

One way to enhance the coverage is the introduction of downlink diversity methods. Downlink diversity will add 3 dB to received power. The UMTS specification supports transmit diversity in the downlink with two transmit antennas since its earliest release. It supports the Alamouti space–time transmit diversity scheme [13] for the downlink, which does not require channel knowledge at the Node B transmitter. This is the "open loop" downlink transmit diversity employing a space–time block coding based transmit diversity (STTD). The STTD encoding is optional in the UMTS radio access network. STTD support is mandatory at the mobile station. Channel coding, interleaving, and spreading are done as without transmit diversity.

UMTS also supports closed-loop transmit diversity schemes [14]. These are known as closed loop modes 1 and 2, respectively. The spread complex-valued signal is fed to both TX antenna branches, and weighted with antenna-specific complex-valued weight factors. The weight factors (actually the corresponding phase adjustments in closed loop mode 1 and phase/amplitude adjustments in closed loop mode 2) are determined by the mobile station and signaled to the UMTS radio access network access point (=cell transceiver) using the uplink Dedicated Physical Control Channel (DPCCH).

For the closed loop mode 1, different orthogonal dedicated pilot symbols in the DPCCH are sent on the two different antennas. For closed loop mode 2, the same dedicated pilot symbols in the DPCCH are sent on both antennas.

It is anticipated that Release 7 of 3GPP will specify MIMO transmission with four transmit antennas at the base station.* A total of eight MIMO proposals for UMTS are listed in [15]. Similarly, the IEEE 802.11n standard will incorporate MIMO transmission using OFDM with adaptive modulation based on channel state information signaled through a feedback control channel.

A tricky problem is interference in MIMO systems with several concurrent users. Each dominant interferer — each MIMO user to all the others in the system or cell — will require (at least) one degree of freedom to eliminate. This reduces the capacity gain of spatial multiplexing such that MIMO *system* capacity will be only insignificantly larger than what can be obtained by smart antennas alone [16,17]. Multi-user detection and cooperation of access points (base stations) in a cellular environment will be necessary to restore capacity to the theoretical limit [18].

---

* More precisely, MIMO is currently a study item within the Third Generation Partnership Project (3GPP) for "Long Term Evolution" of UMTS.

12.2 Deployment

MIMO capacity can show large local variations, depending on the position of the antenna arrays and on the environment. How does one deploy MIMO systems to maximize capacity practically? It is concluded from double-directional MIMO channel measurements [19] that directions of arrival and directions of departure may be discrete. Such environments lend themselves for transmission into limited angles. Even if the directions of departure and arrival are not discrete, it is likely that there are specific sectors from which much power can be received. This favors systems with directional transmission. The measurements in office environments [7,19] clearly demonstrate that there exist at least large angular sectors, either from which no significant power was received or into which no power is transmitted, that would eventually reach an intended receive antenna. How does the terminal or the access point know which are the favored directions? The deployment procedure proposed in [20] relies on measurements in the actual propagation environment.

Transmit diversity schemes are particularly appropriate for microcell scenarios [4] where capacity gains are large and beamforming enhancements are less appropriate due to the large angular spread [21]. Transmit diversity provides a relatively simple capacity upgrade solution in terms of configuring additional hardware. For the Alamouti scheme, two transmit antennas are required on the downlink — either a single cross-planar antenna or a pair of vertically polarized antennas. In terms of power amplifier modules, the operator may be able to share existing power. This is possible when the site is configured with multiple carriers and multi-carrier power amplifiers. This is not possible for single-carrier scenarios, and supplementary power amplifiers must be included.

12.3 Smart Antenna Planning Example

In this section, we discuss a specific planning problem through a simulation study. We discuss the connectivity of users in a 3G cellular network with conventional sectorization and compare it with the connectivity when the network deploys 4-element uniform linear array antennas.

Specifically, we assume a UMTS radio access network configuration including approximately 100 sectors. In order to investigate the system performance, system-level simulations were carried out using a state-of-the-art Monte Carlo simulator. The system-level simulator models the network similarly to most of the commercially available radio network planning tools. Additionally, we included the modeling of smart antenna processing. In this study, we consider two deployment scenarios: In the first scenario, we consider all base stations to be equipped with conventional sector antennas (three sectors per site). In the second scenario, the base stations are equipped with 4-element uniform linear array antennas. In the second scenario, we assume that the baseband signal processing at the base station implements optimum combining.

The typical questions to be answered by the operator are:

- How do smart antennas influence the system performance?
- In which regions are (dis-)connected users concentrated?
- Does the distribution of (dis-)connected users change in the case of smart antenna deployment?
- Does it make sense to deploy smart antennas throughout the entire network or only in certain areas?

For simplicity we assume an idealized service mix with just two different service types, i.e., we assume 40% of the users to be served at a data rate of 12.2 kbit/s and 60% of the users at 64 kbit/s.

The results are shown in Figure 12.6 and Figure 12.7. Figure 12.6 shows the distribution of the connected users for conventional sector antennas (a) and 4-element-ULAs (b). Conversely, Figure 12.7 shows the distribution of disconnected users in case of conventional sector antennas (a) and 4-element-ULAs (b).

This comparison shows that smart antennas are capable of increasing the number of served users. Hence, the network's aggregated throughput can be increased substantially by the use of smart antennas. In the studied scenario, the deployment of 4-element-ULAs is capable of doubling the network throughput. Only a very small number of users remain disconnected in areas of poor coverage.

We note that smart antennas provide a "cluster gain," rather than just a performance gain limited to the smart antenna serving cell. This means that a single cell equipped with a smart antenna will not just improve the coverage and capacity of the serving sector, but will also improve the performance of the neighbouring cells, due to the highly reduced intercell interference power.

Recently, a statistical analysis in live CDMA networks was carried out in North America [22]. The analysis has shown that among all sites in which the load exceeds the capacity threshold, around 51% have a single overloaded sector, while the other sectors operate below the capacity threshold. Another 38% of the sites have two sectors exceeding the capacity threshold, and only 11% of the sites have all three sectors exceeding the capacity threshold. This means that, when deployed per sector, smart antennas allow the operator to address capacity where additional resources are needed, increasing the utilization of every deployed carrier.

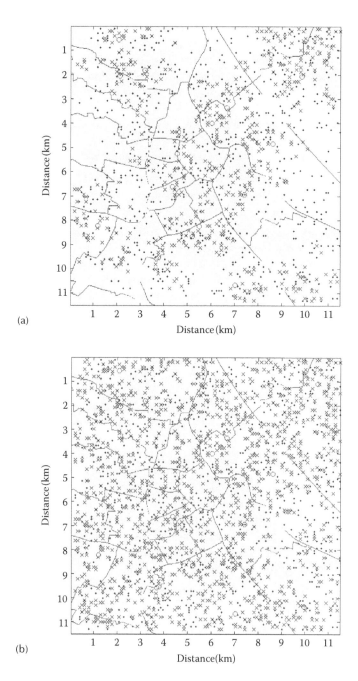

(a)

(b)

FIGURE 12.6
Distribution of connected users in the system area. Each base station is either equipped with a single element sector antenna (a) or a 4-element uniform linear array (ULA, performing optimum combining) (b), respectively. The "." indicates users with a 12.2 kbit/s service, and the "x" shows users with 64 kbit/s service.

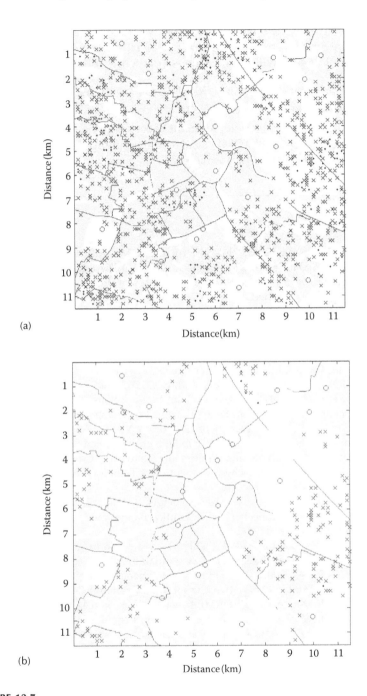

(a)

(b)

FIGURE 12.7
Distribution of disconnected users in the system area. Each base station is either equipped with a single element sector antenna (a), or a 4-element uniform linear array (ULA, performing optimum combining) (b), respectively. The "." indicates users with a 12.2 kbit/s service, and the "x" shows users with 64 kbit/s service.

Considering both the sector-based capacity limitations and the cluster gain of smart antennas, future challenges for smart antenna and MIMO radio network deployment will be to find answers to the following questions:

- How many smart antennas/MIMO systems should be deployed?
- Where should they be deployed?
- How should the mixed deployment of conventional sector antennas, smart antennas, and MIMO systems be configured in order to provide the best system performance?

References

1. W. Weichselberger. 2003. "Spatial Structure of Multiple Antenna Radio Channels — A Signal Processor Viewpoint," Doctoral Dissertation, Vienna University of Technology, Vienna, Austria. Available at: http://www.nt.tuwien.ac.at/mobile/theses_finished.
2. S.N. Diggavi, N. Al-Dhahir, A. Stamoulis, and A.R. Calderbank. 2004. "Great expectations: the value of spatial diversity in wireless networks," *Proceedings of the IEEE*, Vol. 92, No. 2, Feb. 2004, pp. 219–270.
3. A. Paulraj, R. Nabar, and D. Gore. *Introduction to Space-Time Wireless Communications*, Cambridge, U.K., Cambridge University Press.
4. J. Laiho, A. Wacker, and T. Novosad, eds. 2002. *Radio Network Planning and Optimization for UMTS*, New York, John Wiley & Sons.
5. A. Kuchar. 1999. "Real Time Smart Antenna Processing for GSM1800", Doctoral Dissertation, Vienna University of Technology, Vienna, Austria, 1999. Available at: http://www.nt.tuwien.ac.at/mobile/theses_finished.
6. Towards Mobile Broadband Multimedia Networks, Cooperation euroéenne dans le domaine de la recherché Scientific et Technique Action 273. Available at: http://www.lx.it.pt/cost273/.
7. H. Özcelik, M. Herdin, W. Weichselberger, J. Wallace, and E. Bonek. 2003. "Deficiencies of 'Kronecker' MIMO radio channel model," *Electronics Letters*, Vol. 39, No. 16, pp. 1209–1210.
8. Durlacher Research Ltd. 2001. "UMTS-Report — An Investment Perspective," Press Release. Available at: http://www.dad.be/library/pdf/durlacher3.pdf.
9. 3GPP TS 32.642. 2005. Telecommunication management; Configuration Management (CM); UTRAN network resources Integration Reference Point (IRP): Network Resource Model (NRM), Release 6, Version 6.4.0, April 2005.
10. T. Giles, J. Markendahl, J. Zander, P. Zetterberg, P. Karlsson, J. Lind, and G. Malmgren. 2004. "Cost drivers and deployment scenarios for future broadband wireless networks — key research problems and directions for research," *Proc. IEEE Vehicular Technology Conference*, Spring, May 2004.
11. M. Haardt, C.F. Mecklenbräuker, M. Vollmer, and P. Slanina. 2001. Smart Antennas for UTRA TDD, *European Transactions on Telecommunications* (ETT), Special Issue on Smart Antennas, Vol. 12, Issue 5, pp. 393–406, Sept.–Oct. 2001.
12. ETSI TR 101 112. 1998. "Selection procedures for the choice of radio transmission technologies of the universal mobile telecommunications system UMTS (UMTS 30.03), Version 3.2.0," April 1998.

13. 3GPP TS 25.211. 2005. Technical Specification Group Radio Access Network; Physical channels and mapping of transport channels onto physical channels (FDD) Release 6, Version 6.4.0, March 2005.

14. 3GPP TS 25.214. 2005. Technical Specification Group Radio Access Network; Physical layer procedures (FDD), Release 6, Version 6.5.0, March 2005.

15. 3GPP TR 25.876. 2005. Technical Specification Group Radio Access Network; Multiple-Input Multiple Output in UTRA, Release 7, Version 1.8.0, October 2005.

16. S. Catreux, P.F. Driessen, and L.J. Greenstein. 2000. "Simulation results for an interference-limited multiple-input multiple-output cellular system," *IEEE Comm. Lett.*, Vol. 4, No. 11, Nov. 2000, pp. 334–336.

17. S. Catreux, P.F. Driessen, and L. Greenstein. 2001. "Link-optimal space-time processing with multiple transmit and receive antennas," *IEEE Comm. Lett.*, Vol. 5, pp. 334–336.

18. H. Dai, A.F. Molisch, and H.V. Poor. 2004. "Downlink capacity of interference-limited MIMO systems with joint detection," *IEEE Trans. Wireless Comm.*, Vol. 3, pp. 442–453.

19. M. Steinbauer et al. 2001. "The double-directional radio propagation channel," *IEEE Antennas & Propagation Magazine*, Vol. 43, No. 4, August 2001, pp. 51–63.

20. E. Bonek. 2002. "Link-specific MIMO system deployment for ad-hoc networks," *Proc. WWRF Meeting #7, Working Group 4*, Eindhoven, the Netherlands, December 3–4, 2002.

21. S. Andersson et al. 1999. "Adaptive antennas for GSM and TDMA systems." *IEEE Personal Communications*, June 1999, pp. 74–86.

22. T. Crook. 2002. "Why smart antennas & subscriber device based enhancements are important to wireless service providers," Sprint. Available at: http://www.cdg.org/news/events/CDMASeminar/cdg_tech_forum_02/1_sprint_cdg_smart_antenna_rev.pdf.

Index